AF803372

Design of Mechanical Systems Based on Statistics

Advanced Research in Reliability and System Assurance Engineering

Series Editor:
Mangey Ram
Professor, Graphic Era University, Uttarakhand, India

Modeling and Simulation Based Analysis in Reliability Engineering
Edited by Mangey Ram

Reliability Engineering
Theory and Applications
Edited by Ilia Vonta and Mangey Ram

System Reliability Management
Solutions and Technologies
Edited by Adarsh Anand and Mangey Ram

Reliability Engineering
Methods and Applications
Edited by Mangey Ram

Reliability Management and Engineering
Challenges and Future Trends
Edited by Harish Garg and Mangey Ram

Applied Systems Analysis
Science and Art of Solving Real-Life Problems
F. P. Tarasenko

Stochastic Models in Reliability Engineering
Lirong Cui, Ilia Frenkel, and Anatoly Lisnianski

Predictive Analytics
Modeling and Optimization
Vijay Kumar and Mangey Ram

Design of Mechanical Systems Based on Statistics
A Guide to Improving Product Reliability
Seong-woo Woo

For more information about this series, please visit: https://www.routledge.com/Advanced-Research-in-Reliability-and-System-Assurance-Engineering/book-series/CRCARRSAE

Design of Mechanical Systems Based on Statistics

A Guide to Improving Product Reliability

Seong-woo Woo

CRC Press
Taylor & Francis Group
Boca Raton London New York

CRC Press is an imprint of the
Taylor & Francis Group, an **informa** business

First edition published 2021
by CRC Press
6000 Broken Sound Parkway NW, Suite 300, Boca Raton, FL 33487-2742

and by CRC Press
2 Park Square, Milton Park, Abingdon, Oxon, OX14 4RN

© 2021 Taylor & Francis Group, LLC

CRC Press is an imprint of Taylor & Francis Group, LLC

The right of Seong-woo Woo to be identified as author of this work has been asserted by him in accordance with sections 77 and 78 of the Copyright, Designs and Patents Act 1988.

Reasonable efforts have been made to publish reliable data and information, but the author and publisher cannot assume responsibility for the validity of all materials or the consequences of their use. The authors and publishers have attempted to trace the copyright holders of all material reproduced in this publication and apologize to copyright holders if permission to publish in this form has not been obtained. If any copyright material has not been acknowledged please write and let us know so we may rectify in any future reprint.

Except as permitted under U.S. Copyright Law, no part of this book may be reprinted, reproduced, transmitted, or utilized in any form by any electronic, mechanical, or other means, now known or hereafter invented, including photocopying, microfilming, and recording, or in any information storage or retrieval system, without written permission from the publishers.

For permission to photocopy or use material electronically from this work, access www.copyright.com or contact the Copyright Clearance Center, Inc. (CCC), 222 Rosewood Drive, Danvers, MA 01923, 978-750-8400. For works that are not available on CCC please contact mpkbookspermissions@tandf.co.uk

Trademark notice: Product or corporate names may be trademarks or registered trademarks and are used only for identification and explanation without intent to infringe.

Library of Congress Cataloging-in-Publication Data
Names: Woo, Seong-woo, author.
Title: Design of mechanical systems based on statistics : a guide to
improving product reliability / Seong-woo Woo.
Description: First edition. | Boca Raton : CRC Press, 2021. |
Series: Advanced research in reliability and system assurance engineering |
Includes bibliographical references and index.
Identifiers: LCCN 2020045685 (print) | LCCN 2020045686 (ebook) |
ISBN 9780367076269 (hardback) | ISBN 9780429022050 (ebook)
Subjects: LCSH: Machine design–Statistical methods. | Quality of products.|
Reliability (Engineering)
Classification: LCC TJ245.5 .W66 2021 (print) | LCC TJ245.5 (ebook) |
DDC 621.8/15015195–dc23
LC record available at https://lccn.loc.gov/2020045685
LC ebook record available at https://lccn.loc.gov/2020045686

ISBN: 978-0-367-07626-9 (hbk)
ISBN: 978-0-367-74561-5 (pbk)
ISBN: 978-0-429-02205-0 (ebk)

Typeset in Times
by codeMantra

Contents

PART II Parametric Accelerated Life Testing

Preface

End users choose to use a product because of its good performance, ease of use, and trouble-free nature, rather than its technological complexity. The reliability of a product is one of its achievable characteristics, regardless of both the kind of product and the number of parts integrated into it. At the beginning of the twentieth century, new advanced mechanical systems such as airplanes, automobiles, rockets, space shuttles, refrigerators, and appliances were designed for people to lead pleasant lives through the product design processes. A mechanical system transfers (generated) power to accomplish its intended functions, thus producing mechanical advantages by adapting product mechanisms. For instance, an automobile is a wheeled motor vehicle utilized for transit. Its power is generated by the engine and transmitted the wheels through mechanisms such as the transmission and drive train.

As a customer requires new functions of mechanical products, they have been modified and expanded in the last century. The product developing process can generally be summarized as follows: (1) define the problems and their specifications; (2) develop the product design, prototype, and testing; and (3) production. Mechanical products with multiple modules should also satisfy the customer's requirements such as reliability, higher performance, and cost-reduction. Therefore, only global companies with high techs and finance can survive competitive situations in the market. Due to the increase of (intended) functions and parts, the products have become more complex, which might increase the risk of product failure. Any design faults in a product might be reproduced and modified through a proper statistical method or reliability testing in the design process.

However, as the current statistical methodologies including Design Of Experiment (DOE) and Taguchi's robust design only pursuit product optimization without considering the concepts of product design, they require huge computations to find an optimum solution but without results, because they haven't figured out failure mechanics. That is, if there are design faults in a product that cause an inadequacy of strength (or stiffness) when subjected to repeated loads, the product will fail before its expected lifetime due to fatigue failure. Mechanical fatigue accounts for about 90% of all structural failures, which was estimated by Battelle in 1982 to cost about 1.5 billion in the United States. From an economics standpoint, if a product failure occurs in its expected product lifetime, the company's profit will decrease because of the regulation of global market such as Product Liability Law. The brand image of the product will deteriorate, and finally, it will be kicked out of marketplace. Therefore, product reliability has become a critical element (to consider) in system engineering.

Since the 1970s, there have been lots of disasters due to defects of technology. It has been recognized that huge gaps exist between reliability engineering and mechanical system design. Therefore, engineers have started to consider why product recalls happen frequently with low possibility. To elicit this problem, they should know what mechanical product is. That is, the mechanical systems comprising several different modules transfer power to carry out tasks by obtaining mechanical advantages from properly chosen product mechanisms. In this process, the mechanical products will be subjected to repetitive stress. Product material can also be imperfect, i.e., it may have a small crack or pre-existing defect on the surface of a component. After repeated stresses, mechanical failure such as fatigue might happen in the field. To avoid failures, a mechanical system should be robustly designed to endure the operating conditions made by the customers who buy and utilize. Otherwise, the mechanical system will abruptly fail before achieving its expected lifetime, and thus, it will no longer satisfy the conventional specifications of product working.

In the last century, the studies of product reliability have been deepened to prevent product recalls. Even though there are numerous concepts, methodologies, and books on product reliability, a new methodology of reliability design is required in the areas of the mechanical systems, because product recalls for new products still happen in marketplace. The most effective way to prevent the design mistakes of a newly developed product is to develop a parametric accelerated life testing (ALT) as a reliability methodology in the established design process in parallel with the reliability-embedded design process. That way, before the product is launched, the manufacturer can modify the design flaws.

Product reliability can be defined as the ability of a system or module to function under specified conditions for a stated period of time. To develop a new reliability methodology, we have to combine the design concepts of the mechanical systems with statistical concepts including probability. This parametric ALT as a reliability methodology therefore comprises: (1) parametric ALT plans, (2) load analysis for accelerated testing, (3) parametric ALTs with corrective plans, and (4) assessment of whether the last design(s) of product satisfies the target BX lifetime. So, we would explain a generalized life-stress model with an effort concept, accelerated factor, and sample size equation in this book.

The main purpose of writing this book, therefore, is to explain a parametric ALT embedded on the product development process. It consists of two parts and nine chapters; that is, Part I: data and statistics in mechanical system and Part II: parametric ALT for mechanical system. Part I, from the standpoint of statistics, will describe how to deal with the data obtained from field failures after a mechanical system is designed and manufactured. First of all, after we define the mechanical system of a product such as airplane, automobile, refrigerator, and construction machine, we will discuss how to check the data of product lifetime. Though each product has different design principles – thermodynamics, mechanism, fluid mechanics, etc. – we will recognize that the failure mechanisms – fatigue, vibration, fracture, and erosion – due to the design problems and loads are similar. That is why they use the same materials and designs in products. And if there are design faults in products when subjected to repetitive stress, the products will fail in its lifetime. The trend of product life can be described with the bathtub curve.

In Chapter 1, after defining the mechanical system, we will briefly review the design principles and their implementation in the product design process of a mechanical system. When neglecting some design points in the developing process, many manufacturers experienced the product recalls. We also explain the reliability evaluation tools that are developed throughout the history. Chapter 2 will discuss what reliability data in the mechanical system is. Because it is very hard to systematically obtain proper data from field, companies should rely on the data from proper reliability testing, but this has limitations because there is no systematic methodology. Chapter 3 will deal with a fundamental knowledge of statistics including probability and statistical distribution. Chapter 4 will discuss descriptive statistics for lifetime testing and analysis of mechanical product. We also discuss numerous concepts of reliability engineering such as bathtub, mean time to failure (MTTF), and failure rate. Combining the conventional statistics with design principles of mechanical system such as strength and stiffness, it might be possible to figure out the fundamental concepts of parametric ALT such as the general life stress model based on failure mechanism and load analysis, acceleration, and sample size equation.

Part II will discuss the basic concepts of the parametric ALT. Chapter 5 will deal with the structure (including mechanisms) of mechanical products and their modeling used by Newtonian, D'Alembert's principle, Lagrangian, or bond graph. The mechanical products convert power to force (effort) and velocity (flow) to complete a mission by adapting its mechanisms and mechanical advantages. Therefore, mechanical products are subjected to repetitive stress. If the products are not designed to have a proper strength (or stiffness), it will be a root cause of mechanical (fatigue) failures. Chapter 6 will discuss the design of mechanical systems – strength and stiffness. Chapter 7 will discuss what mechanical system's failures, especially fatigue, are. A system's failure may start from the presence of a minute material defect when repeatedly and randomly subjected to variable tensile and compression loads. That is, if there are design faults in a mechanical system that cause an inadequacy of strength (or stiffness), the mechanical system will fail in its lifetime when subjected to repetitive loads.

Chapters 8 and 9 will deal with parametric ALT as a new reliability methodology and its case studies that will assess the design of mechanical systems. Currently, various schemes have been suggested to properly determine the sample size for assessing the product lifetime. Here, the Weibayes model was used to derive sample size equation. A new sample size equation connected with accelerated factor (AF) could attain the mission cycle for parametric ALT from the target BX lifetime on the test plans. So, the new sample size equation enables the parametric ALT to quickly assess the targeted lifetime of a product as an engineer uncovers the design faults affecting the reliability for lifetime target. Finally, the author recognizes whether the targeted product lifetime is achieved for its mission cycle.

Author Biography

Seong-woo Woo has a BS and an MS in Mechanical Engineering, and he received his PhD in Mechanical Engineering from Texas A&M. His major interests are in energy systems, such as HVAC and its heat transfer, optimal design and control of refrigerators, reliability design of mechanical components, and failure analysis of thermal components in marketplace using non-destructive methods, like scanning electron microscopy and X-ray. Most notably, he has developed a parametric accelerating life testing (ALT) as a new reliability methodology, which would find if there is a design fault in a mechanical system that is subjected to repetitive stress. Using this parametric ALT, engineers can find the design faults before the product is launched and thus can avoid failures.

From 1992 to 1997, he worked in the Agency for Defense Development, Chinhae, South Korea, where he was the researcher in charge of the Development of Naval Weapon System. From 2000 to 2010, he worked as a senior reliability engineer in Side-by-Side Refrigerator Division, Digital Appliance, SAMSUNG Electronics, where he focused on enhancing the lifetime of refrigerators using a parametric ALT. He is now working as an associate professor in the Mechanical Department at Addis Ababa Science & Technology University, Ethiopia.

Author Biography

Part I

Data and Statistics in Mechanical System

1 Introduction to Reliability Design of Mechanical System

1.1 INTRODUCTION

As customers require high-tech functioned products, numerous technologies implemented in products are continuously emerging and vanishing in the marketplace so that modern human life is diversified and wealthy. Consequently, only a few of global companies with high technologies could survive the unlimited competition in the marketplace. For example, there are many mechanical systems, such as automobile, airplane, refrigerator, and bicycle, developed in the last century. Because customers continually require new functions, the mechanical products have been modifying and expanding so that they become multi-modules embedded on a variety of functions. If a product successfully mounted in high-techs sells well, its manufacturer will survive against the competition in the global marketplace.

Mechanical products should also satisfy the customer requirements such as high response, high control on performance, low noise, precision control for a wide frequency range, long lifetime, compact and highly portable weight, high reliability, low price, advanced hardware design, contamination resistance, and energy efficiency. However, engineers wonder if a product can satisfy these requirements as the development periods of the product are continually shortening in the current developing processes. On the other hand, reliability of the product from the marketplace is more highly required. However, there are no reliability methodologies that can prove the robustness of product design. That is why product recall frequently happens. For instance, as the development period of automobiles with all requirements satisfied has recently been shortened from 65 to 24 months, the reliability increases from 0.9 to 0.99. To meet two objectives – quality and developing time– the global manufacturers need to have a reliability methodology closely bound to the product developing process (Figure 1.1).

When we explain the quality costs, we think the relationship between product life cycles and quality costs due to product failure. That is, we recognize that the faster the reliability concepts – parametric accelerated life testing (ALT) – in the design process are applied, the better the product revenue can be obtained as the failure cost decreases. For instance, if a cost of $1 is needed to correct a design flaw in the previous design phase, the quality cost would increase to $10 after the last engineering phase, $100 at the early production, and $1,000 at the mass product phase.

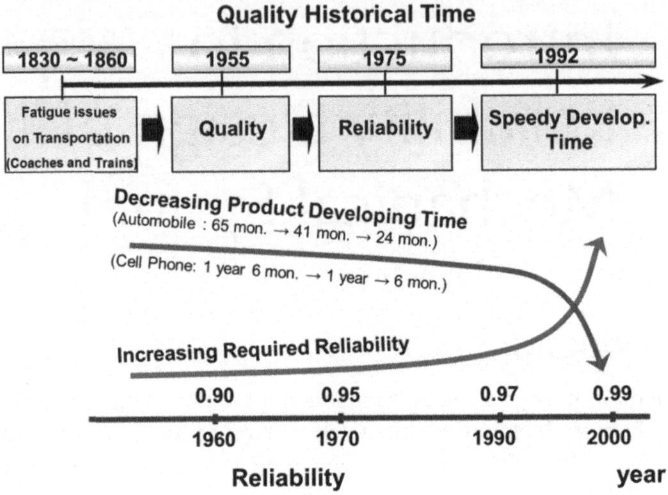

FIGURE 1.1 Historical trend for product quality.

Most product failures such as fatigue generally come from the negligence (or design fault) of a new design. When an 'improved' design or a new feature is introduced, there are design parameters that the designer does not expect from the experience. In other words, new designs can not only offer tremendous advantages from a standpoint of revenue, but they can also pose potential design problems under customer usage conditions.

Consequently, after engineers find inherent design faults by a proper parametric ALT and correct them, manufacturers could safely implement new designs into products because such an approach reduces quality costs due to the frequent failures in the marketplace. We also recognize that the design cost in the product development takes only 5% of the total product cost. However, when a product recall happens, its cost influence takes 70% of the total product cost. To prevent it, we require a new reliability methodology that can evaluate the product design in the design process (Figure 1.2).

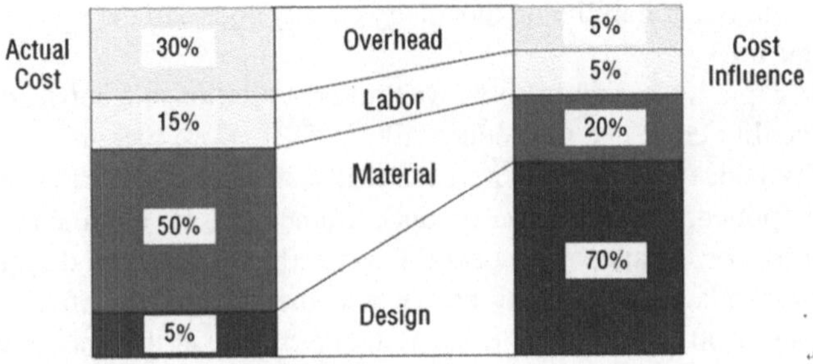

FIGURE 1.2 Leverage in product design.

This book deals with parametric ALT as a new reliability methodology that can assess the design of mechanical products in the developing process. It comprises two parts – part I: data and statistics in mechanical systems and part II: parametric ALT for the design of mechanical systems. From the standpoint of statistics, part I will describe how to deal with the lifetime data obtained from the field after mechanical systems are designed and launched. First of all, we will explain what a mechanical system, such as automobile, airplane, refrigerator, and construction machine, is. Though each mechanical product has different design principles – thermodynamics, mechanism, statistics, dynamics, fluid mechanics, strength of mechanics, etc. – we can find that their failure mechanisms – fatigue, fracture, and erosion – due to the design problems are similar. That is, if there are design flaws when a product is subjected to repeated stresses, it will fail in product lifetime.

Chapter 1 defines the mechanical system. After briefly reviewing the design principles of mechanical products, we explain the historical reliability concepts for evaluating the product lifetime. Finally, we stress on the necessity of a new reliability methodology for assessing the product design because of the incompleteness of current concepts. Chapter 2, from the standpoint of statistics, will discuss what (reliability) data in the mechanical system are. As the customers start to use products, the failure data start to emerge. Based on the field data, we can estimate the product lifetime and describe it on bathtub curves. However, because it is very hard to obtain a complete data, companies independently try to get it from reliability testing. However, there is no reliability methodology. Chapter 3 deals with probability and its distributions such as binomial, Poisson, exponential, and Weibull, thus providing a clear understanding of the reliability testing. Chapter 4, as a kind of descriptive statistics, will discuss how to determine the lifetime from product testing (or field data). It requires numerous statistical concepts such as bathtub, Weibull distribution, failure rate, and lifetime measure such as MTTF (or BX life). BX life is defined as the time at which X percent of the items in a population will have failed. The mean time to failure (MTTF) equates to a B60 life. We have to develop the life stress model and sample size equation. To do this, we need to merge the design concepts of mechanical products.

Part II will focus on parametric ALT as a new reliability methodology combined with the design concepts of mechanical systems, based on probability and statistics. A mechanical system transfers power, which implements mechanical advantages by adapting product mechanisms. In this process, a mechanical product is subjected to repetitive stress. If there are design faults that can cause an inadequacy of strength (or stiffness), the product will fail in its lifetime. To assess its robustness, we will derive the time-to-failure and sample size equation which is a core concept of this methodology. Chapter 5 deals with the structure including the mechanism of mechanical products and their modeling technics such as Newtonian, D'Alembert's principle, Lagrangian, and bond-graph. The mechanical systems transfer power that modulates forces (effort) and velocity (flow). Chapter 6 will discuss how to design mechanical systems by introducing the basic concepts such as strength and stiffness as we discuss with the elasticity and vibration. Chapter 7 will discuss what mechanical system failure – especially, fatigue – is. Chapters 8 and 9 deal with parametric ALT and its case studies.

1.2 MEANING OF TERMS 'QUALITY' AND 'RELIABILITY'

Customers choose a product because of its better performance, ease of usability, lack of problem, and technological simplicity. From the standpoint of a customer, the reliability of the product is one of the acceptable attributes, regardless of both the kind of product and the number of components connected to the product. When a global company manufactures a product with a good quality – performance and reliability – its brand image will be well-known in the marketplace. The term 'quality of product' in daily usage is vaguely defined to signify the inherent level of the product's excellence. In the field, quality is defined as 'conformance to specification at the beginning of product use'. Presuming that the product specifications adequately capture customer needs, the product quality degree can be exactly measured.

On the other hand, as we can see in Table 1.1, the recalls come out in the marketplace when customers start to use a new product. From the investigation of the root causes of product recalls, we recognize that they result from the design faults missed in the developing process. Unless the company ensures the quality of product lifetime for new (intended) functions, it will be expelled from the global marketplace. We therefore seem to friendly use product reliability in our everyday life. In other words, we can figure out the concept of 'reliability' from the following question: how many of these samples satisfy their specifications at the end of one-year assurance period?

We sometimes confuse between the meaning of terms 'quality' and 'reliability'. Quality is related to the product's performance at a particular period – especially, at the start of the product's use. As consumers use product, we can know its performance at certain time through the common specifications in company whether product quality meets. On the other hand, reliability is related to the product's performance for its lifetime due to its design. As repeatedly mentioned, the product unreliability comes from design defects for product functions at a certain period. If a product is out of specifications, we can consider it as a quality defect. Therefore, product reliability explains the ability of a product's design to sustain its performance in different environmental or usage conditions, without showing poor quality at any point of its lifetime that is explained in product specifications.

1.3 WHAT IS A MECHANICAL PRODUCT?

1.3.1 Introduction

Mechanical products transfer power to fulfill a job that requires forces and movement, which produce mechanical advantages by adapting product mechanisms. The term 'mechanical' is originated from the Latin 'machina', which can be explained as relating to machinery or tools. A mechanical product usually consists of (1) a power source and actuators that produce forces and movement, (2) structured mechanisms that shape the actuator input to attain a specific implementation of output forces and movement, and (3) a controller with sensors. A man uses a stake to raise a rock, which is a simple instance of a mechanism. As the man applies a little effort at

TABLE 1.1
Quality Defects and Failures

	Quality Faults	Failure
Notion	Out of common specifications	Physical failure (e.g. fatigue)
Indicator	Defect rate, ppm	Lifetime, failure rate
Measure	Percent, ppm	Percent/year, year
Field	Manufacturing	Design
Probability distribution	Normal distribution	Exponential/Weibull
	$f(x)=\frac{1}{\sigma\sqrt{2\pi}}e^{-\frac{(x-\mu)^2}{2\sigma^2}}$	$F(t)=1-R(t)=1-e^{-\lambda t}$
	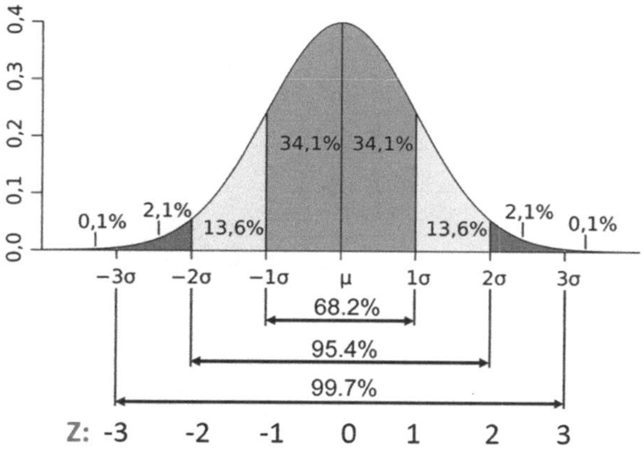	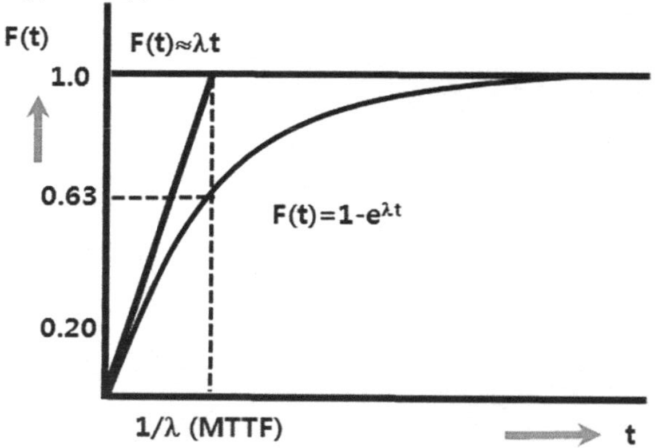

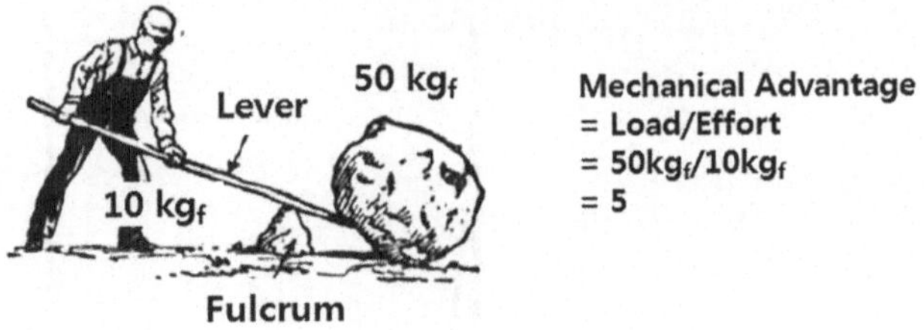

FIGURE 1.3 Man raising a stone by lever.

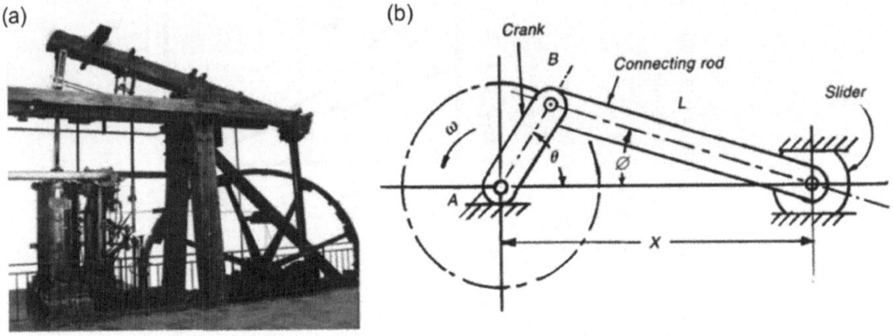

FIGURE 1.4 (a) James Watt's Industrial Steam Engine and (b) slider-crank mechanism (Boulton & Watt, 1788).

one end of the lever, the rock on the other end moves. This effort gain is known as Mechanical Advantage (Figure 1.3).

To understand the fundamental design of mechanical products, we can use the mechanism of Boulton and Watt steam engine as a typical example, which is explained as a slider-crank mechanism where the steam expands to drive the piston and transfers the reciprocating movement to a rotary movement. That is, walking beam, coupler, and crank transform the linear movement of the engine piston into rotation of the output pulley. Lastly, the pulley rotation operates the flyball governor, which can run and control the valve for the steam input to the piston cylinder (Figure 1.4).

1.3.2 Refrigerator

To store fresh (or frozen) food, in a refrigerator chilled air is supplied from the evaporator to both the freezer and the refrigerator. To achieve this, a vapor-compression refrigeration cycle is employed in refrigerators. To produce a continuous cooling effect, it operates at two pressure levels (high and low). Major components of a refrigerator are (1) compressor, (2) condenser, (3) expansion valve, and (4) evaporator (Figure 1.5).

Refrigeration cycle is a process of lowering the temperature of a substance less than that of its surrounding. The refrigerant enters the compressor at a low pressure (state 1). It then runs away from the compressor and arrives at the condenser at an

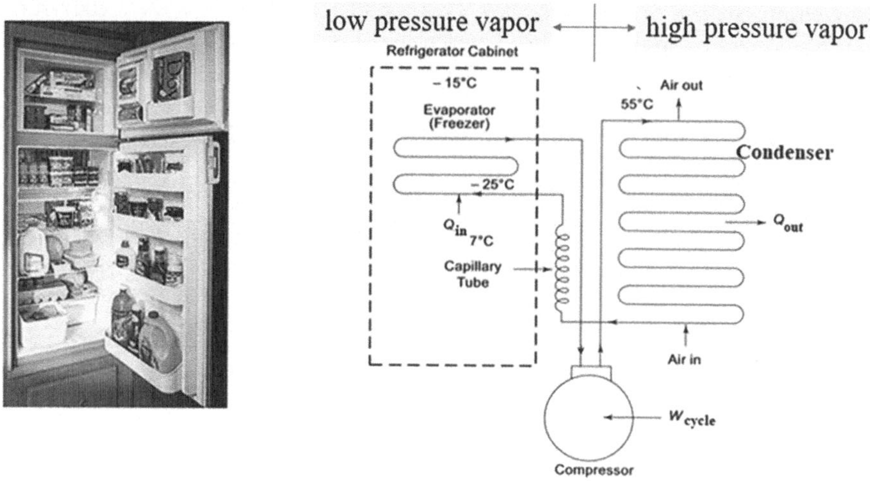

FIGURE 1.5 Refrigerator and its diagram.

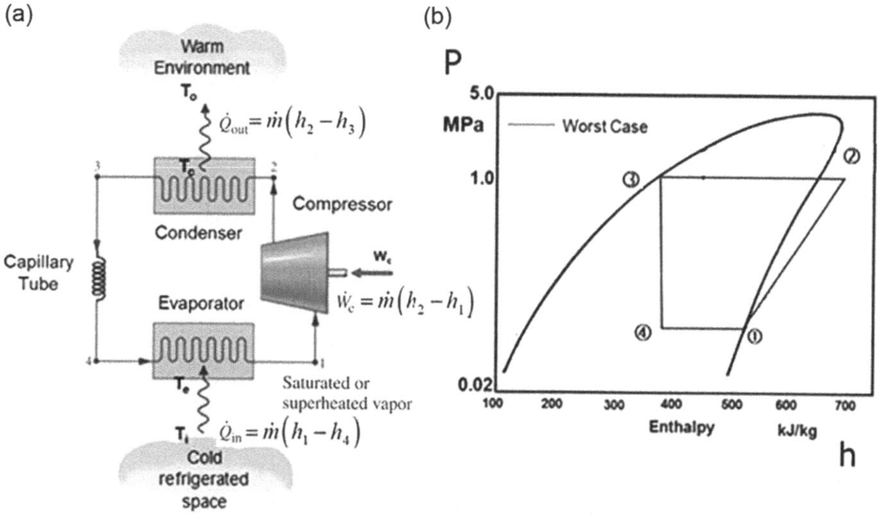

FIGURE 1.6 Vapor-compression refrigeration cycle: (a) refrigeration cycle and (b) *p-h* diagram.

elevated pressure (state 2). As heat is transferred to the surroundings, the refrigerant is condensed (state 3). The pressure of the liquid is decreased as it runs through the expansion valves (state 3). The residual liquid at a low pressure and temperature is vaporized in the evaporator (state 4) (Figure 1.6).

A refrigeration system utilized only for cooling is defined as a refrigerator. Generally, its capacity is defined as the amount of heat lowered than the surrounding environment, which is expressed in tons. It is equivalent to the heat required for melting one ton of ice in one day. One ton of refrigeration is the steady-state heat transfer rate required to melt 1 ton of ice at 32°F in 24 hours. That is, it can be expressed as

$$1 \text{ ton} = 144 \text{ Btu/lb} \times 2,000 = 288,000 \text{ Btu/day}$$

$$= 12,000 \text{ Btu/hour } (= 3.516 \text{ kW}) \quad (1.1)$$

For example, if a system has the capacity of 1,200,000 Btu/hour, we know that it has 100 tons (= 1,200,000/12,000). Cooling Coefficient of Performance (COP_C), refrigeration cycle efficiency, is expressed as

$$COP_C = \frac{\dot{Q}_{in}}{W_{cycle}} = \frac{\dot{m}(h_1 - h_4)}{\dot{m}(h_2 - h_1)} = \frac{(h_1 - h_4)}{(h_2 - h_1)} \quad (1.2)$$

where refrigeration effect: $q_{in} = h_1 - h_4$ and refrigeration capacity: $\dot{Q}_{in} = \dot{m}(h_1 - h_4)$.

On the other hand, there is another mechanical product, namely, Room Air Conditioner (RAC), that is operated by a vapor-compression refrigeration cycle. For example, assume that a room is continually maintained at a temperature of 25°C. In the air conditioner, the air in the room is drawn by a fan and is passed through a *cooling coil*, the surface of which is continued, say, at a temperature of 10°C. By passing through the coil, the air is cooled down (e.g., to 15°C) before being released into the room. After picking up the room heat, the air is passed again through the cooling coil at 25°C (Figure 1.7).

Refrigeration utilizes refrigerant as a working fluid that experiences two phases of change: (1) boiling (evaporator) and (2) condensing (condenser). Therefore, a refrigerant should be a fluid with thermodynamic characteristics that allow to boil at low temperatures. Examples of typical refrigerants are ammonia, *R*12, *R*22, and carbon dioxide. However, most refrigerants are halogenated hydrocarbons. The naming agreement adopted by the American Society of Heating, Refrigerating and Air-conditioning Engineers (ASHRAE) is

$$R(a-1)(b+1)d = C_a H_b Cl_c F_d \quad (1.3)$$

where $c = 2(a-1) - b - d$.

For example, chemical formulation of *R*22 (*R*022) can be found as

$$R22 = C_1 H_1 Cl_1 F_2 \text{ (chlorodifluoromethane)}$$

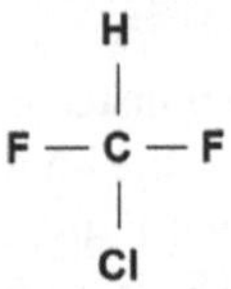

(1.4)

where $a-1=0 \rightarrow a=1$, $b+1=2 \rightarrow b=1$, $d=2$, $c=2(a+1)-b-d=2(1+1)-1-2=1$.

There are largely two refrigeration systems: (1) vapor-compression system (VCS) and (2) vapor absorption refrigeration system (VARS). VARS is one of the oldest

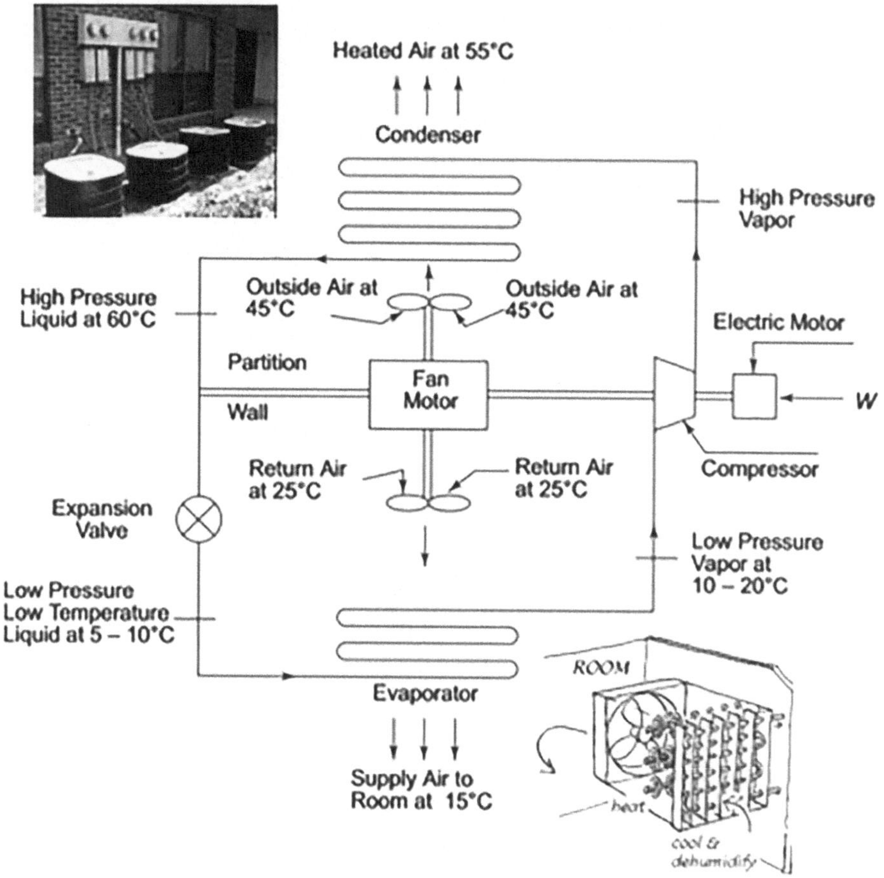

FIGURE 1.7 Schematic diagram of a typical window-type air conditioner.

methods for generating refrigeration, and it utilizes heat energy instead of mechanical energy. In the vapor absorption system, an absorber, a pump, and a generator substitute the compressor in VCS. These components in VARS carry out the same function as that of a compressor in VCS. That is, the absorber absorbs the refrigerant vapor by a proper absorbent and thus produces a rich solution of the refrigerant. Then, the pump pumps the strong solution and thus elevates the pressure of the condenser. The generator distills the vapor from the strong solution and make the solution weak, and again the cycle continues (Figure 1.8).

As customers' expectations keep increasing, manufacturers would continually develop newly designed refrigerators with new functions and features. For this (intended) function, a new mechanical module is designed and expanded to meet customers' requirements. The components in a new module should work under the span of conditions subjected to it by consumers. In this process, a new module might potentially have failures in the field, which could adversely influence the brand name of the company. Hence, a reliability testing should be conducted to identify any

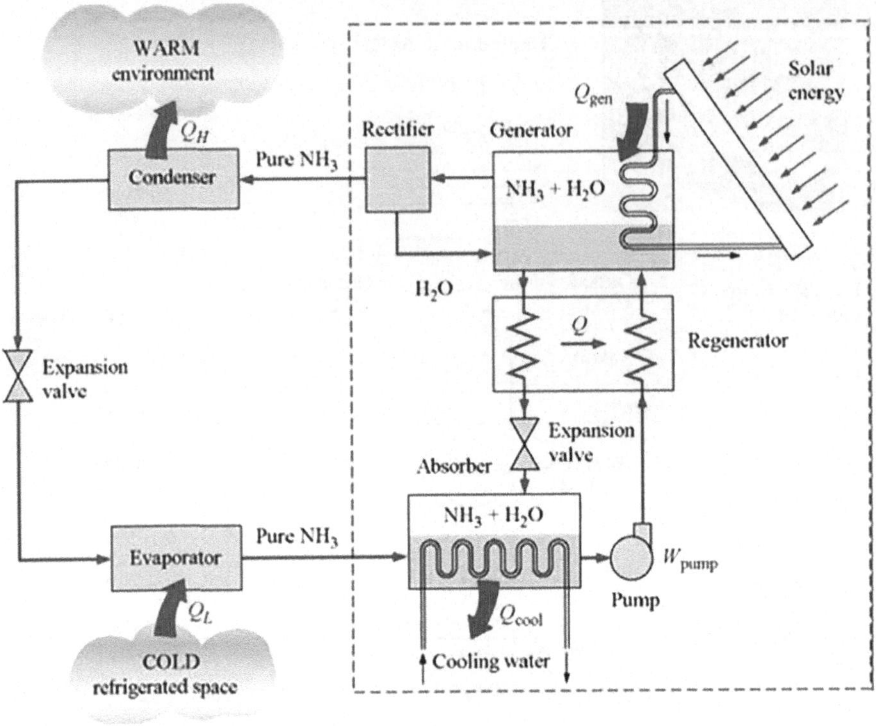

FIGURE 1.8 A simple schematic of a vapor absorption refrigeration system.

defective configurations in the new module before the company launches the product. A refrigerator is comprised of different modules such as cabinet and doors, internal fixtures (shelves and drawers), heat exchanger (condenser and evaporator), generating parts (motor or compressor), sensors and controls, and water supply device. A refrigerator might have as many as 2,000 components. The lifetime of refrigerator is targeted to have a B20 life of at least 10 years (Figure 1.9).

1.3.3 Automobiles

An automobile used for transportation is a wheeled motor vehicle. The term 'automobile' comes from the Ancient Greek term autós, meaning 'self', and the Latin term mobilis, meaning 'movable'. Over the past decades, extra features and controls have been added on vehicles, thus making them steadily more complicated but also more reliable and easier to run. These include air conditioning, rear reversing cameras, and navigation systems.

The power of an automobile is generated by an internal combustion engine that operates in the Carnot cycle, a theoretical ideal thermodynamic cycle. Then, the power is transmitted to each wheel, thus obtaining a mechanical advantage through mechanisms such as transmission and drive systems. An automobile consists of several different modules such as engine, transmission, drive, electrical, and body parts.

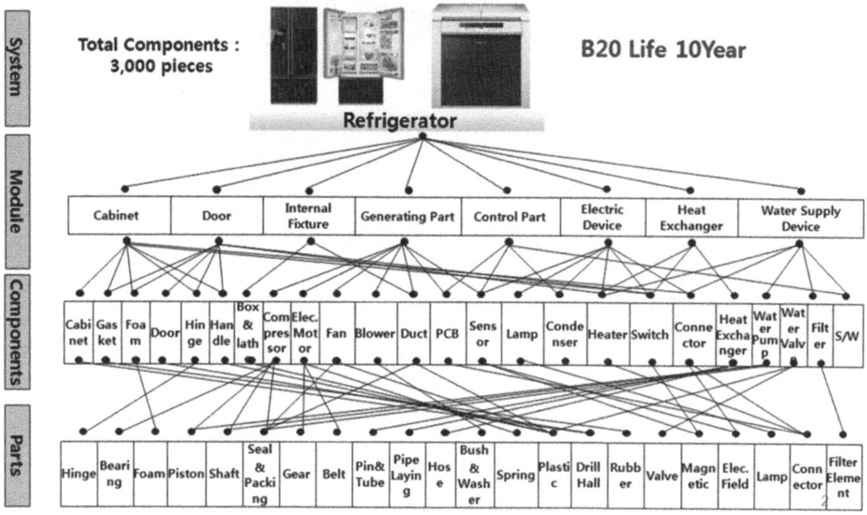

FIGURE 1.9 Structure of refrigeration system with multi-modules.

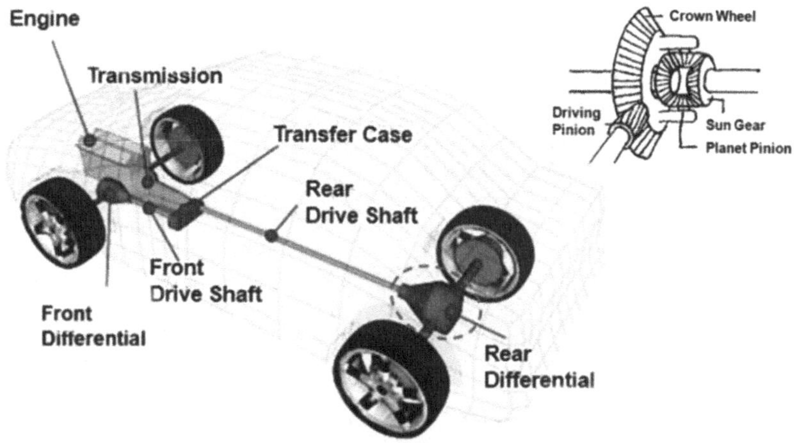

FIGURE 1.10 Automobile structure.

The total number of components in an automobile may be as many as 20,000. Hence, the total failure rate of an automobile in its lifetime is the sum of the failure rates of each module. A car's lifetime is expected to have a B40 life of at least 12 years (Figures 1.10 and 1.11).

1.3.4 Airplane

Airliners are designed to move a cargo payload or passengers. It is propelled forward by a thrust force from an engine. Especially, as the aircraft moves forward,

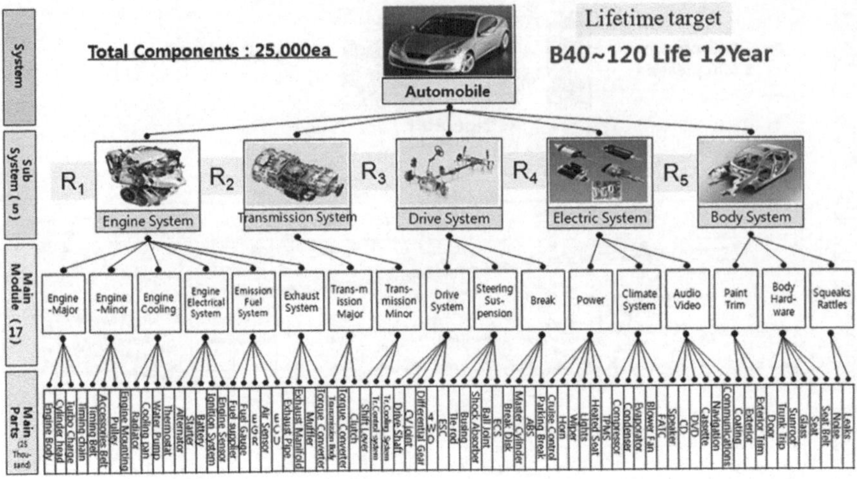

FIGURE 1.11 Breakdown of automobile with multi-modules.

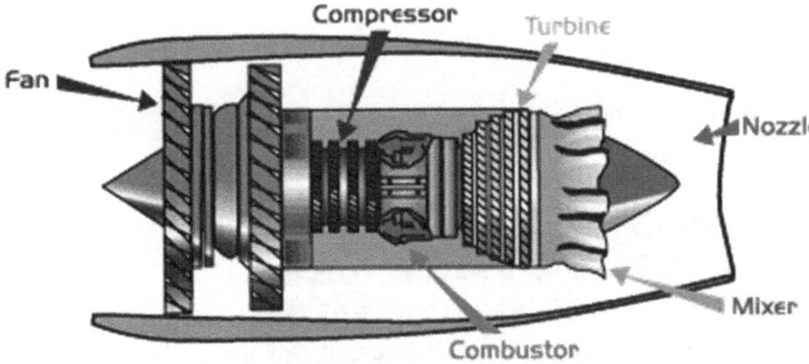

FIGURE 1.12 Jet-engine structure.

the wings deflect the air downward, thus generating a lifting force to carry it in flight that is depended on a variety of wing configurations. The term 'airplane' is derived from the French aéroplane, which comes from the Greek word ἀήρ (air), and either the Latin word 'planus', which means 'level', or the Greek word πλάνος (planos), which means 'wandering'. It also means the wing that moves through the air.

An aircraft utilizes an onboard propulsion from the mechanical power produced by an aircraft engine, which is either a propeller or jet propulsion. Especially, jet engines supply airplane thrust by taking in the air, compressing the air, and injecting fuel into the hot-compressed air mixture in a combustion chamber, and the accelerated exhaust emits rearwards through a turbine. Modern gas turbine engines are operated by the Brayton cycle that comprises a gas compressor, a combustion chamber, and an expansion turbine (Figures 1.12 and 1.13).

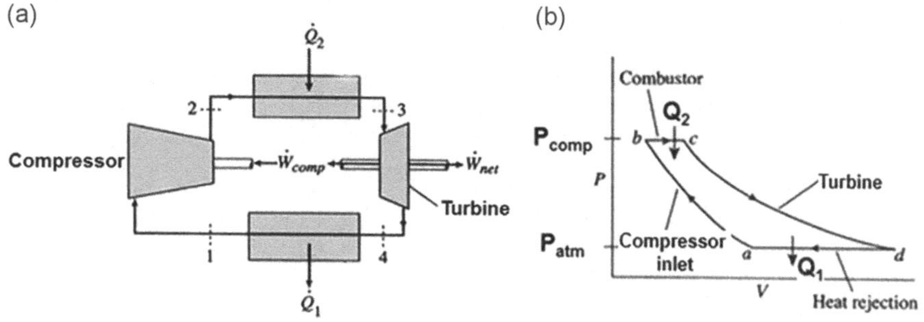

FIGURE 1.13 Brayton cycle: (a) Brayton cycle and (b) *P-V* diagram.

The mechanical structure of an aircraft is composed of approximately 200,000 components which include engine systems, aviation control systems, power delivery systems, door systems, and machine systems. Airplanes are built with multiple modules that have their own structures and mechanisms (Figure 1.14).

1.3.5 Heavy Machinery

A heavy machinery is designed for performing construction tasks and is operated by a hydraulic power, which functions through the mechanical advantage of a simple machine. By properly using hydraulic circuits, the machine transmits a hydraulic fluid throughout various hydraulic motors and hydraulic cylinders and becomes pressurized in hoses and tubes of circuits. The hierarchical configuration of a construction machine, such as an excavator, consists of an engine device, a track system, an electric device, an upper appearance system, a driving system, a main control valve unit, a hydraulic operation machine system, a cooling system, and other miscellaneous parts. Multiple modules in a product have their own mechanisms and structures. A typical appliance contains over 5,000 parts (Figure 1.15).

1.3.6 Machine Tools

A machine tool is usually a machine for forming or machining metal or other rigid materials by boring, cutting, shearing, grinding, etc. It utilizes some sorts of cutting or shaping tools. All machine tools have some ways of constraining the work piece and supply a guided movement of the parts of the machine. Therefore, the relative movement between the cutting tool (which is called the 'tool path') and the work piece is constrained or controlled by the machine to at least some extent, rather than being totally 'freehand' or 'offhand'.

The hierarchical configuration of machine tools comprises an automatic tool or pallet altering device, a drive unit, a spindle unit, a tilting index table, a hydro-power unit, a turret head, a computer numerical control (CNC) controller, a cooler unit, etc. The machine tools possess over 1,000 parts. Its reliability design will focus on the modules. The machine tools can effortlessly compute the module reliability because of the serial connection system (Figure 1.16).

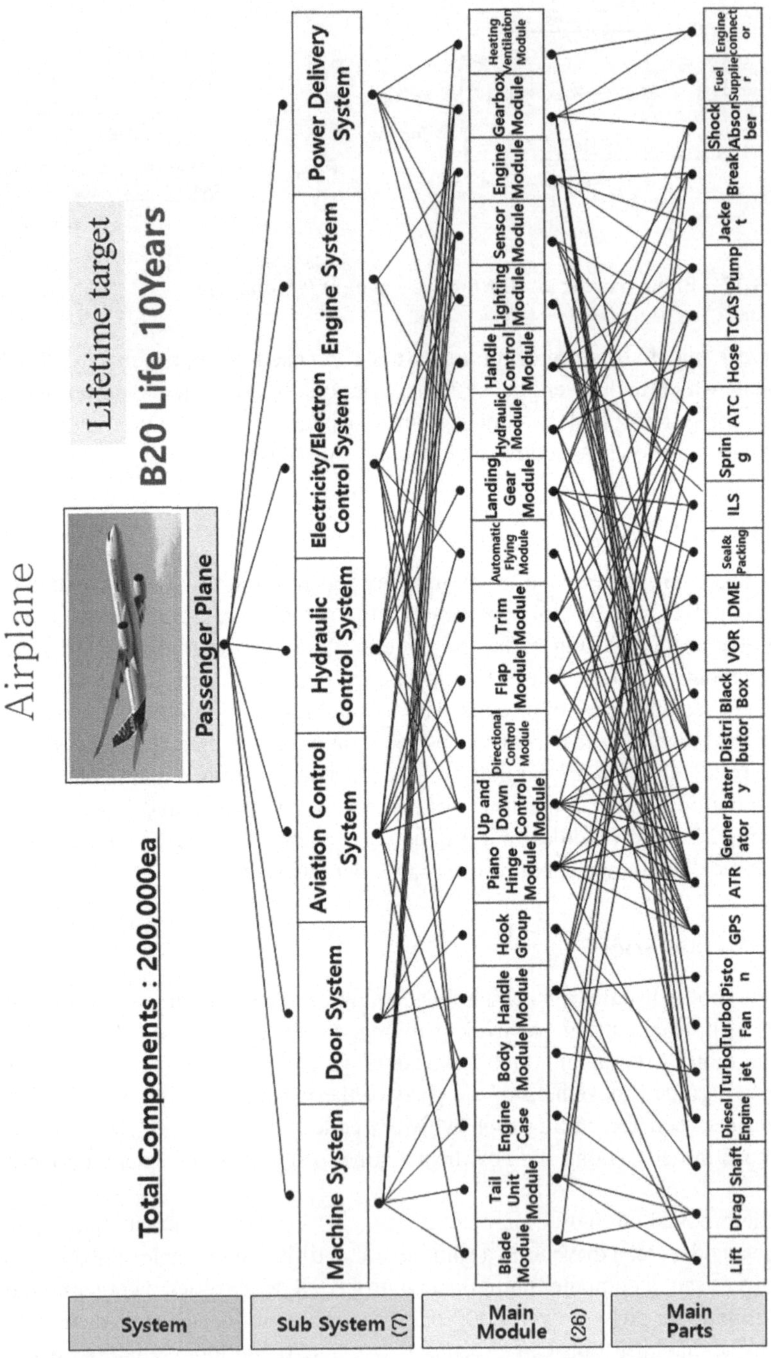

FIGURE 1.14 Structure of airplane with multi-modules.

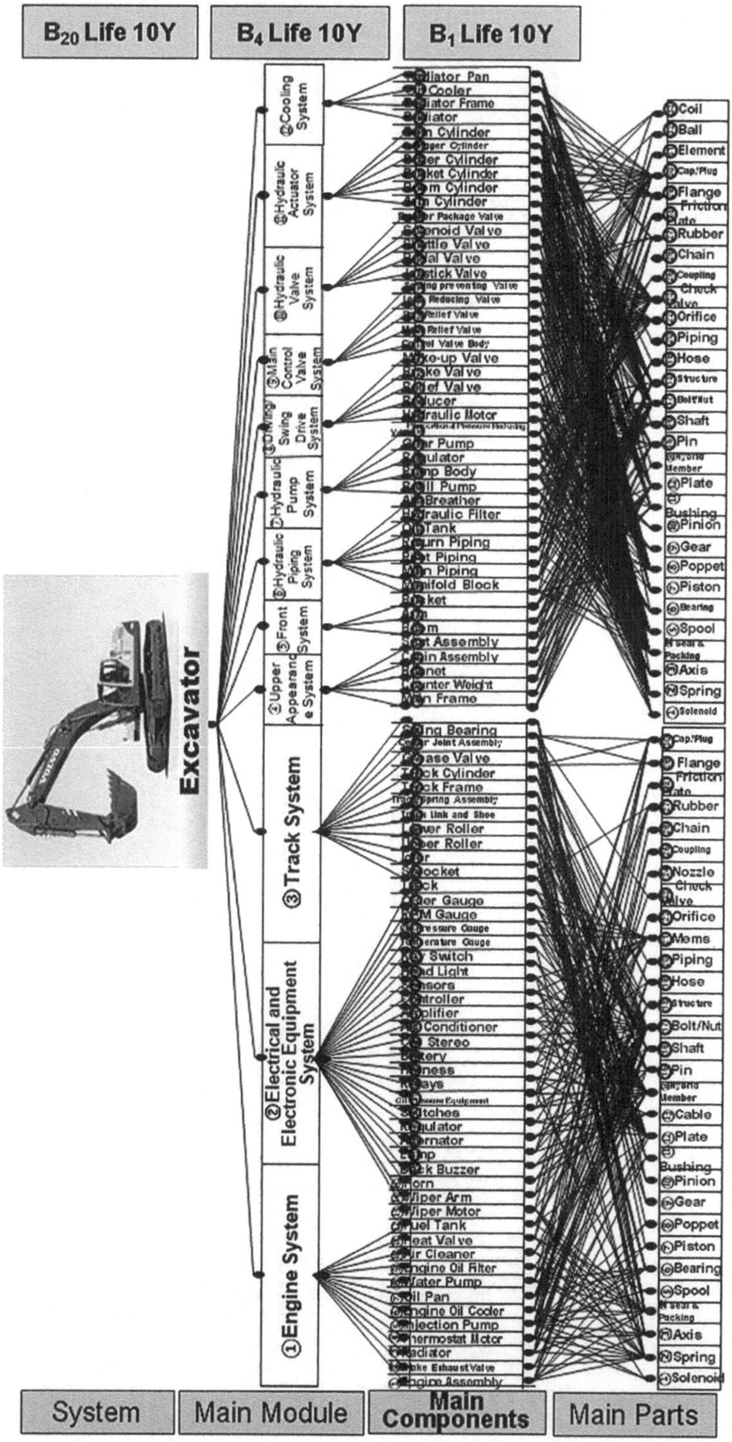

FIGURE 1.15 Structure of excavator with multi-modules.

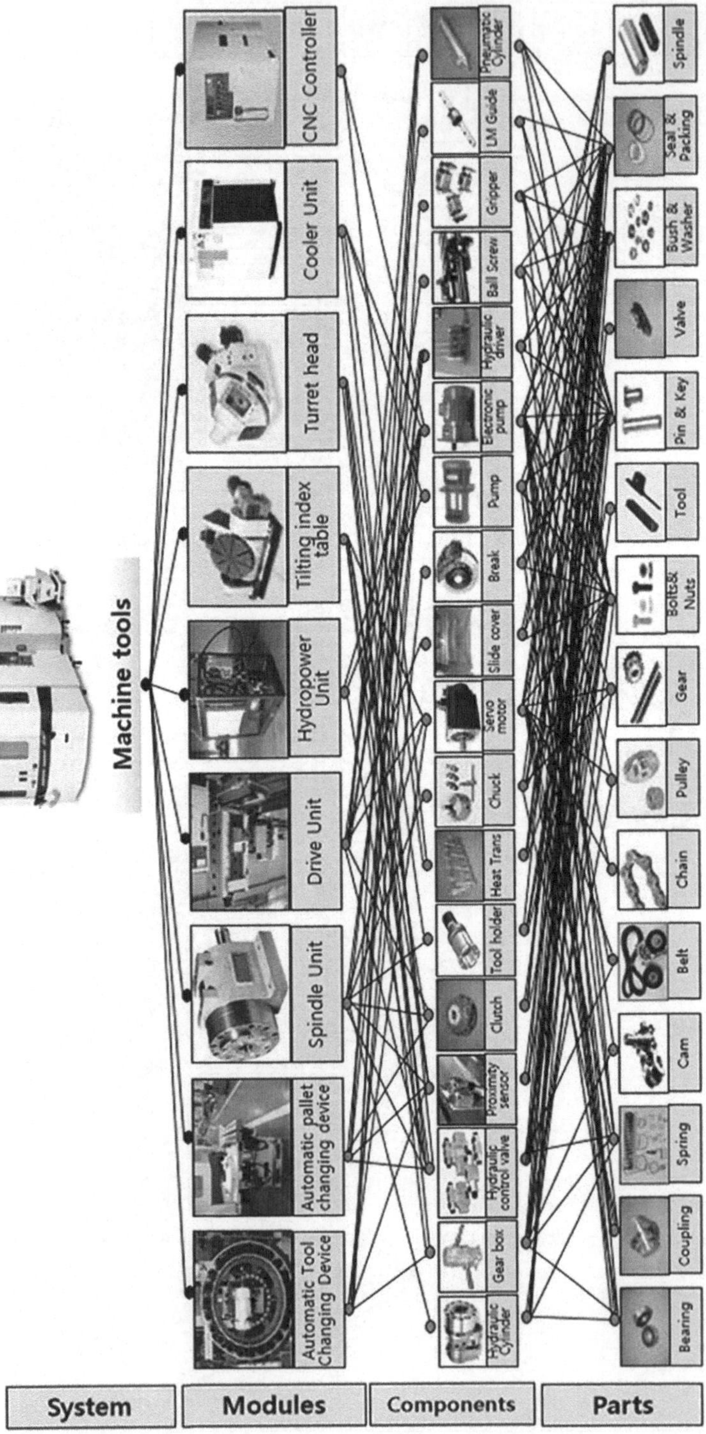

FIGURE 1.16 Structure of machine tools with multi-modules.

1.3.7 Product Design Process

The design process of a new intended function in mechanical products is formulation of a plan to help a mechanical engineer in producing a newly designed product. An engineer starts to develop a new feature of mechanical products based on the user requirements. To solve the design problems with the limited conditions, an engineer requires the inventive thoughts to develop new features that consist of research works, mechanisms, and structures.

Most mechanical systems enclose a complete process: (1) produce the customer requirements from the marketplace, (2) make conceptual and physical designs including new mechanisms, (3) make prototype, (4) demonstrate the product, and (5) assess the last product (Figures 1.17 and 1.18, Table 1.2).

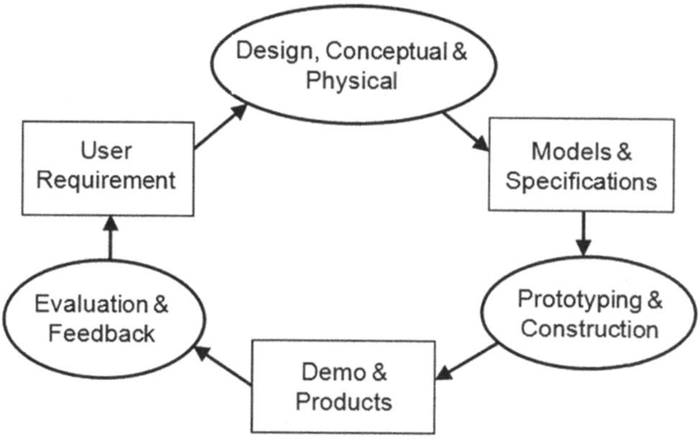

FIGURE 1.17 Iterative design overview of mechanical product.

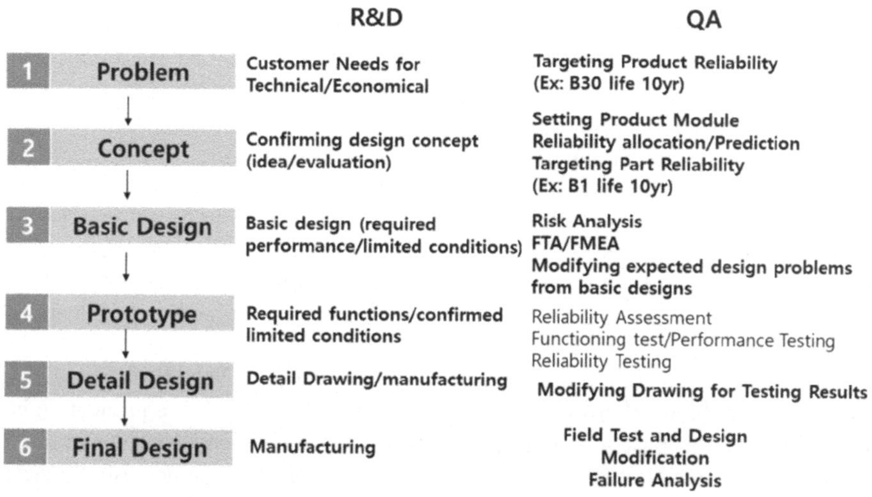

FIGURE 1.18 Design process.

TABLE 1.2
Development Tools in the Design Phase

Design Phase	Contents	Tools
Problem define	Market investigation, analysis of customer's requirement	QFD, questionnaire, patent analysis
Concept	Generalization, qualitative, trial and error	Morphology analysis, brainstorming, sketching
Detail	Analysis, quantitative, repetitive engineering calculation, optimization, numerical calculation, experiment, drawing, model	CAE, CAD, optimization, RP/3D printing, design of experiment (DOE)

Affirmed by Quality Assurance (QA) and Research & Development (R&D), a company with an established developing process often implements the customer requirements as structures including mechanisms into products. As a customer uses a product, the product is subjected to mechanical stress. If improperly designed, it will collapse in its lifetime and fail to perform its intended functions. However, there is no design methodology to assure if a mechanical system is properly designed or has enough strength and stiffness to bear its own loads. As a result, the company often faces recalls for its newly designed products. Therefore, the company requires parametric ALT as a new reliability methodology embedded in the product developing process. It comprises (1) product reliability target/allocation/prediction and (2) identifying the design problems of the problematic parts by parametric systems if lifetime target is satisfied.

1.3.7.1 Product Design – (Intended) Functions

We would establish the (intended) functions that meet the customers' requirements in a product. These intended functions are the functionalities that the product has to perform the voices of the customer and explain the company specifications. For a television, the intended function is to view programs that composes moving images. To establish superiority over other manufacturers, it is required as the fundamental advantages like the superiority of picture and sound for TV. A product like TV comprises multiple units (or modules) that can be put together as a subassembly. As a product puts an input, it gets intended functions that need to be realized as an output in response. Intended functions are implemented in the product design process, and their performance will be measured by specifications (Figure 1.19).

Intended functions might implement through the design development process. It starts to define the problems that need to be realized: (1) What is the problem to be developed? (2) How have others approached it? (3) What are the product constraints? Product design would start from the failure experiences. Companies usually state their past mistakes in the internal documents that specify the design requirements.

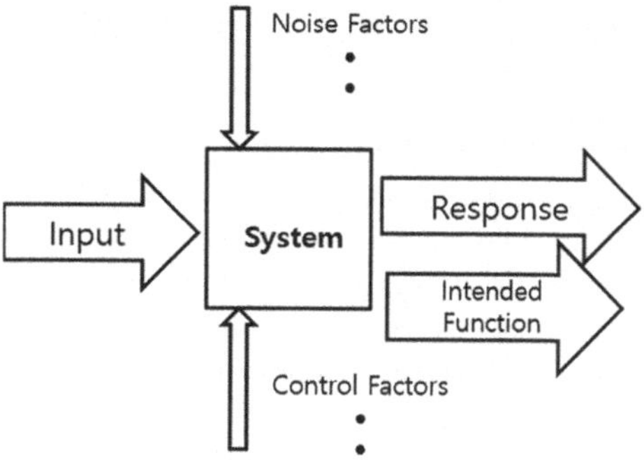

FIGURE 1.19 Intended function in parameter diagram.

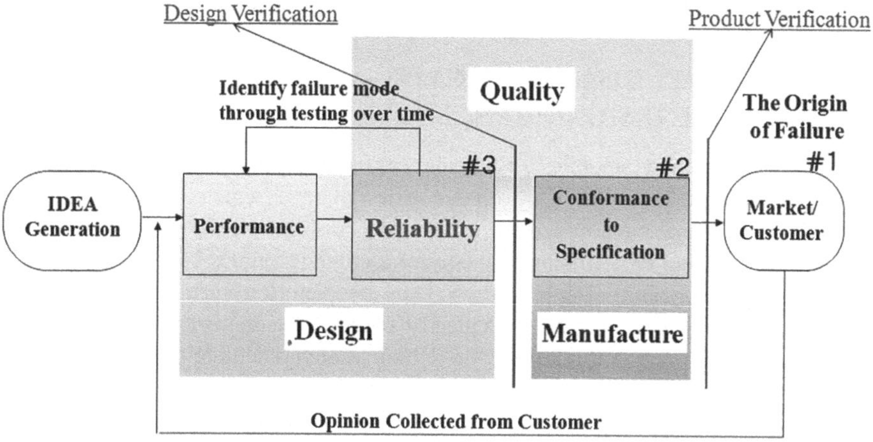

FIGURE 1.20 Realizing intended functions in the product design process.

A product's quality in serving its intended functions is an important requisite to guarantee its continued success in the marketplace. On the other hand, recall of a product due to its poor quality negatively impacts the brand name loyalty and customer satisfaction.

A prototype is an initial operating product of an intended function that fulfils the customer requirements. The product design process means multiple repetitions like redesigns of a company's solution before setting on the last design. The final product can define the performance criteria in evaluating the specifications: (a) intended functions (or fundamental advantage), (b) specified product life, and (c) the failure rate of product under operating/environmental states.

As seen in Figure 1.20, performances are to consider each of the key characteristics that consist of a variety of specifications. Reliability is to recognize the failure mode through testing over time or lifetime. A manufacturer decides whether the product complies with specifications of the production process. After production, a new concept of product decides the feedback of customers in field.

1.3.7.2 Limitations of the Traditional Design Process

The traditional developing process of products usually focuses on implementing the customer requirements, and a company produces its structure affirmed by R&D and QA. However, there is no design process that confirms whether a new architecture of mechanical products has a good quality – enough strength and stiffness. Consequently, the company often recalls a new product. According to the reliability concept, the company requires a new reliability-embedded developing process with a reliability methodology discussed in Chapters 8 and 9.

It comprises (1) product reliability target/allocation/prediction, (2) parametric accelerated life testing with accelerated factor and sample size equation, (3) searching out the design problems of the problematic parts and modifying them, and (4) proving the effectiveness through the analysis of the field failure data.

1.4 RELIABILITY DISASTERS AND ITS ASSESSMENT SIGNIFICANCE

A disaster – like an automobile recall, a nuclear plant accident, and an oil spill – is a failure of a product's intended function due to design flaws, which causes catastrophic people, and environmental or economic impacts. This disaster also has no predicting power of the society or community to control its own resources because people will recognize its significance after revelation. Thus, people often learn lessons from their past mistakes so that they can be prevented by recognizing their root causes early. For example, in 1912, we can think the RMS Titanic that was sank in the ocean injuring more than 2,200 people on-board. As the crew saw the approaching icebergs very late, they couldn't turn the ship quickly, which led the ship to hit on a floating iceberg at its right-hand side. It only took 2 hours and 40 min to sink, which costed the lives of more than 1,500 people. The reason behind this incident is the ship metal iron was too fragile to bear the iceberg impact due to the cold seawaters of Atlantic. There were no alternative plans for rescue, and also the ship sank fast within few hours (Figure 1.21).

The Titanic sinking, from standpoint of material, was primarily caused by the increased brittleness of the steel used to construct the hull of the ship due to the impact of cold seawater of the North Atlantic. Though the impact of the iceberg on the ship is small, it hence caused a large amount of damage. Consequently, the bolts holding on to the steel plates fractured, which led to the breakdown of the ship hull. Immediately after the incident of RMS Titanic, every ship was mandated to have a proper emergency evacuation plan.

Another amazing story of reliability disaster is that of the V-1 missile developed by a Germany engineering group during World War II (WWII). In spite of attempts to supply careful attention and high-quality parts, the first ten V-1 missiles ended up in a total failure. All the first developed missiles either exploded on the launching pad or finally landed 'too soon'. One of the famous mathematicians, Robert Lusser, analyzed the missile system on V-1. He derived the product

FIGURE 1.21 RMS Titanic of the White Star Line sinking around 2.20 AM, Monday morning, April 15, 1912. (Wikipedia.)

probability law of series parts – systems functioning only if all the components are working. That is, system reliability is equal to the reliabilities of the individual parts.

After WWII, the product continued to evolve as a hybrid system with multiple functions, composed of many parts like airplane, automobile, refrigerator, television sets, electronic computers, etc. Moreover, as elaborated control and devising for system safety also became more requiring and adding in their system, the chance for reliability disaster increased in market due to their complexities. As the lifetime of a mechanical product was decided by its problematic parts, it was important to assess them in the design process before the release of the final product. However, the results were not satisfactory because a systematic reliability methodology could not been developed.

For example, today commonly a Boeing 747 jumbo jet airplane is made of approximately 4.5 million parts including fasteners, multiple modules, and subsystems. An automobile is made of more than 22,000 parts, multiple modules, and subsystems. In 1935, a farm tractor was made of 1,200 parts, but in 1990, the number increased to around 2,900. Even for relatively simpler mechanical products such as bikes, there has been a significant increase in complexity with respect to parts. Consequently, the design of mechanical products such as automobiles depends on these parts. If one of them is problematic (or wrongly designed) in the field, there will be a massive recall of all mechanical systems (Figure 1.22).

As the product complexity increases when a new product structure is continually adapted to satisfy the product performance and reduce the cost, there are other chances for the inherent problematic design of the parts. According to a study carried out by the U.S. Navy concerning parts failure, 43% of the product failures were attributed to design, 30% to operation and maintenance, 20% to manufacturing, and

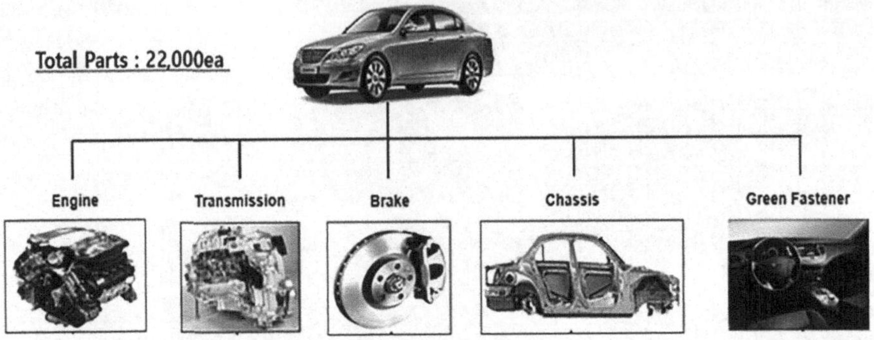

FIGURE 1.22 Structure of automobile composed of multi-modules.

7% to miscellaneous factors. While the design cost occupies only 5% of the actual cost, the cost effect due to a recall takes 70% of the actual cost.

The root cause of most structural failures is generally the negligence during application of a new design or material. When an 'improved' design is introduced in the marketplace, there are some factors that the product designer does not expect. That is, new materials (or a new design) can not only supply big advantages, but they also brings potential problems. Consequently, a new design should be changed into customer service only after a proper testing like parametric ALT. Such an approach will decrease the chance of failures and prevent them.

Thus, we list few typical examples of reliability disasters in the following:

- Space Shuttle Challenger: This disaster happened in 1986, in which all the crew lost their lives. The main cause of this disaster was design flaws – rubber O-rings under chilled winter environment in southern Texas. At that time, engineers from the booster manufacturer detected a possible problem with the O-ring seals and advised that the launch time should be detained. However, they had no evidence data to convince National Aeronautics and Space Administration (NASA) officials. Many Americans are crumpled in pride because of Space Shuttle Challenger's failure.
- Nuclear Reactor Explosion in Chernobyl: This disaster happened in 1986, in the former Soviet Union, in which 31 people died, and still this area is polluted due to the presence of radioactivity materials. The disaster was the result of design flaws such as a problematic switch in the reactor design.
- Point Pleasant Bridge Collapse: Located on the West Virginia/Ohio border, the bridge fell in 1967. This disaster resulted in the loss of 46 lives and its basic cause was the metal fatigue of a critical eye bar in the suspension bridge (Figure 1.23).
- Because welded designs are faster and cheaper than the riveted ones, earlier Liberty ships were fabricated by all-welded hulls. As measured by the Charpy impact tests, the steel used to build the Liberty ships also had poor

FIGURE 1.23 Point pleasant bridge disaster. (Wikipedia.)

FIGURE 1.24 Early liberty ships disaster. (Wikipedia.)

toughness. The Liberty ships suffered hull and deck cracks. During WWII, there were nearly 1,500 examples of notable brittle fractures. Including three of the 2,710 Liberties, twelve ships suddenly broke in half, including the SS John P. Gaines that sank on November 24, 1943 costing the lives of ten people (Figure 1.24).

- A Leak at McKee Refinery in 2007: Propane gas leaked from the McKee Refinery's Propane Deasphalting Unit in Sunray, Texas. Three labors were

FIGURE 1.25 Tanker Kurdistan disasters. (Wikipedia.)

hurt with critical burns, and the refinery was closed for 2 months. Gas prices soared 9 cents per gallon in the west.

- In 1979, the tanker Kurdistan broke completely in half while sailing off North Atlantic Ocean, Cape Breton Island. The contact between the chilled salt water and the hull with warm crude oil in the tanker created critical thermal stresses. The shipwreck happened because of a wrongly welded bilge keel. The structural detail was penetrated by a severe stress concentration on the weld which led to a crack (Figure 1.25).
- Seongsu Bridge Collapse: This disaster happened on October 21, 1994. Because of the bad welding of the steel trusses, the bridge fell down. In this accident, 32 people lost their lives and 17 people were injured. Consequently, the bridge was presumed to repair, but it had to be entirely redesigned and rebuilt. The new design was finished on August 15, 1997, and is identical to the original one (Table 1.3).

1.4.1 Versailles Rail Crash in 1842

- The Versailles rail crash happened in 1842 on the railway between Paris and Versailles. Following King Louis Philippe I's celebrations at the Versailles Palace, a train coming back to Paris derailed at Meudon. As the main train broke an axle, the carriages behind piled into it and caught fire. With about 200 deaths including that of the explorer Jules Dumont d'Urville, it was recorded as one of the earliest and deadliest railway accidents in the world. Because most of the passengers died were wearing the seat belt, the accident led to abandon the practice of locking passengers in their carriages.

TABLE 1.3
Summary of Seongsu Bridge Collapse

	Reliability Disaster
Phenomenon	
Structure	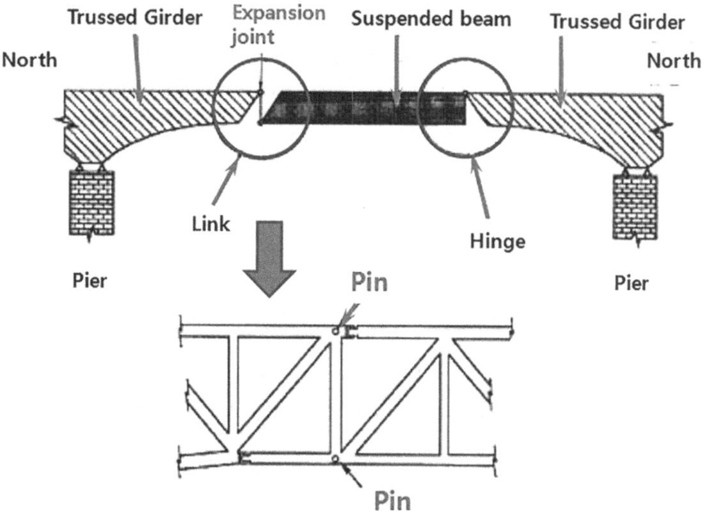
Root cause	Bad welding of the steel trusses suspended beneath the concrete slab roadway

It began the study of metal fatigue subjected to repeated loads like S-N curves (Figure 1.26).

- Root Cause: Metal fatigue on the rail was badly understood, and the accident is connected to the start of systematic research into the failure problem.

FIGURE 1.26 Versailles rail accident (1842) from Wikipedia.

1.4.2 Collapse of Tacoma Narrows Bridge in 1940

The dramatic Tacoma Narrows disaster on November 7, 1940 is mentioned as an instance of forced resonance of a mechanical system. The Tacoma Narrows Bridge was a pair of twin suspension bridges that connected the city of Tacoma to the Kitsap Peninsula. It went past State Route 16 over the strait and was collapsed by a wind-induced natural frequency. The collapse of the bridge had no loss of human life. As recorded on the Tacoma Narrows Bridge film collection, it has still been famous to architecture, engineering, and physics/engineering students as a cautionary story (Figure 1.27).

As the natural frequency of mechanical structures matches with the frequency of the external excitation like wind (vortices), resonance happens. Due to unconstrained deflections and the subsequent product failures, vibration testing has become a standard procedure in the design and development of most mechanical systems.

- Root Cause: Without any final conclusions, three feasible failure causes are presumed:
  1. Aerodynamic instability by self-induced vibrations in the bridge structure
  2. Periodical eddy vortices in bridge, matched with the natural frequency
  3. By wind velocity of the bridge, random turbulence effects like the random fluctuations.

1.4.3 Collapse of De Havilland DH 106 Comet in 1953

As the first commercial jet engine airplane, the de Havilland DH 106 Comet was replaced with the propeller airplane. People can have a transatlantic flight with this plane. The earliest original Comet plane had an aerodynamic design with four

FIGURE 1.27 Collapse of Tacoma Narrows Bridge (1940) from Wikipedia.

turbojet engines in two wings, an aerodynamic fuselage, and big square-type windows. It flew on July 27, 1949. Since 1952, it supplied a quiet and comfortable passenger cabin.

After a year in 1953, the Comets started to experience design problems and three airplanes broke up while flying. Due to metal fatigue in its airframe, a design flaw in the corners of the square windows, which was subjected to repeated stresses, was discovered in the Comet. As a result, the Comet was redesigned with oval windows, structural reinforcement, and other changes (Figures 1.28 and 1.29).

- Root Cause: Tiny cracks near the fixing nails of big square windows → repeated stress such as pressurization and decompression in airplane → spreading cracks → air blast by the broken window of airplane.

1.4.4 Company and M Company Rotary Compressor Recall in 1981

In 1981, market share and profits of G Company appliance division were declining. For instance, manufacture of a refrigerator compressor required 65-min work of labor in comparison to 25 min for that of competitors in Italy and Japan. Moreover, labor costs of G Company were higher than that of Japanese companies. The alternative was to buy compressors with a better design model from Italy or Japan. By 1983, G Company decided to design a new rotary compressor in-house along with a commitment for a new $120 million factory. G Company and a rival M Company had invented the rotary compressor technology that had been

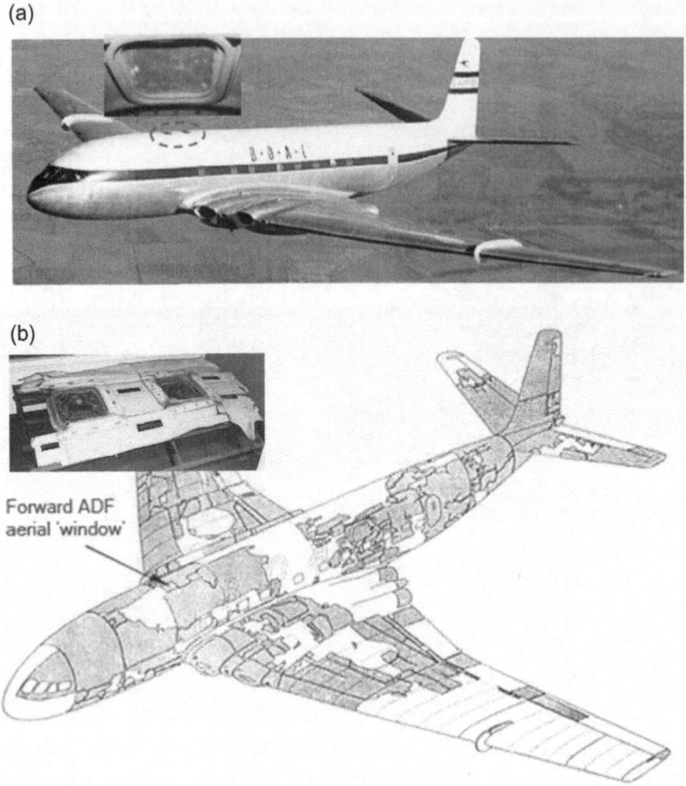

FIGURE 1.28 De Havilland DH 106 Comet (1954) from Wikipedia: (a) De Havilland DH 106 Comet and (b) Recovered parts of the wreckage of De Havilland DH 106 Comet.

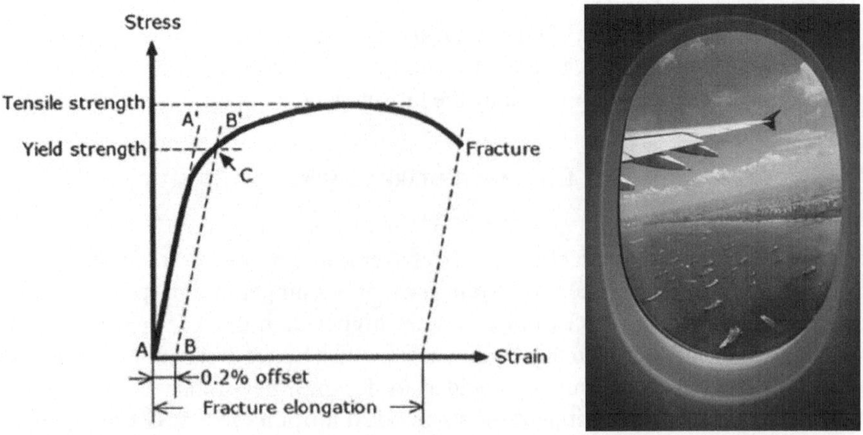

FIGURE 1.29 Stress-strain curve and the oval window in airplane.

TABLE 1.4
G Company and M Company Rotary Compressor Recall Summary

	G Company	M Company
Product	Household refrigerator	
Unit	Rotary Comp (sealed refrigerant compressor)	
Production date	1986.3	1985.1
Issued date	1987.7	1991.1
Failed cost	$450 million	$560 million
Failed amount	1.1 million	1 million
Failure mechanism	Abnormal wear out (sintered iron)	Wear out (lubrication at high temperature)
	Oil reaction/sludge imbedding	Oil reaction/sludge imbedding
User environment	Worst case	Worst case
After disaster	Withdraw Comp BIZ	Lock out factory

used in air conditioners for many years. However, this was a start for compressor recall (Table 1.4).

- Root Cause: Unexpected wear out of sintered iron under harsh operating conditions.

Because of its higher energy efficiency and reduced one-third in size of the conventional reciprocating compressors, a rotary compressor had the less weighted part. The rotary compressors occupied less space, thus supplying more room inside the refrigerator and better satisfying customer requirements. The rotary compressor for the refrigerator was almost similar to that operated in air conditioners.

However, the coolant in a refrigerator flows only one-tenth as fast and the unit operates about four times longer in 1 year than that of an air conditioner. Because this powdered metal material could exhibit much closer tolerances and reduce the machining costs, two small parts inside the compressor were made out of powdered metal rather than the hardened steel and cast iron used in air conditioners. The design engineers did not consider the critical failure in the rotary compressor until the noise claims of domestic house in 1987.

When a rotary compressor suddenly locked in 1987, M Company and G Company faced massive recalls for the rotary compressor. As the (generated) oil sludge stuck in the capillary tube in the refrigeration system, the cooling capacity of the refrigerator abruptly decreased. In the compressor development process, reproducing this failure mode and preventing the blocking of this tube were critical to improve the reliability of the refrigerator. However, reliability testing methods such as the parametric ALT was not used at that time.

1.4.5 Firestone and Ford Tire in 2000

In the start of the 2000s, the Ford Tire and Firestone faced extraordinary tire failures on the Ford Explorer furnished with Firestone tires. The Ford Motor Company had a

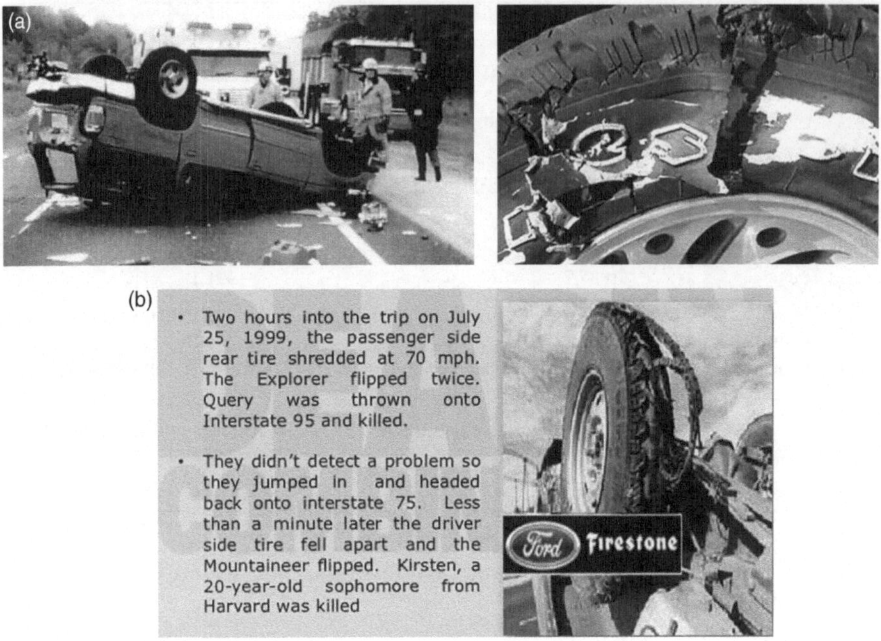

FIGURE 1.30 Ford and Firestone tire controversy from Wikipedia: (a) Ford explorer and Firestone tires and (b) Timeline of Firestone fallout.

close relationship with Firestone. As Firestone became a subsidiary of Japanese tire manufacturer Bridgestone in 1988, they drifted apart. In May 2000, the U.S. National Highway Traffic Safety Administration (NHTSA) contacted Ford and requested the high chance of Firestone tire failure on Ford Explorer's model. Soon Ford caught that it had extreme failure rates from 15-inch Firestone tire models (Figure 1.30).

Including 2.8 million Firestone Wilderness AT tires, millions of Firestone tires were recalled. Because there had been over 3,000 catastrophic injuries and 240 deaths, a large number of lawsuits have been filed against both Ford and Firestone. Most of the accidents due to separating a kind of tire tread when Ford corners on cloverleaf interchange in high speed.

- Root Cause: Eliminate tire air (minor design change) → tire heat up → damaged tire → interaction of steel and rubber tire → separation of tread.

1.4.6 Toshiba Satellite Notebook and Battery Overheating Problem in 2007

As more than 100 cases were filed against about 41,000 Toshiba laptops due to melting issues and minor injuries, Toshiba notebook fell into massive recalls in 2007. The root cause was overheat which produced a burning effect to consumers. Heat generated as processors and batteries run. Therefore, laptops were designed to provide adequate airflow with a fan and eliminate overheat from the case. However, to satisfy the requirements of

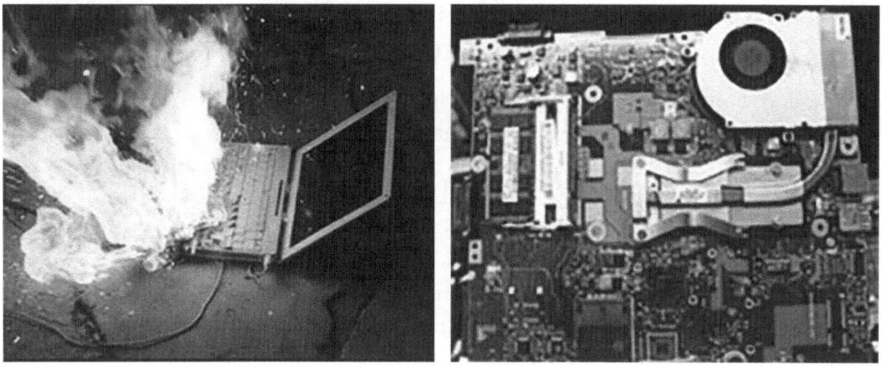

FIGURE 1.31 Toshiba Satellite T130 notebook and battery-overheating problem.

less weight, and slim and compact design, notebooks will urge to push heat-generating components including the micro-process into a compact space (Figure 1.31).

- Root Cause: urging so much processing power and battery into such a compact space (design problem).

1.4.7 Toyota Motor Recalls in 2009

Due to the pedal entrapment/floor mat problem in about 5.2 million vehicles, the recalls of Toyota Motor occurred in 2009, and especially, in 2.3 million vehicles, it is caused by the accelerator pedal problem. As Toyota widened the recalls to include 1.8 million vehicles in Europe and 75,000 in China, the total recalls of Toyota cars in the world were 9 million. The NHTSA concluded that pedal misapplication was responsible for most of the accidents (Figure 1.32).

- Root Cause: the pedal entrapment/floor mat problem.

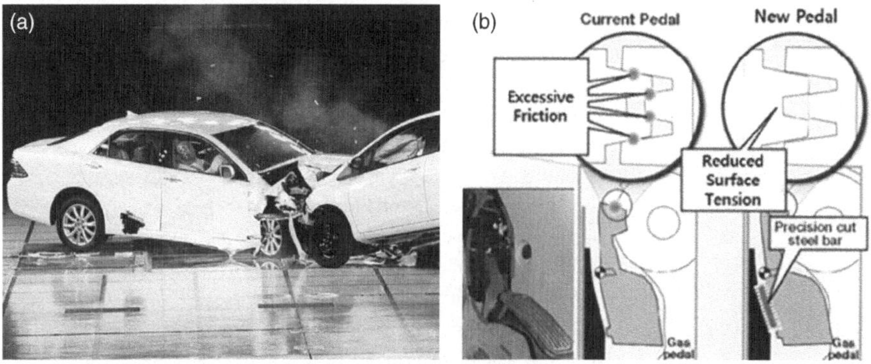

FIGURE 1.32 Recalls of automobiles by Toyota Motor Corporation: (a) Toyota crash and (b) pedal design problem.

1.5 HISTORICAL REVIEW OF DEVELOPMENT OF RELIABILITY METHODOLOGIES

The modern concept of reliability started in 1816. Reliability in statistics at that time was defined as the measurement consistency to define a test. That is, a test is reliable if the identical result is repeated. As an attribute of a product, reliability was a general concept that had been recognized. We summarize the historical milestones in the early twentieth century in Table 1.5.

In 1842, there were rail accidents that often happened in France Versailles. August Wöhler studied the root causes of fracture in railroad axles and began the earliest systematic studies of S-N curve (or Wöhler's curve) [1,2]. To prevent the railroad accidents, S-N curves of materials can be utilized to minimize the fatigue incidents by dropping the critical stress in a part.

During World War I (WWI), Griffith studied fracture mechanics to tell the failure of brittle materials. He suggested the first law of thermodynamics to formulate a fracture theory based on a simple energy balance. As a flaw in material is unstable for (dynamic) loading, fracture starts to occur when the strain-energy change is enough sufficient to endure the surface energy of the material. He also suggested that

TABLE 1.5
History Summary of Reliability Technology

~1950	–	Wilhelm Albert publishes the first article on fatigue (1837)
		A. Wöhler summarized fatigue test results on rail-road axles (1870)
		O.H. Basquin proposes a log-log relationship for S-N curves (1901)
		John Ambrose Fleming invented vacuum tubes in 1904
		Griffith's theory of fracture (1921)
		A.M. Miner introduces a linear damage hypothesis (1945)
WWII	Germany	V-I and V-II Rocket development (R. Lusser's law)
WWII	US	Reliability of the electron power tube (aircraft electronic devices failure in the WWII)
1954	Japan	Surveys and studies on the electron power tube reliability in the Vacuum Committee of the Institute of Electrical Engineers
1952–1957	US	US DOD formed the Advisory Group on the Reliability of Electronic Equipment (AGREE)
		AGREE suggested vacuum tube to follow the bathtub curve
1954	US	First National Symposium on Reliability and Quality Control, New York
1950s	US	Several conferences began to focus on various reliability topics (e.g., 1955 Holm Conference on Electrical Contacts)
1961	Italy	The Rome Air Development Center (RADC) introduced a PoF program
1962	US	Launched the Apollo program (FMEA and FTA)
		First reliability and maintainability conference
1962	US	First Symposium on Physics of Failure in Electronics, Chicago
1965	IEC	Reliability and Maintainability Technical Committee, TC 56, Tokyo
1968	–	Tatsuo Endo introduces the rain-flow cycle count algorithm
1971	Japan	First reliability and maintainability symposium

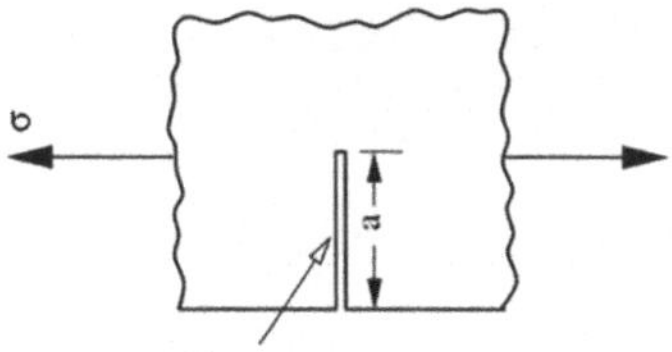

FIGURE 1.33 An edge crack (flaw) of length a in a material.

the low fracture strength found in tests was due to the existence of microscopic flaws in the bulk material that can be still effective (Figure 1.33) [3].

$$\sigma_f \sqrt{a} \approx C \tag{1.5}$$

where σ_f is the failure stress.

Failure happens when the free energy pertains to a peak value at a critical crack length.

$$C = \sqrt{\frac{2E\gamma}{\pi}} \tag{1.6}$$

where E is Young's modulus of the material and γ is the surface energy density of the material.

Vacuum tubes, invented in 1904 by John Ambrose Fleming, were a critical part of electronic instruments such as television, diffusion of radio, radar, sound recording and reproduction, sound reinforcement, analog and digital computers, large telephone networks, and industrial process control. As of the vacuum tube was invented, it made modern technologies of the product applicable. Use of radio with vacuum tubes began in the public in 1916, and the reliability concept by the problematic vacuum tubes started to develop (Figure 1.34).

In 1895, Karl Pearson stated the concept of 'negative exponential distribution'. His exponential distribution had lots of amazing properties that were applicable in the 1950s and 1960s. In other words, one property of the serial system is the ability to add on failure rates of different parts in products. Simply adding it was more applicable at the time when using mechanical and later electric systems:

$$R(t) = R_1(t) \cdot R_2(t) \cdots R_n(t) \tag{1.7}$$

$$R(t) = e^{-\lambda_1 t} \cdot e^{-\lambda_2 t} \cdots e^{-\lambda_n t} \tag{1.8}$$

$$R(t) = e^{-(\lambda_1 + \lambda_2 + \cdots + \lambda_n)t} \tag{1.9}$$

where R is the reliability function, λ is the failure rate, and t is the use time.

As automobiles came into common use in the early 1920s, Walter A. Shewhart at Bell Laboratories introduced product improvement by the statistical quality control.

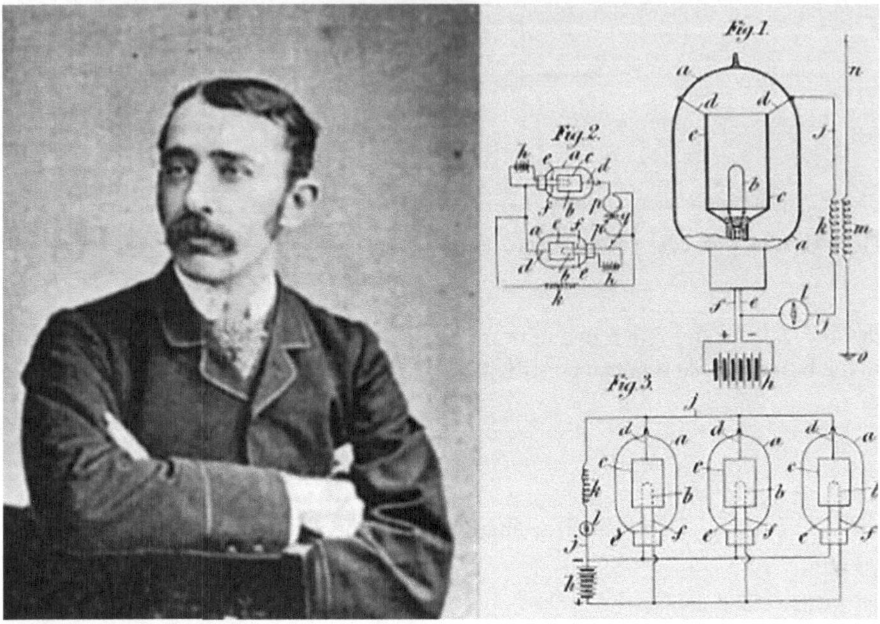

FIGURE 1.34 British engineer John Ambrose Fleming and his vacuum tubes patents [4].

In 1924, he introduced the control chart and the concept of statistical control. Statistics as a measurement tool was combined with the development of reliability concepts. While designers were responsible for product quality and reliability, technicians took care of the failures. In the 1930s, quality and process measures in automobile were still growing.

In a military short lecture, W. Edwards Deming in the 1940s mentioned the management's responsibility for quality. He stated that quality problems occur actually due to design errors, not due to workers' mistakes [5]. For example, an inceptive reliability concept was applied to the spark transmitter telegraph due to the uncomplicated design. It was a battery-powered system connected with simple transmitters by wire. The leading failure mode was a broken wire or insufficient voltage. After WWI, this system was changed with considerable ungraded transmitters based on vacuum tubes.

Before WWII, there were still no ideas in reliability engineering. However, many new electronic products such as vacuum tube portable radios, electronic switches, electronic detonators, and radar were introduced into the army during WWII. As the war started, it was found that half of the airborne electronic apparatus in store was failed in lifetime and unable to satisfy the army requirements (the Air Core and Navy). Reliability effort for this time had to do with a new (metal) material testing. The root causes for the failures were only the products' fatigue or fracture. For example, M.A. Miner published a seminal paper titled 'Cumulative Damage in Fatigue' in 1945 in *ASME Journal*. B [6]. In 1948, Epstein published a paper titled 'Statistical Aspects of Fracture Problems' in the *Journal of Applied Physics* [7].

During WWII, under Wernher von Braun developing the V-1 missile, one of the Germany research groups was collaborating for better designs. After the end of WWII, he reported that first V-1 missiles were totally non-success. Although they attempt to supply high-quality components and pay careful attention in details, all first missiles either exploded on the pad or landed 'too soon'. By developing the reliability theory, they had to find solutions.

During WWII, Germany applied the basic reliability concepts to upgrade the reliability of their V1 and V2 rockets with multi-modules. To finish the assignment of VI and VII rockets, German engineers had to make better the reliability of VI and VII rockets. Robert Lusser, a mathematician, was called in as one of the consultants. His work was to examine the missile system, and he shortly elaborated the product probability law for series parts. This theorem is to define systems functioning only if all parts are working and is valid under special assumptions. That is, he defines that the reliability of a series system is equal to the product of the part reliability. In other words, a series system is 'weaker than that of its weakest link', as the product reliability of a series of parts can be less than the lowest-reliability part. If the system consists of a large number of parts, the system reliability may therefore be rather low, even though each part has high reliabilities. The product lifetime, therefore, is decided by the weakest chain (or module) and extended from its design improvement. After WWII, the Department of Defense in the United States severely acknowledged the inevitability for reliability of its military apparatus. It became the theoretical basis of MIL-HDBK-217 and MIL-STD-756. However, the reliability engineering was yet completely undeveloped because of the restrictions on the understanding of the design, failure mechanism, and reliability testing.

In the beginning of the 1950s, the principal military application areas of reliability design were the vacuum tube in radar systems, military equipment, or other electronics because these systems were troublesome as in WWII. The vacuum tube computers that had a 1024-bit memory were invented to take a huge space and use kilowatts of power, though very ineffective in the modern computer. During WWII, lots of airplanes were wrecked during bombarding. The vacuum tubes installed in these airplanes had been demonstrated as the troublesome components, though the part price was not expensive. After the war, half of the electronic equipment for shipboard was unsuccessful in achieving its lifetime. The root cause of most failures also was the vacuum tubes. Failure modes of vacuum tubes in sockets were irregular working problems. The action plans for a problematic electronic system were to break down the system, eliminating the tubes, and re-mounting them at a proper period.

In the modernization process, because the army should think about the cost matters, the operation and logistics costs for the vacuum tubes would become huge. To work out the problem of vacuum tubes, Institute of Electrical and Electronic Engineers (IEEE) in 1948 organized the Reliability Society. In 1948, Z.W. Birnbaum had established the Laboratory of Statistical Research at the University of Washington, which sufficed to utilize the concept of statistics. In 1951, Air Development Center (RADC) was found in New York and Rome to study reliability problems with the Air Force Rome.

In 1950, a study group in army was commenced, which was thereafter called the Advisory Group on the Reliability of Electronic Equipment (AGREE). By 1959, this group in its result report proposed the following three suggestions for the reliable systems such as vacuum tube: (1) there was a necessity to develop reliable parts for supplier, (2) the army should initiate quality and reliability specifications for component suppliers, and (3) real field data should be gathered on parts to find out the root causes of problems.

A close meaning of modern product lifetime originated in 1957 AGREE Commission Report. Task Group 1 in AGREE has evolved minimum-acceptable units for the reliability of various types of military electronic apparatus, demonstrated in terms of Mean Time Between Failures, though describing lifetime for electronic components was currently improper. The last report of AGREE committee proposed that the reliability of the products such as vacuum tube followed the bathtub curve. Consequently, reliability of parts is usually expressed in 'the bathtub curve', which shows trends as early failure, useful life, and wear-out failure. Now the cumulative distribution function corresponding to a bathtub curve is replaced with a Weibull chart in reliability engineering (Figure 1.35).

AGREE committee also suggested to officially testing products with a statistical confidence level. It would perform the environmental tests under extreme temperature and vibration conditions, which changed as Military Standard 781. The AGREE report at first defined the reliability as 'the probability of a product performing without failure a specified function under given conditions for a specified period of time'.

In the start of the 1950s, a conference on electrical contacts and connectors was commenced to investigate the reliability physics – failure mechanisms and reliability topics. In 1955, Joseph Naresky in RADC published 'Reliability Factors for Ground Electronic Equipment' [8]. The proceeding entitled as 'Transaction on Reliability and Quality Control in Electronics' was presented in a conference, which was integrated with an IEEE Reliability conference and grew the Reliability and Maintainability Symposium.

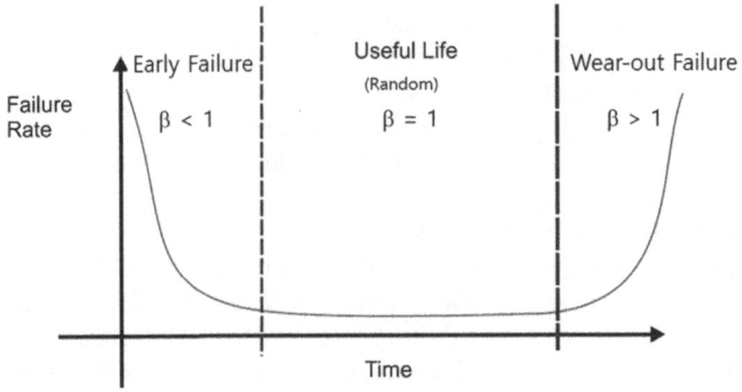

FIGURE 1.35 Bathtub curve for vacuum tube radio systems.

In the 1950s, as television usage was started at American homes, more vacuum tubes were already used. Due to the failure of one or more vacuum tubes, repair problems usually occurred. One of the key switching devices in TV was vacuum tube, which controls electric current through a vacuum in a sealed container – cathode ray tube. Representative reliability issues developed in the tubes with oxide cathodes are as follows: (1) decreased ability to release electrons, (2) a stress-related fracture of the tungsten wire, (3) air leakage into the tube, and (4) glowing plate – a sign of an overloaded tube. Most vacuum tubes used in radio systems that followed a bathtub-type curve were easy-to-develop, replaceable electronic modules – Standard Electronic Modules (SEMs) and then reinstate a failed system.

In a 1957 report titled 'Predicting Reliability', Robert Lusser in Redstone Arsenal stated that 60% of the failures of one army missile system were due to its parts. He also emphasized that contemporary quality methods for electronic parts were improper and new ideas for electric parts should be executed. Aeronautical Radio INC (ARINC) planned to upgrade the suppliers of vacuum tubes and reduced early mortality removals by a factor of four. This period of 10 years ended up with Radio Corporation of America publishing information in TR1100 on the failure rates of some army parts. RADC utilized these ideas as the fundamentals for Military Handbook 217. Over the next several decades, Birnbaum suggested on Chebychev's inequalities, reliability of complex systems, non-parametric statistics, competing risk, cumulative damage models, survival distributions, and mortality rates.

Walodie Weibull in Sweden investigated on the fatigue of materials and immediately suggested a Weibull distribution. Weibull in 1939 proposed an easy mathematical probability distribution, which could show a wide span of failure attributes by altering two parameters. Though his failure distribution would not be applicable to every failure mechanism, it is a useful tool to examine the lots of the reliability problems. With seven case studies, his most famous papers in 1951 were suggested to the American Society of Mechanical Engineers (ASME) on the Weibull distribution. Between 1955 and 1963, he conducted studies on the creep and fatigue mechanisms of materials. He also suggested the Weibull distribution based on the weakest chain model of failures in materials. In a Wright Air Development Center Report 59-400 for the US military, he produced 'Statistical Evaluation of Data from Fatigue and Creep Rupture Tests: Fundamental Concepts and General Methods' in 1959 [9].

While working as a consultant for the US Air Force Materials Laboratory in 1961, he issued one book on materials and fatigue testing [10]. In 1972, the ASME granted their gold medal to Weibull. King Carl XVI Gustaf of Sweden personally awarded the Great Gold Medal from the Royal Swedish Academy of Engineering Sciences to him in 1978.

As his analysis methods and applications were spreading worldwide, many people began to utilize the Weibull chart. In the late 1950s, Dorian Shainin wrote down the earliest booklet on Weibull, while Leonard Johnson in General Motors in 1964 could make better the plotting techniques by proposing median ranks and beta Binomial confidence bounds. Professor Gumbel stated that the Weibull distribution is a Type III Smallest Extreme Value distribution such as Equations (1.10) and (1.11).

For the Weibull function, Dr. Robert Abernethy suggested many implementations, investigation methods, and rectifications [11].

$$F(x) = \exp\left(-\left(\frac{a-x}{b}\right)^c\right) \tag{1.10}$$

where $x \leq a$, $b > 0$, $c > 0$,

$$K = A\exp\left(-\frac{E_a}{RT}\right) \tag{1.11}$$

where k is the rate constant of a chemical reaction, T is the absolute temperature, A is the pre-factor, E_a is the activation energy, and R is the universal gas constant.

While visiting an Institute in Columbia as a professor for the Study of Fatigue and Reliability in 1963, Weibull collaborated with professors Gumbel and Freudenthal. As a consultant for the US Air Force Materials Laboratory, he issued the associated reports and a book on materials and fatigue testing until 1970.

After WWII, as manufacturers produced more complicated products composed of an ever-increasing number of parts (radio, television, computers, etc.), the development continued throughout the world. As introduction of automatic equipment in manufacturing is increasing, the need for complex control and safety systems is also becoming more crucial.

In the 1960s, as the transistor in 1947 and transistor radio in 1954 invented, a lot of failure occurred in the field. To improve it, Physics of Failure (PoF) for electronic parts began, which was the most well-liked electronic transmission device for the 1960–1970s. People enjoyed music everywhere using pocket-size radios. These tools had some difficulties such as transistor failure, capacitor problems, and electromechanical faults. To prevent these failures, PoF was used as a structured method to the design and development of reliable products. By understanding of the root causes of the failures, the products can be made with better performance (Figure 1.36).

In one Electronics Conference financed by Illinois Institute of Technology (IIT), RADC earnestly functioned the PoF. In the 1960s, NASA was established to achieve America's forceful dedication to space exploration. Their key efforts were to make the reliability of parts and systems better that could function satisfactorily to finish the space commissions. RADC issued the document 'Quality and Reliability Assurance Procedures for Monolithic Microcircuits'. Semiconductors were popular in compact movable transistor radios. Next, cheap price germanium and silicon diodes could satisfy the requirements. Through the Institute of Radio Engineers (IRE) Dr. Frank M. Gryna issued a Reliability Training Book.

In this time, as the nuclear power industry and airplanes, helicopters, missiles, and submarine in the army utilized, the reliability issues of various technologies mounted in their product started PoF. The effects of the Electro-Magnetic Compatibility system were studied at RADC in the 1960s (Figure 1.37).

As one of the milestones, the success of the Arrhenius model for semiconductors was proved in the Proceedings of the Seventh National Symposium of Reliability and Quality Control in 1962. G.A. Dodson and B.T. Howard in Bell Labs published the

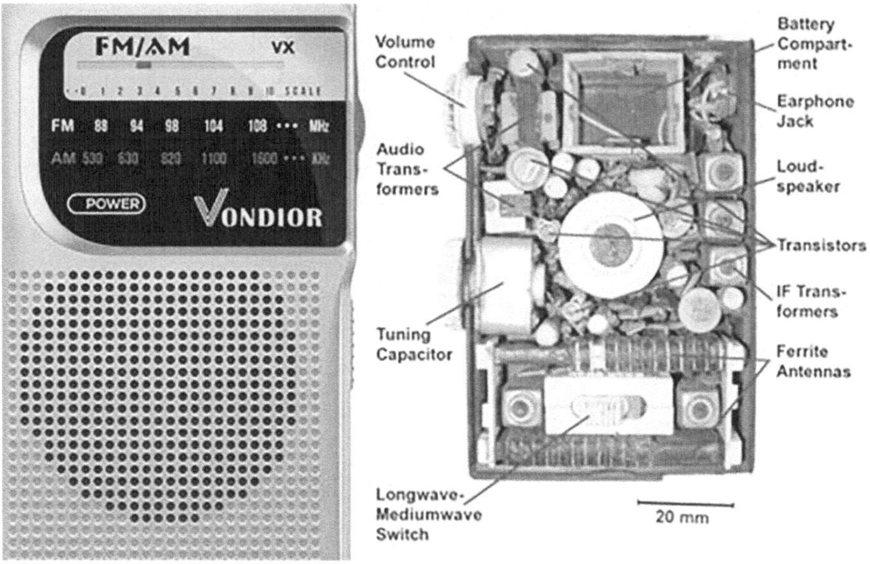

FIGURE 1.36 A transistor radio with multiple parts. (Wikipedia.)

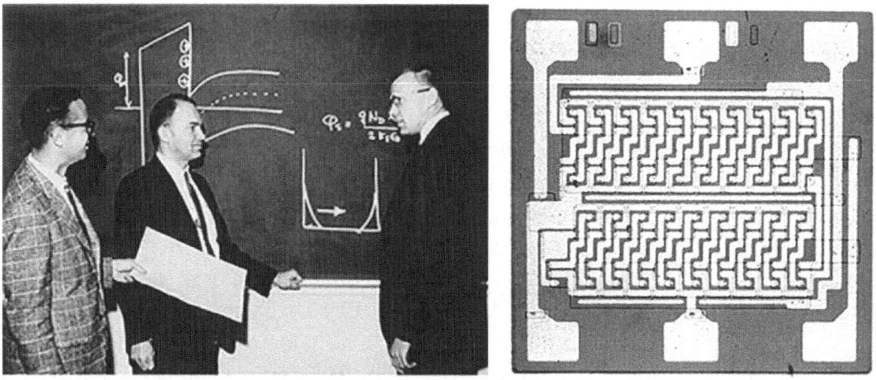

FIGURE 1.37 Andy Grove, Bruce Deal, and Ed Snow at the Fairchild Palo Alto R & D laboratory and first commercial Metal Oxide Semiconductor (MOS) IC in 1964. (Wikipedia.)

paper entitled 'High Stress Aging to Failure of Semiconductor Devices' [11]. This conference also published many other papers and improved the reliability of other electronic parts. It then was renamed as the Reliability Physics Symposium (RPS) in 1967. When applied to Integrated Circuits (ICs), Shurtleff and Workman in the late 1960s published a paper on step stress testing that sets up its limits.

When the metals are pressurized at high current densities, electro-migration in electronic part is one of the critical failure mechanisms that can be applicable to the transportation of mass in metals. On the physics of electro-migration, J.R. Black issued his work in 1967. As the quantity of free-charge carriers is growing with temperature, silicon in semiconductor has started to dominate reliability activities in a variety of industries. The U.S. Army Material Command published a Reliability Handbook (AMCP 702-3) in 1968. On the other hand, Shooman's Probabilistic Reliability was also published to describe statistical methods.

Automotive business published a FMEA handbook for technical development of suppliers to investigate the failure mode of electronic parts, not yet issued as a Military standard. As many commercial satellites were launched, International Telecommunications Satellite Organization (INTELSAT) that was supplying international broadcast services between the United States and Europe in 1965 also intensified the reliability study for communications. World experts took part in reliability conferences. As Apollo was landing on a moon, people acknowledged how far reliability had progressed in the new decade.

$$f(t) = \frac{1}{2\mu^2\gamma^2\sqrt{\pi}}\left(\frac{t^2-\mu^2}{\sqrt{\frac{t}{u}}-\sqrt{\frac{u}{t}}}\right)\exp\left[-\frac{1}{\gamma^2}\left(\frac{t}{\mu}+\frac{\mu}{t}-2\right)\right] \tag{1.12}$$

where γ is a shape parameter and μ is a scale parameter.

As seen in Equation (1.12), Birnbaum and Saunders in 1969 provided a life distribution model that could be obtained from a physical fatigue procedure where crack growth leads to failure. Since one of the leading methods to select a life distribution model is to obtain it from a physical/statistical argument that is compatible with the failure mechanism, the Birnbaum-Saunders fatigue life distribution is worth taking into consideration.

After the microcomputer was designed in the 1970s, RAM dimensions were rapidly growing. ICs were substituted by vacuum tubes. The variations of ICs such as NMOS, Bipolar, and CMOS rapidly increased. In the middle of the 1970s, Electrostatic Discharge (ESD) and Electrical Over Stress (EOS) were issued by some papers and finally turned the debating issues of a conference in the end of the decade.

In the same way, in International Reliability Physics Symposium (IRPS), studies for passive parts such as resistor, capacitor, and inductor passed to a Capacitor and Resistor Technology Symposium (CARTS). The forward-thinking papers on gold aluminum inter-metalics, accelerated testing, and the use of Scanning Electron Microscopes (SEM) were in a few climax of the decade.

Based on field data, Hakim and Reich in the middle of 1970s issued a paper on the assessment of plastic-encapsulated transistors and ICs. In addition, two most significant reliability papers were published: one on soft-errors from alpha particles first issued by Woods and May, and the other on accelerated testing of ICs with activation energies calculated for a variety of failure mechanisms by D.S. Peck. Bellcore in the end of the decade gathered commercial field data and became the basis of the Bellcore reliability prediction methodology utilized extensively with MIL-STD-217F.

Toward the end of the 1950s and the beginning of the 1960s, interest of the United States was focused on space research and intercontinental ballistic missiles, chiefly joined to the Gemini and Mercury programs. In the competition with Russia, the launching of a manned spacecraft was critical. During the Apollo space program, the spacecraft and its parts reliably functioned the entire road to the moon. In coming back to the Navy, all agreements should hold specifications for reliability instead of exactly performance requirements. As Military Standard 1629 on FMEA was issued in 1974, NASA could jump at designing and developing spacecrafts such as the space shuttle. Their accent was on risk management through the reliability, utilization of statistics, system safety, QA, human factors, and software assurance. As technology rapidly advanced, reliability had expanded into a number of new areas.

Highlighting random vibration and temperature cycling became Environmental stress screening (ESS) testing, finally published as a Navy document P-9492 in 1979 and issued on Random Vibration with Tustin in 1984. The bygone quality procedures were changed with the Navy Best Manufacturing Practice program. An association for engineers working with reliability questions was soon established. IEEE Transactions on reliability came out in 1963, and many books ón the subject were issued in the 1960s.

In the 1970s, concerns on safety and risk features of the construction and operation of nuclear power plants expanded in the other areas of the world as well as in the United States. In the United States, a huge research commission financed by the multi-million dollar project, so-called Rasmussen report, was established to examine the problem. Despite its ineffectiveness, this report points out the first safety analysis of the complex system as a nuclear power plant.

Comparable efforts in the business of Asia and Europe have also been made in the analysis of risk and reliability issues. Particularly, the offshore oil industry of Norway happens. As in the North Sea, the offshore gas and oil development is moving forward into deeper waters, growing numbers of faraway-operated subsea production systems are operated. In many regards comparable to the reliability of spacecrafts, the reliability of subsea systems is critical because a moderate reliability cannot be recompensated by extensive maintenance.

During this decade, as the failure rates of many electronic components including mechanical components were dropped by a factor of 10, engineers questioned on the bathtub curve. For such a situation, the traditional failure rate typified by the bathtub curve can be reduced to resemble the failure rate represented by a flat, straight line with the shape parameter β (Figure 1.38).

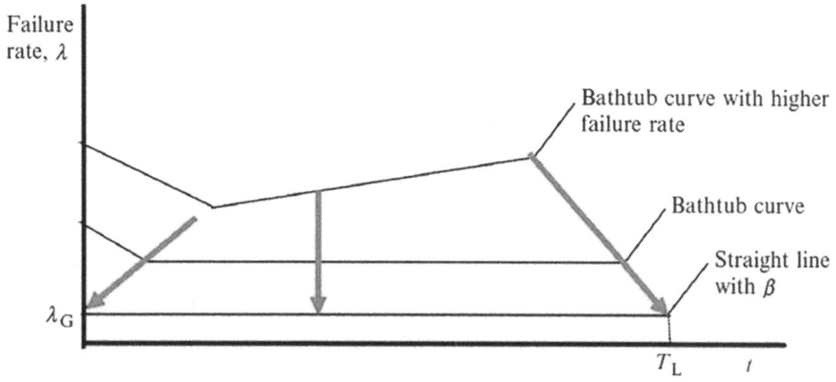

FIGURE 1.38 Bathtub curve and straight line with slope β.

However, a new methodology for reliability is still required to find the problematic parts because recall before production still happens. Later, it will deal with a new reliability methodology – parametric ALT – that can estimate product lifetime, as discussed in Chapters 8 and 9.

REFERENCES

1. Wöhler, A., 1855, Theorie rechteckiger eiserner Brückenbalken mit Gitterwänden und mit Blechwänden, *Zeitschrift für Bauwesen*, 5, 121–166.
2. Wöhler, A., 1870, Über die Festigkeitsversuche mit Eisen und Stahl, *Zeitschrift für Bauwesen*, 20, 73–106.
3. Griffith, AA, 1921, The phenomena of rupture and flow in solids, *Philosophical Transactions of the Royal Society of London A*, 221, 163–198.
4. Fleming, J.A., U.S. Patent 803,684, 17 Nov. 1815.
5. Deming, W.E., Stephan, F., 1940, On a least squares adjustment of a sampled frequency table when the expected marginal totals are known, *Annals of Mathematical Statistics*, 11 (4), 427–444.
6. Miner, M.A., 1945, Cumulative damage in fatigue, *Journal of Applied Mechanics*, 12 (3), 59–64.
7. Epstein, B., 1948, Statistical aspects of fracture problems, *Journal of Applied Physics*, 19 (2), 140–147.
8. Naresky, J.J., 1962, Foreword. *Proceedings of First Annual Symposium on the Physics of Failure in Electronics*, September 26–27, Chicago, IL.
9. Weibull, W., 1959, *Statistical Evaluation of Data from Fatigue and Creep Rupture Tests, Part I: Fundamental Concepts and General Methods.* Wright Air Development Center Technical Report 59-400, Sweden, September.
10. Weibull, W., 1961, *Fatigue testing and analysis of results*, London: Pergamon Press.
11. Abernethy, R., 2002, *The new Weibull handbook*, 4th ed. Self-published ISBN 0-9653062-1-6.

2 Data and Statistics in Mechanical System

2.1 INTRODUCTION

Statistics in mechanical engineering is a system of methods for gathering, examining, and explaining the data from field (or testing) when customers use products, and drawing to the conclusions based on the data. It also includes the presentation of test data by collection, processing, and interpretation (descriptive statistics). Especially, inference statistics, to be developed later in the chapter, is a basis for decision-making to determine whether lifetime target for a new product design is achieved. To assess the test data collected from selected product samples, statistics in mechanical engineering should be combined with design principles. Statistics requires more than just putting numbers in a table and its graphical presentation.

In other words, statistical knowledge therefore should be connected to the fundamental design principles of mechanical engineering – thermodynamics, mechanism, strength of mechanics, and load analysis. A mechanical engineer requires to have the knowledge of the fundamentals of probability and statistics, besides the design principles of mechanical product, materials, and loading. For example, an automobile engine generates the rotating motion through the crank-slider mechanism that was invented by James Watt and operated by the Carnot cycle. As the (generated) power successively transfers to transmission and driving system that can increase (adjusted) torque, automotive will move forward. Consequently, the automobile is subjected to repetitive stresses. If a mechanical system has design faults, we see that the product will fail in its lifetime in the marketplace.

However, because fatigue failures of mechanical products due to design faults rarely occur in the field, we have to know the statistics and failure mechanism to estimate the lifetime of multi-module products. To clearly understand it, we should develop new reliability methodologies – generalized failure model and sample size equation – which combine statistics with the engineering principles. We can improve the reliability (especially, lifetime) of mechanical systems, such as automobiles, by assessing and modifying their design.

There are many traditional statistical methods that can be used to evaluate the robust design of mechanical products – for example, Design of Experiments (DOEs) [1] and Taguchi's method [2,3]. Especially, DOE is a structured method to decide the connection between parameters influencing a process and the yield of that process. DOEs are performed for many factors that impact on the product lifetime (or designs). The usefulness of a factor can be assessed through analysis of variance (ANOVA). DOE substantially could not reflect the design principles of mechanical design and the failure mechanics. That is, if there are design faults that can cause an inadequacy of strength (or stiffness) when a mechanical structure (or mechanism) is

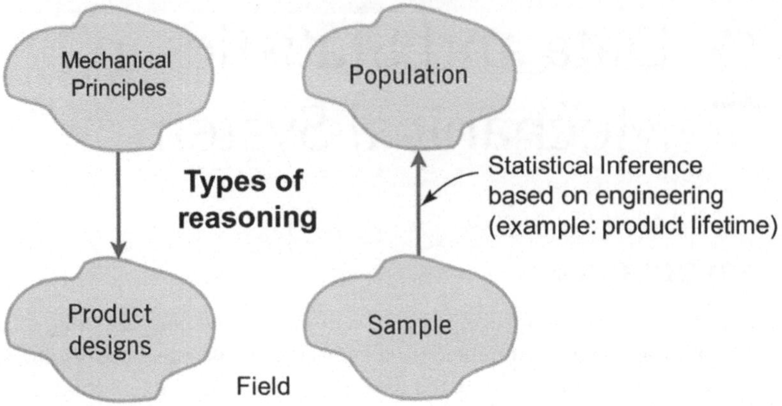

FIGURE 2.1 Statistical inference based on the design principles of mechanical engineering.

subjected to repeated stresses, there is a possibility for the product to fail in its lifetime. Though DOE neglects the design principles such as the relationship between loading and design faults, it pursues only product optimization. So it requires a lot of time to reach solutions and is still controversial.

To determine whether the final design of a product satisfies the lifetime targets based on mechanical engineering principles, the current statistical methodology needs to be developed with the following research items: (1) test planning based on multi-modules product; (2) defining lifetime index such as BX life; (3) how to perform reliability testing from a complete understanding of product loads, failure mechanics, and design; (4) descriptive statistics: summarizing and exploring data from testing; and (5) statistical inference (to be developed): affirming whether product satisfies the lifetime targets – B1 life of 10 years. Before discussing with the debating issues further, we will briefly look over the methodology issues – sampling, data, and its distribution – in classical statistics (Figure 2.1).

Example 2.1

Comparison between automobile performances

From eight automobiles produced in 2015 (Type A), the time to travel 60 km/hour is measured as followed: 18.7, 19.2, 16.2, 19.0, 13.5, 12.3, 17.6, and 13.9. Half of these types of automobiles reach 60 km/hour within 17 second, and 25% of them reach in more than 19 second. The mean can be calculated from the test data of eight samples. That is,

$$M = \frac{18.7+19.2+16.2+19.0+13.5+12.3+17.6+13.9}{8} = 16.3$$

So the manufacturer can announce that the automobiles produced (type A) can reach 60 km/hour within 17 second on average. Based on this, the company will be evaluated for the next year.

If the speed data of automobiles are represented as a dot graph, it can be as follows:

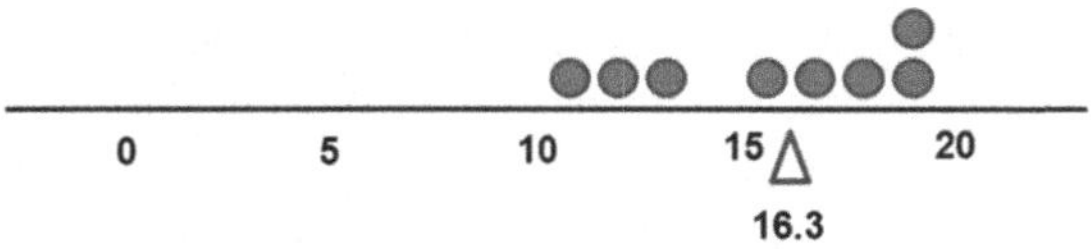

By the dot graph, we can figure out how the data are spread. Once all data are compared, we can locate the mean value.

2.2 TERMINOLOGIES USED IN STATISTICS

We can define statistical terminologies such as population, sample, parameter, and statistics as follows (Figure 2.2):

- A *population* is a group of all things or items under analysis in a statistical study. Population can be characterized as the set of persons or things in which an engineer is interested and from which the engineer wants to draw conclusions. For example, if an engineer is about to design an automobile engine, the whole set of population will be related to the engine.
- A *sample* is the subset of the population that is really being noticed or learned. The reasons to choose a sample from the population are as follows: (1) population investigation is impossible because of time and economics, (2) investigation results should be provided in time, and (3) possibility of inaccuracy in total sampling because of investigation error.

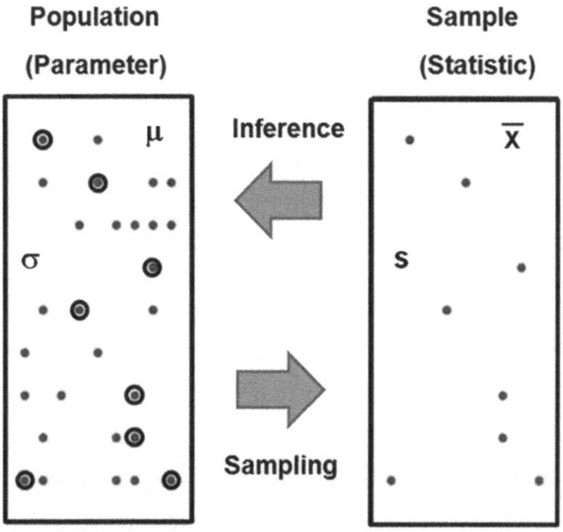

FIGURE 2.2 Population and sampling.

- A *random sampling* is a kind of method to select a well-representative sample from a population. In order for a sample to be precise, it should be random with a similar possibility of every representative of the population having a chance of being selected. This can lessen bias.
- A *parameter* is a quantitative attribute of the populace in guessing or testing such as a population mean or proportion. For instance, if you want to find out which classes freshmen choose at a college, you can query everyone (via email) and obtain a parameter such as mean weight (or height) of freshmen.
- A *statistics* is a quantitative attribute of a sample that helps to guess or test the population parameter such as a sample mean or proportion.

2.3 A BRIEF HISTORY OF STATISTICS

The term 'statistics' originates from the Latin term 'statisticum collegium' which means 'political state' or 'government'. Statistics was utilized by rulers to gain information on agriculture, land, commerce, and populations of their states to evaluate the aspects of their governments. By 1700 and 1800, astronomy introduced probability models and statistical theories as physics advanced. Especially, theories of many gambling were well understood and systematized by Fermat, Pascal, Huygens, and Leibniz, which is a basis of modern probability in the area of mathematics.

As earliest statistics and probability theory were developed in the nineteenth century, statistical reasoning and probability models utilized by social scientists also proceeded with the modern sciences of experimental sociology and psychology. It was applied to thermodynamics and statistical mechanics by physical scientists in the last century. In the early 1810s, Laplace suggested the central limit theorem. He also read Gauss's work and made the revolutionary connection between the least squares estimation and central limit theorem. The Gauss–Laplace combination is considered as one of the crucial milestones in the history of science. Karl Pearson (1857–1936) established *P*-value, Pearson's correlation coefficient, Pearson's chi-square test, and principal component analysis, among many other things.

At the beginning of the twentieth century, Ronald A. Fisher (1890–1962) implemented statistical procedures in the design of scientific experiments – for example, ANOVA. Fisher also summed up his statistical effort in Statistical Methods and Scientific Inference (1956). Based on small sets of data, William S. Gosset established the methods for decision-making. In the manufacture of mechanical products, there is a possibility of recalls from the marketplace. A new reliability methodology might be developed in the future based on the probability and its statistics (Figure 2.3).

2.4 STATISTICS FOR THE DESIGN OF MECHANICAL PRODUCT

As mentioned, statistics is a science of gathering, organizing, examining, and making inference from (testing or field) data. It is a particularly practical area of mathematics that is not only studied theoretically by advanced mathematicians but also

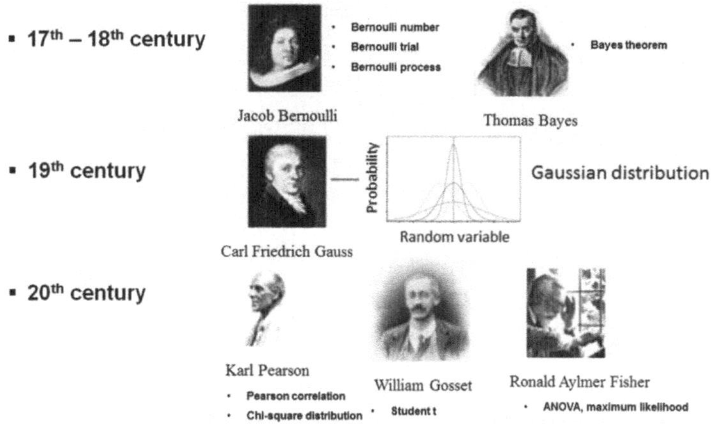

FIGURE 2.3 A brief history of statistics.

one that is used by researchers in many fields. Statistical methods and their analyses are usually used to communicate research findings and to hold hypotheses and give integrity to research methodology and conclusions.

There are two major areas in statistics: descriptive and inferential statistics, which can be expanded in the analysis of the mechanical system and its design principles. Descriptive statistics is a kind of methods for summarizing and organizing data. It comprises the building of charts, graphs, and tables, and the calculation of various descriptive measures such as averages, measures of variation, and percentiles. On the other hand, inferential statistics is related to using sample data to make an inference about a population of data (Figure 2.4).

Inferential statistics are suggested through compound mathematical computations, which allow engineers to deduce tendencies from the study of a sample taken from a large population. They utilize inferential statistics to analyze the relations between variables within a sample and then generalize or predict how those variables will relate to a bigger population. For example, we can suggest a new reliability methodology such as parametric ALT, which will be discussed in Chapter 8. It, therefore, can be expanded on generalizations or predictions about a product design such as the determination of lifetime for a newly designed mechanical product if a proper lifetime model, based on design concepts, is combined with statistical theories.

2.4.1 Confidence Interval

Because it is not possible to analyze each member of a population of products, engineers select a typical subset of products from the population, named as a statistical sample. From this statistical analysis, there can be generalizations about the population, such as product lifetime, when certain results are obtained from the sample by hypothesis or mathematical formulations. There are some terminologies that can be used in inferential statistics:

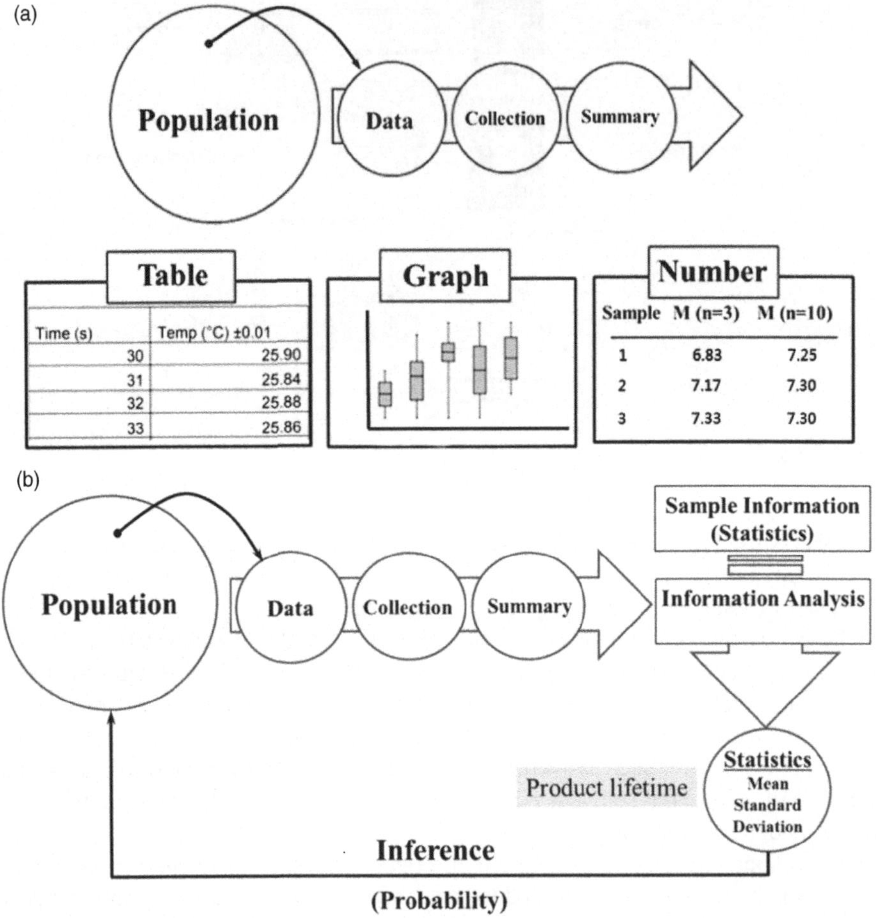

FIGURE 2.4 Differences between descriptive and inferential statistics: (a) descriptive statistics and (b) inferential statistics.

- In statistics, the aim of choosing a random sample from a population is to approximate the population mean. Because the analysis data are observed from the lifetimes of a particular sample unit, there is uncertainty in the results due to the limited sample size. An issue that always arises is how well the sample statistics evaluates the underlying population value.
- A confidence interval (CI) explains this issue because it supplies a range of values that are likely to hold the population parameter of interest. That is, it stands for the frequency of feasible CIs that hold the true value of the unspecified population parameter (Figure 2.5).
- A CI supplies a range of values for undefined parameters of the population by computing a statistical sample. This is expressed as an interval and the degree of confidence that the parameter is in the interval.

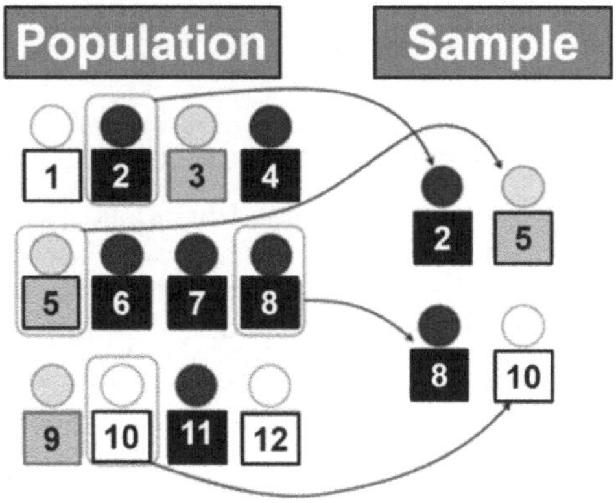

FIGURE 2.5 Concept of confidence interval.

- Hypothesis testing is where scientists claim the population by examining a statistical sample. By design, there is some uncertainty in this process. This can be expressed as a level of significance.
- A CI is distinguished as the probability that a random value lies within a certain range. CI is expressed as a percentage. For example, a 90% CI suggests that in 90 out of 100 cases, the observed value falls in this certain interval. After a particular sample is chosen, the population parameter is either or not in the interval achieved. The desired level of confidence is set by the researcher. A 90% CI means a significance level of $\alpha = 10\%$. The confidence level also relies on the product field.
  - 60% CI is widely used as international standard organization.
  - For IEC and GM, 50% CI is used.
  - MIL standard adapts 60%–90% CI.

2.4.2 Data Type

There are two kinds of data in statistics: categorical and numerical. Nominal data in the categorical category are used for labeling variables if it is associated to a certain group. Few examples are urban vs. suburban vs. rural, male vs. female, and aspirin vs. placebo. On the other hand, ordinal data in the categorical category can be placed into some kind of meaningful order but without any sign about the size of the interval. Examples are runners that come in the first, second, and third places in a race (Figure 2.6).

Numerical data are defined with digits as opposed to letters or words. For example, the height of a building or the weight of a desk is numerical data. Numerical data can be broken down into two different classes: discrete and continuous data. Discrete data can choose only integer values, whereas continuous data can choose any value. Most of the failure data obtained from the reliability test can therefore be classified as continuous data.

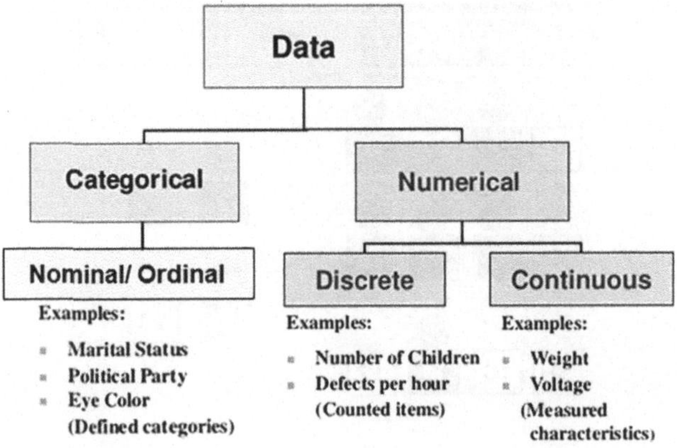

FIGURE 2.6 Data types.

2.4.3 Data in Mechanical System

Data are a set of numerical values of qualitative or quantitative variables that lead to a certain conclusion. It is the task of assigning a number to a characteristic of an event or object. The data of a company extensively come from internal testing or field data. The patterns of (internal) testing include research tests, prototype tests, reliability growth tests, qualification tests, environmental tests, production assessments, tests on purchased items, production acceptance tests, and tests on failed or malfunctioning items.

If used properly, the data will include information on the following: component (or assembly) tested, environment (type of test, vibration, temperature, etc.), test length, and failure description. When using inferential statistics to assess particular data, engineers conduct a hypothesis testing to determine whether they can make a general assumption with their results to a bigger population. These suggest the probability that the analysis results of the sample are representative of the population as a whole. For example, if the lifetime is targeted to have B1 life of 10 years, we can obtain the mission cycles from sample size equation in case of the given samples. If there is no product failure for mission cycles, we can guarantee the lifetime target – B1 life of 10 years (Chapter 8).

The part or assembly test data describe what was really tested. This includes any information necessary to a particular detail such as the revision or version of the item, the source of the item, and the type of the device. Accelerated testing is run to reveal design flaws under severe use conditions and harsh environments. The test period describes the total length of the period of all successful tests, as well as the times to failure of any failed parts or units. There are two kinds of failures: catastrophic failure and degradation failure. Product recall usually comes from the catastrophic failure which happens suddenly in a unit's lifetime due to a faulty design.

For a company, field failures are identified through customer complaints, warranty returns, field representative information, and distributor/dealer information. Unit failures from the field are the most precious source of data to improve the product design. They must also be reproduced and modified for the same types of failures by reliability testing such as parametric ALT. Finally, we can obtain the lifetime data from testing or field.

2.4.4 Central Tendency and Dispersion

All collected data might be expressed by a measure of central tendency and measure of dispersion in descriptive statistics. Different patterns of data are summarized by distinct measures of central tendency and dispersion (Figure 2.7).

The central tendency means where the observed data are located densely. There are three statistics that are commonly used to represent the center of a collected set of measurements (or data):

- Mean (also called the average or arithmetic mean): the sum of all the data in a data set divided by the number of data. Therefore, it is the average of all of the data and can only describe continuous data. Nominal data do not have a mean, and describing ordinal data with a mean is confusing. If a data set has a few extreme values, it will alter the mean enough to make it an unreliable measure of central tendency. For lifetime, the central tendency of product lifetime can be described as Mean Time To Failure.

  If n observed data are x_1, x_2, x_3, …, x_n, the mean of the variable x is defined as follows:

$$\bar{x} = \frac{1}{n}\left(x_1 + x_2 + \cdots + x_n\right) = \frac{\sum_{i=1}^{n} x_i}{n} \tag{2.1}$$

- Median (the middle of a data set): when all the numbers have been put in order from the least to the greatest, the median is the mid-point in the data. The median is not influenced by extreme values because it only responds to the number of observations, not the magnitude of the observations. Ordinal

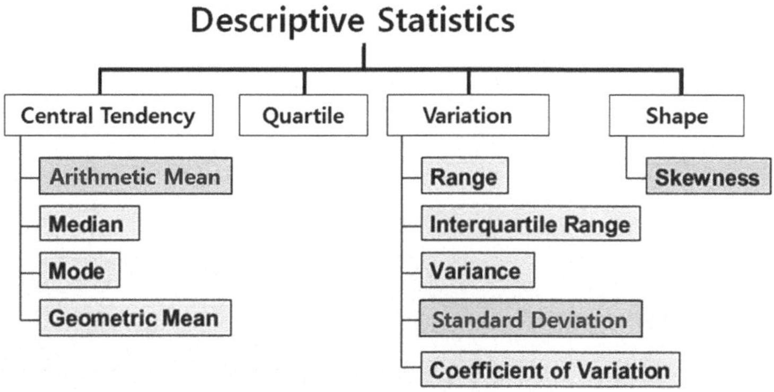

FIGURE 2.7 Measures summary of center and variation.

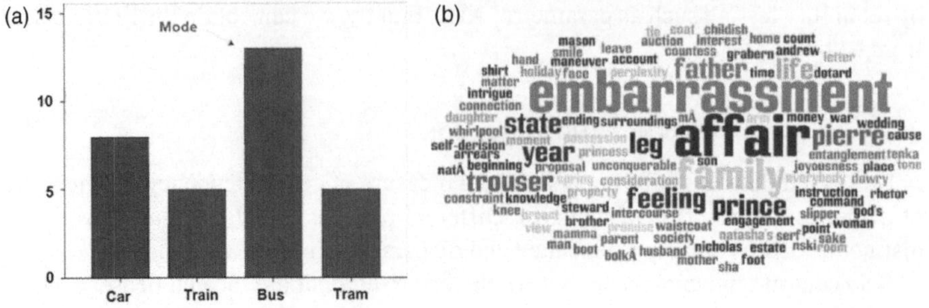

FIGURE 2.8 Mode in this particular data set: (a) most common form of transport and (b) word cloud.

data are best described by the median. Continuous data with extreme values can also be expressed by the median.

- Mode: The value that emerges most often in the data set. It is often used to express nominal data. It is not effected by extreme values.

The mode is the variable that happens with the biggest frequency in collected data. It can be differently defined that a mode for quantitative data is the observed value, whereas a mode for qualitative data is the name itself. As seen in Figure 2.8, we know that the mode in this particular data set is the bus for the most common form of transport and the word 'Will' in the sentence.

Example 2.2

Choose the data (1, 2, 3, 4, 5, 6, 7, 8, 9). What is the mean, median, and mode?

Mean: $\bar{x} = \dfrac{1+2+3+4+5+6+7+8+9}{9} = 5$

Median: 5

Mode: 5 (in this case, since all numbers are represented only once, it is typical to pick the median as the mode).

Example 2.3

Choose the data (1, 2, 3, 4, 5, 6, 7, 8, 500). What is the mean, median, and mode?

Mean: $\bar{x} = \dfrac{1+2+3+4+5+6+7+8+500}{9} = 59.6$

Median: 5

Mode: 5.

2.4.5 Data Distributions

For each pattern of data, you can get the frequency with which each value shows and plots it on a graph such as histogram. This supplies you a data distribution. All of these terms apply to the bell-shaped curve that we are all familiar with. We recognize that many natural phenomena fall into a normal distribution. The mean, mode, and median for a normal distribution are all the same (Figure 2.9).

Choose the failure data and draw histogram. We can get the skewed right (or left) histogram that is referred to as non-parametric distributions such as Weibull. If not all data are normally distributed, we know that these extreme outlier data tend to 'pull' the mean in a certain direction away from the true midpoint of the data. When a failure behavior is represented graphically, basic probability concepts are mean, median, mode, and standard deviation.

2.4.6 Standard of Variability

To completely describe all collected data, we have to report not only a measure of central tendency of the data but also the measure of dispersion of that data. Measures of dispersion, therefore, tell how much data are spread. Look at the following frequency graph to see why (Figure 2.10), notice that the central tendency of data (mean) is the same, but the dispersion of data (Standard Deviation) is obviously somehow different.

We will explain three of the most frequently utilized standards of variation: range, inter-quartile range (IQR), and standard deviation.

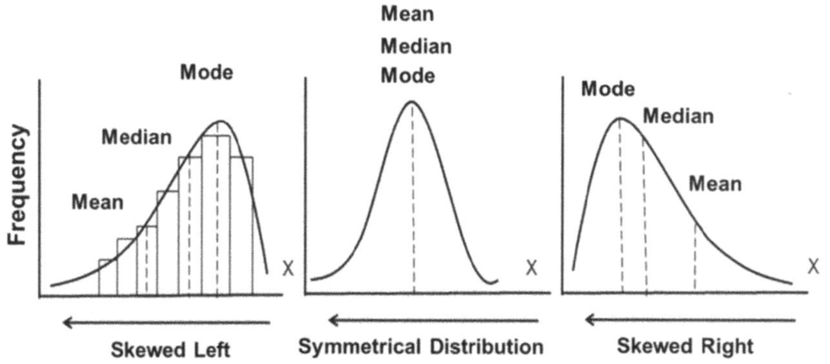

FIGURE 2.9 Mean, median, and mode for skewed left/right, and symmetric distribution.

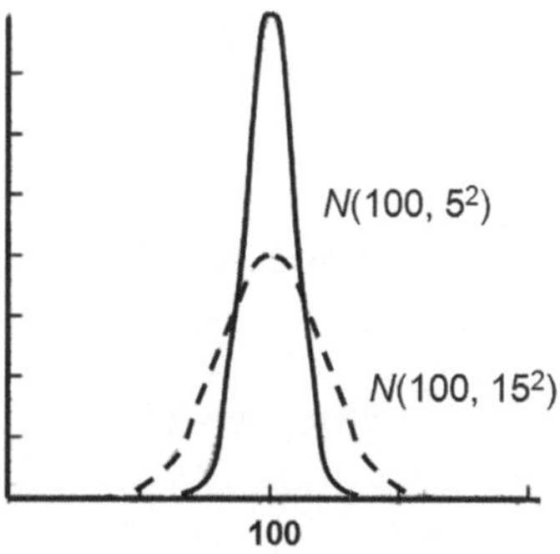

FIGURE 2.10 Normal distribution with the same central tendency but different variations.

- Range: The range of the variable is the differentiation between its minimum and maximum values in a data set. That is, Range = Max − Min. The sample range of the variable is quite easy to calculate and affected by outliers. However, in utilizing the range, a lot of information is disregarded because the other observed values are disregarded (Figure 2.11).
- IQR: The most usually utilized percentiles are quartiles. The quartiles of the variable divide the observed values into quarters. The three quartiles, Q_1, Q_2, and Q_3, can be defined as: the first quartile Q_1 is at location $(n+1)/4$, the second quartile Q_2 (the median) is at location $(n+1)/2$, and the third quartile Q_3 is at location $3(n+1)/4$. So the inter-quartile range, denoted as IQR, is the difference between the first and third quartiles of the variable, that is, IQR = $Q3 - Q1$. The IQR gives the range of the middle 50% of the observed values (Figure 2.12).

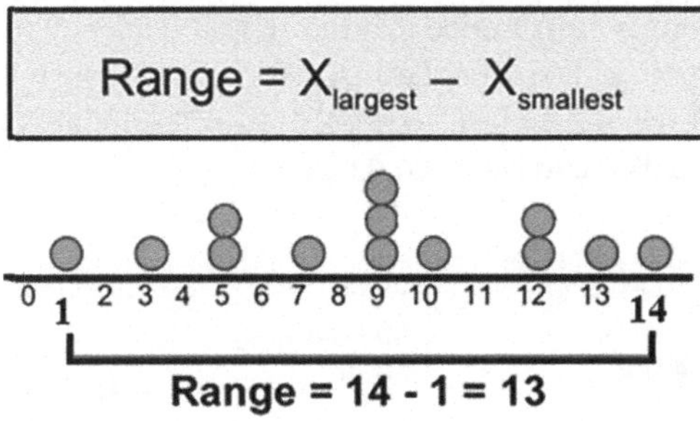

FIGURE 2.11 Range concept.

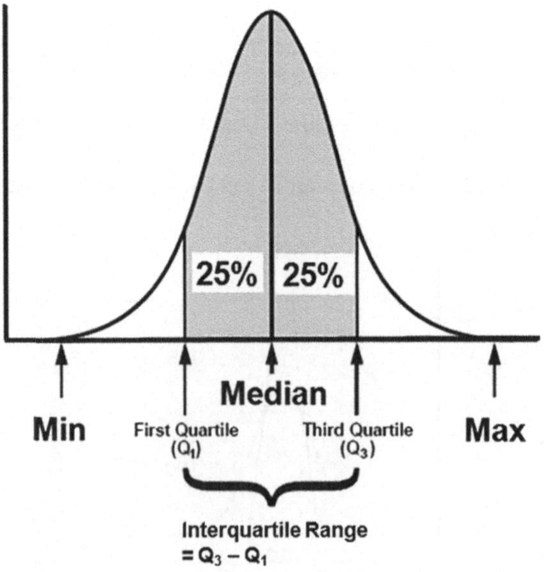

FIGURE 2.12 Inter-quartile range (IQR).

Example 2.4

Seven members in a bike race had the following completing times in minutes: 28, 22, 26, 29, 21, 23, and 24.

What is the interquartile range?

Solution: First of all, rearrange the data in the order of magnitude (smallest first).

21, 22, 23, 24, 26, 28, 29.

Median is 24, $Q_1=22$, $Q_3=28$. So $\text{IQR}=Q_3-Q_1=28-22=6$.

Example 2.5

Eight members in a bike race had the following finishing times in minutes: 28, 22, 26, 29, 21, 23, 24, and 50.

What is the interquartile range?

Solution: First of all, rearrange the data in the order of magnitude (smallest first).

21, 22, 23, 24, 26, 28, 29, 50

Median is $(24+26)/2=25$, $Q_1=23$, $Q_3=28$. So, $\text{IQR}=Q_3-Q_1=28-23=5$.

- Standard Deviation (SD): The sample standard deviation is the most often utilized standard of variability. A unit of standard has to do with variance around an average with continuous data. SD can only be utilized with normally distributed data.

  For a variable x, the sample standard deviation, designated by s_x, is

$$s_x = \sqrt{\frac{\sum_{i=1}^{n}(x_i-\bar{x})^2}{n-1}} = \sqrt{\frac{\sum_{i=1}^{n}x_i^2-n\bar{x}^2}{n-1}}. \tag{2.2}$$

  Since the standard deviation is defined as the sample mean $\bar{x}$ of the variable x, it is a preferred standard of variation when the mean is utilized as the standard of center. Note that the standard deviation is always a positive number, i.e., $s_x \geq 0$.

  In a formula of the standard deviation, the sum of the squared deviations from the mean is called the sum of squared deviations and supplies a standard of total deviation from the mean for all the observed values of the variable. Once the sum of squared deviations is divided by $n-1$, we can get

$$s_x^2 = \frac{\sum_{i=1}^{n}(x_i-\bar{x})^2}{n-1} \tag{2.3}$$

  which is called the sample variance.

  68% of all data fall in 1 SD of a mean, and 95% of all data fall in 2 SD of a mean (Figure 2.13).

So if we obtain an average of 80 and a standard deviation of 10 for heart rate in a population, then 68% of all people will have a heart rate between 70 and 90, and 95% of all people will have a heart rate between 60 and 100.

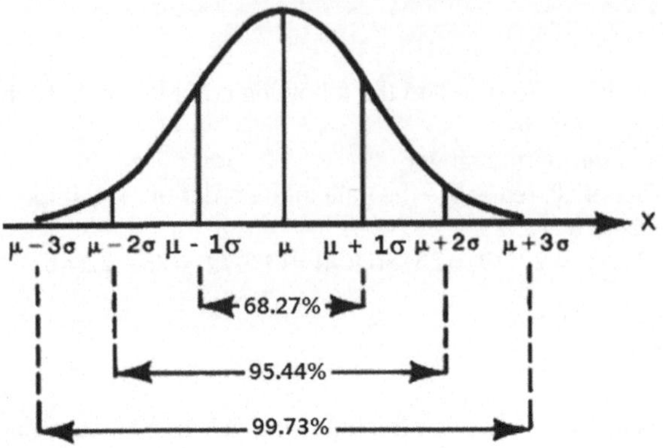

FIGURE 2.13 Standard deviation in normal distribution.

2.5 DESCRIBING DATA IN MECHANICAL ENGINEERING

To reach defensible conclusions from an experiment's outcomes, mechanical engineers would analyze data after a new product is tested under customer usage (or environment conditions) for its lifetime. To design a product, an engineer has to figure out the meaning and interpretation of numerical and diagrammatic statements requiring data. At that time, we rely on the fundamental tabular and diagrammatic techniques for displaying the central trends and spread of data. We can find another trends, not typical histogram that is expressed as data and its number. A lot of product failures occur initially. As time goes on, the number of failures will decrease (See Table 2.1). We know that the curve is a right-skewed shape. The basic question is how to effectively describe its trends. The final answer is the bathtub curve. That is, if the failure rate is represented in time, we can find another trend of the product life – bathtub curve.

On the bathtub curve, we can graphically represent the product life. That is, it comprises three periods: (1) an early mortality period with a decreasing failure rate,

TABLE 2.1
Quality Defects and Failures

Time	Number of Failure	$\lambda(t)$
1	5	5/100
2	10	10/95
3	35	35/85
4	30	30/50
5	15	15/20
6	2	2/5
7	2	2/3
8	1	1/1
Total	100	

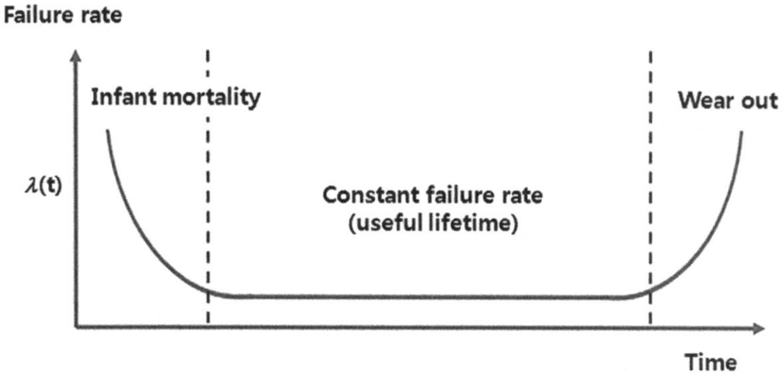

FIGURE 2.14 Bathtub curve.

(2) followed by a normal life period (also known as 'useful life') with a low and stable failure rate, and (3) concluding with a wear-out period that shows an increasing failure rate. Bathtub therefore supplies an overview of how early mortality, normal life failures, and wear-out modes combine to produce the general product failure distributions (Figure 2.14).

REFERENCES

1. Montgomery, D., 2013, *Design and analysis of experiments*, 8th ed., Hoboken, NJ: John Wiley.
2. Taguchi, G., 1978, Off-line and on-line quality control systems. *Proceedings of the International Conference on Quality Control*, Tokyo, Japan.
3. Taguchi, G., Shih-Chung, T., 1992, *Introduction to quality engineering: bringing quality engineering upstream*, New York: American Society of Mechanical Engineering.

3 Probability and Its Distribution in Statistics

3.1 INTRODUCTION

As seen in Figure 3.1, firstly, when two dices are rolled, it is impossible always to get sixteen as a result. On the other hand, we certainly know that the sun will rise in the east every day. In between two different cases, we can find the possibility. That is, we know that there is a chance for both cases to happen, e.g., possibility of getting head or tail when a coin is tossed. Lots of events happening in natural phenomena or in our modern life occur either randomly or repeatedly with a possibility, uncertainty, or impossibility. The typical examples are as follows:

1. A structural failure that occurs suddenly in a newly designed product due to repeated loads or overloading in daily consumer usage in the field.
2. Possibility of failure in future in a machine in a factory that manufactures products continuously due to defective assembly in its design process.
3. The probability of getting heads when a coin is tossed will be fairly 1/2.
4. Every Sunday consumers order pizza, but they don't know how long it will take to reach their home.
5. The chance of a great earthquake to happen on the San Andreas Fault in the following 30 years is about 21%.
6. The chance of humankind to extinct by 2100 is approximately 50%.

The random phenomena mentioned above will occur repeatedly in our daily life. The modern technology cannot predict their outcomes with certainty in advance, but we can recognize them from a set of all the possible outcomes of test data.

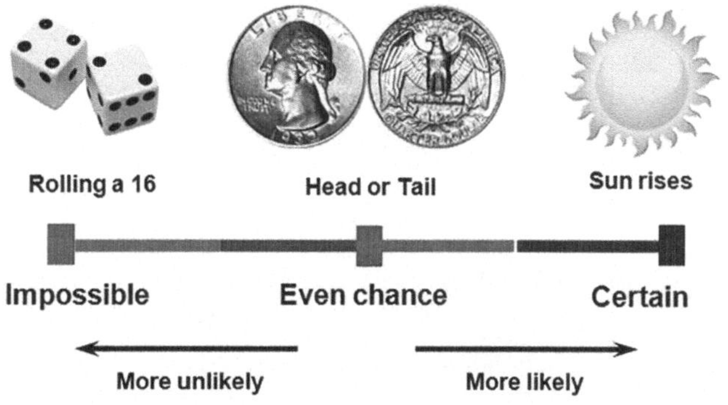

FIGURE 3.1 Probability line.

Analysis using these test data to find the probability of an even is called 'statistical experiment'. A statistical experiment is a procedure that after drawing a sample from a population, we test it to make a decision in some measurable space. The sample values can be defined as random variable, and the decisions made are to be functions of this random variable.

All statistical experiments have commonly three properties: (1) the experiment might have more than one outcome, (2) each outcome can be stated in advance, and (3) the outcome of the experiment relies on chance. For example, tossing a coin is a kind of statistical experiment. There is more than one possible outcome that might be tail or head. And there is an element of chance since the outcome is unknown. A statistical experiment can provide a random outcome. The set of all possible outcomes is defined as the sample space. An event is a subset of outcomes from the sample space. Examples of events are getting tail when tossing a coin or getting '5' when rolling a dice.

Furthermore, to improve the lifetime of a mechanical product from test data, conventional engineering statistics is required to be more than just the tabulation of numbers and its graphical presentation of data. It, therefore, should combine the statistical concepts discussed in Part I with the fundamental principles of mechanical engineering – mechanism design and load analysis discussed in Chapter 5. Therefore, we can obtain the solutions for (1) experimental design, (2) description: summarizing and exploring data, (3) inference: assessing whether a system satisfies the lifetime targets – B1 life 10 years (Chapter 8).

3.2 FUNDAMENTALS OF PROBABILITY

Choosing a particular event $A_i \in S$ provides an observation. Definite set theory concepts have special terminology and meaning when utilized in the conditions of probability (Figure 3.2).

1. A subset A of S is called an event.
2. For any event A, $B \in S$:
    - The complement event A^c is the event that is not in A, or not in A'.
    - Union $A \cup B$: an outcome if it is either in A or in B.
    - Intersection $A \cap B$: an outcome if it is in both A and B.

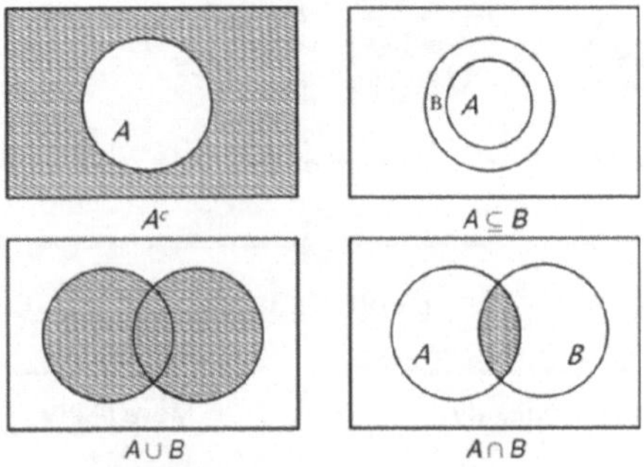

FIGURE 3.2 Set theoretic operations by the Venn diagrams.

- $A \subseteq B$: the occurrence of A means the occurrence of B.
- $A \cap B = \phi$: A and B are mutually exclusive or disjoint events.

If the number of possible outcomes is to be counted, the sample space is $S = \{A_1, A_2, A_3, \ldots\}$.

In probability, the sample space composes a finite number of equally likely (or mutually exclusive) outcomes, $S = \{A_1, A_2, A_3, \ldots\}$. The probability can classically be defined as follows:

$$P(A) = \frac{\text{number of outcomes in } A}{\text{number of outcomes in } S} = \frac{|A|}{N} \tag{3.1}$$

Example 3.1

If an equitable coin is tossed twice, what is the chance of obtaining the head at least once?

Solution: The sample space of this experiment is
S = {HH, HT, TH, TT}.

The event A is given by
A = {at least one head} = {HH, HT, TH}.

The chance of A is the summation of the chances of its basic events. Thus, we obtain
$P(A) = P(HH) + P(HT) + P(TH) = 1/4 + 1/4 + 1/4 = 3/4$

This classical definition for probability in Equation (3.1) is instinctively logical, but it makes several practical presumptions about the existence of the limit and the fact that the ratio merges to a single value for all possible orders of experimental outcomes. That is, if the trial number n increases infinitely, the relative frequency $\frac{r}{n}$, where r is the frequency for event A, will approach $P(A)$. For example, the probability of tail in a single flip of a fair coin is $P(A) = \lim_{n \to \infty} \frac{r}{n} = \frac{1}{2}$ as $n \to \infty$.

Rather than make these kinds of presumptions, modern probability theorists prefer to explain a more fundamental set of axioms about the essence of the probability measure P. Using the following axioms, we can obtain more complicate results, including the intuitive meaning of probability as a fraction of occurrence.

The probability of event A is a number $P(A)$ assigned to A that satisfies the following axiom conditions:

1. $0 \leq P(A) \leq 1$ for sample space S and event E.
2. $P(S) = 1$, $P(\phi) = 0$.
3. For any sequence of events $A_1, A_2, \ldots, A_n$ that are mutually exclusive, satisfying as following:

$$P(A_1 \cup A_2 \cup \cdots \cup A_n) = P(A_1) + P(A_2) + \cdots + P(A_n) \tag{3.2}$$

The first axiom explains that the chance of an event is a number between 0 and 1. The second axiom explains that the event expressed by the whole sample space has the probability of 1. The third axiom explains how to integrate the probabilities of mutually exclusive events.

Example 3.2

Every Sunday a consumer orders pizza. The delivery time varies from 10 to 30 min with the same opportunity in a day. What is the probability of delivering between 20 and 25 min?

Solution: The sample space of this experiment is all real numbers between 10 and 30 min, {(10, 30)}. The delivering time between 20 and 25 min is {(20, 25)}. The probability is (25 – 20)/(30 – 10) = 1/4.

3.2.1 Counting Techniques in Probabilities

For a discrete sample space, engineers can obtain probabilities as counting outcomes in the sample space and the corresponding events. If the number of possible events in the sample space are not much, probability can be easily calculated. Most of the cases are difficult to count them because of their complexity. There are several basic counting methods – permutation, multiplication rule, and combination – that can effectively be used to count the outcomes of complex cases.

3.2.2 Multiplication Rule

If A_1 is a test with n_1 possible outcomes and A_2 is a test with n_2 possible outcomes, the test that performs A_1 first and then A_2 can have $n_1 \times n_2$ possible outcomes.

Example 3.3

Discover the feasible number of outcomes in a sequence of two tosses of a fair coin. The number of possible outcomes is $2 \cdot 2 = 4$. This is obvious from the following tree drawing.

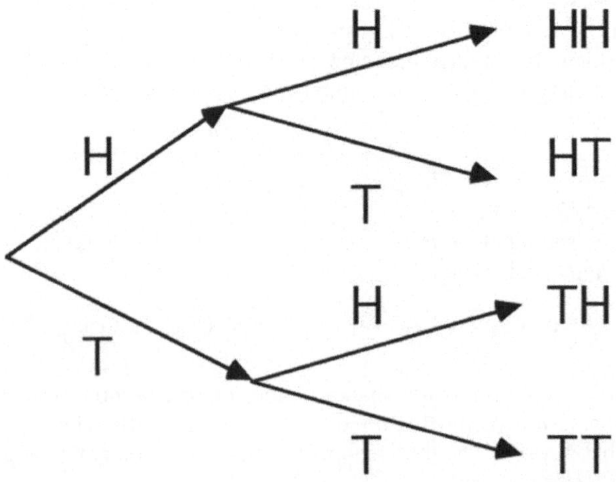

Example 3.4

How many unlike license plates are feasible if Kentucky utilizes three letters followed by three digits?

$(26)^3 (10)^3 = (17,576) (1,000) = 17,576,000.$

3.2.3 Permutation

From a set of n elements, the number of ways for getting an ordered subset of k elements is given by

$$_{n}P_{k} = n(n-1)(n-2)\cdots(n-r+1) = \frac{n!}{(n-r)!} \tag{3.3}$$

The numeral of permutations of n discrete objects is n factorial, usually expressed as $n!$, which means that the product of all positive integers is less than or equal to n. That is,

$$_{n}P_{n} = n(n-1)(n-2)\cdots 1 = n! \tag{3.4}$$

Example 3.5

In a conference meeting, the committee tries to arrange four panel seats (A, B, C, and D) in a row. Find out all cases that four people are dispatched. What is the probability of the case that A in the left most seat is dispatched?

Solution: Total cases in the sample space are calculated as follows:

(Case dispatched in the left most seat) × (Except left, case dispatched in the second seat) × (Except two left, case dispatched in the third seat) × (Except three left, case dispatched in the right seat)

$= 4 \times 3 \times 2 \times 1 = 4! = 24.$

Cases that dispatched A in the left most seat exclude A seat, and arranges the other three seats as second and third. So they are $3 \times 2 \times 1 = 3!$ Therefore, the probability for dispatching A in the left most seat is calculated as follows:

$3!/4! = 6/24 = 0.25.$

3.2.4 Combination

In permutation, the order is important. But combination is a choice of entries from a collection, so that the sequence of choice does not have any importance. For instance, given three fruits, say an apple, an orange, and a pear, there are three combinations of two that might be took out from this set: an apple and a pear; an apple and an orange; or a pear and an orange. In other words, it is the numbers of method of picking k unordered outcomes from n possibilities. It also is familiar as the binomial coefficient or choice number that can read 'n choose k',

$$_{n}C_{k} = \begin{pmatrix} n \\ k \end{pmatrix} = \frac{_{n}P_{r}}{k!} = \frac{n!}{(n-k)!} \tag{3.5}$$

Example 3.6

How many different groups of one physicist and two chemists can be established from three physicists and four chemists?

$$ {}_4C_2 \times {}_3C_1 = \begin{pmatrix} 4 \\ 2 \end{pmatrix}\begin{pmatrix} 3 \\ 1 \end{pmatrix} = (6)(3) = 18 $$

Thus, 18 different committees can be formulated.

3.2.5 Addition Theorem of Probability

The chance of event A or B can be obtained by adding on the chances of both events A and B and subtracting any intersection of the two events. That is,

$$ P(A \cup B) = P(A) + P(B) - P(A \cap B) \tag{3.6} $$

If $A \cap B = \phi$, $P(A \cup B) = P(A) + P(B)$. That is, events A and B are mutually exclusive events. And more instinctively, if we take the region of the Venn diagram below, we will add on the regions of figures A and B and take away one overlapping region, which stands for the intersection of A and B (Figure 3.3).

Example 3.7

In a semester, there are 40 sophomores in the statistics department. Of them, 25 students take economics as a selective course. On the other hand, 30 students take business and 20 students take both courses. If we meet a sophomore in the department, find out the probability that he or she has taken economics or business.

Solution: The event in which a student takes business is A. On the contrary, the event in which a student takes economics is B.

$P(A) = 25/40$, $P(B) = 30/40$, and $P(A \cap B) = 20/40$. So, the probability that they take economics or business is

$P(A \cup B) = P(A) + P(B) - P(A \cap B) = 25/40 + 30/40 - 20/40 = 35/40$.

3.2.6 Conditional Probability

The conditional probability of an event B is the chance that the event will happen given the knowledge that an event A has already happened. This chance is written as $P(B|A)$, notation for the probability of B given A. In the case where events A and B are

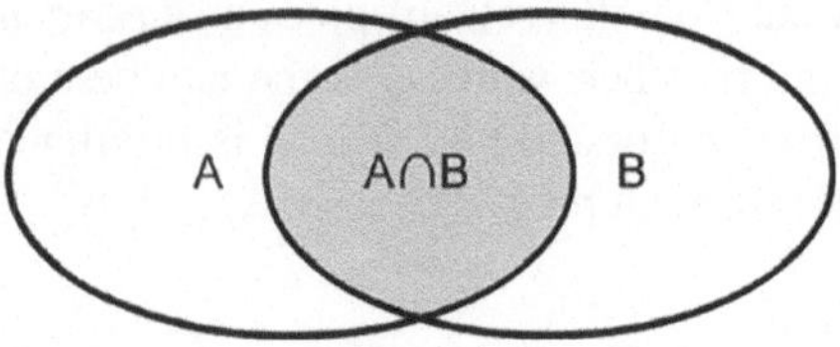

FIGURE 3.3 Addition theorem of probability.

independent (where event A has no impact on the chance of event B), the conditional probability of event B given event A is simply the chance of event B, that is, $P(B)$. If events A and B are dependent on each other, the chance of the intersection of A and B (the chance that both events happen) is expressed by

$$P(A \cup B) = P(A)P(B \mid A) \tag{3.7}$$

From this meaning, the conditional chance $P(B|A)$ is simply obtained by dividing by $P(A)$:

$$P(B \mid A) = \frac{P(A \cap B)}{P(A)} \tag{3.8}$$

Example 3.8

As seen in a card play (Figure 3.4), assume a player needs to draw two cards of the same suit to win. Of the 52 cards, there are 13 cards in each suit. Assume first the player draws a heart. Now the player wishes to draw a second heart. Since one heart has already been drawn, there are now 12 hearts remaining in a deck of 51 cards. So, the conditional chance P(Draw second heart|First card a heart) = 12/51.

Example 3.9

Assume a person applying to a college decides that he has an 80% chance of being received, and he knows that dormitory housing will only be supplied for 60% of all of the accepted students. The possibility of the student being approved and getting dormitory housing is defined by

P(Accepted and Dormitory Housing) = P(Dormitory Housing|Accepted) P(Accepted) = (0.60)*(0.80) = 0.48.

3.2.7 Multiplication Theorem of Chance

If $P(A) > 0$ and $P(B) > 0$, then

$$P(A \text{ and } B) = P(A \cap B) = P(A)P(B \mid A) = P(B)P(A \mid B) \tag{3.9}$$

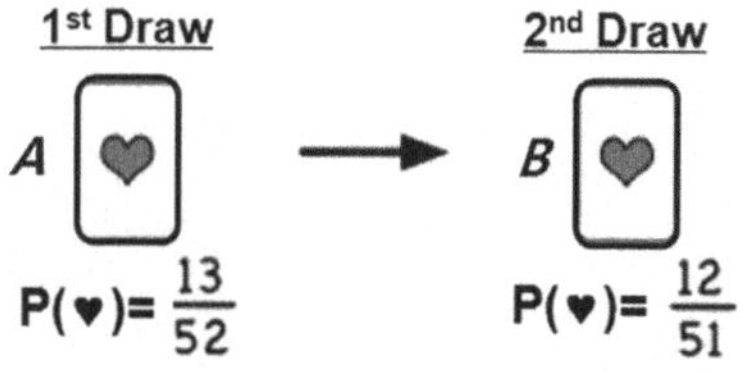

FIGURE 3.4 Card drawing – example of conditional probability.

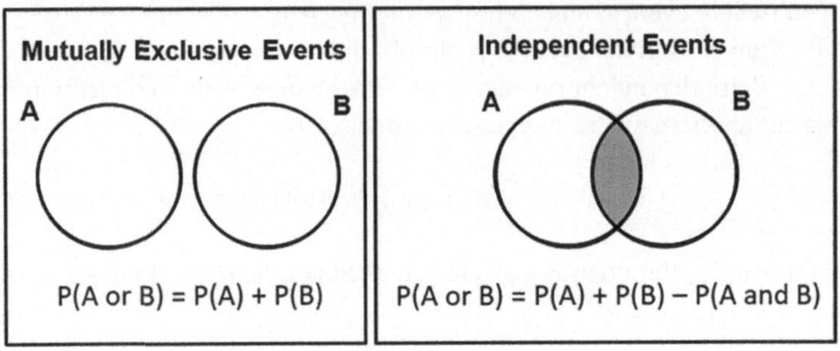

FIGURE 3.5 Probability mutually exclusive and independent events.

The chance of events A and B to happen can be calculated by finding the chance of event A to occur and multiplying it by the chance of event B to happen, given that event A has already occurred. Events A and B are said to be independent if there is no impact of one event's occurrence on that of the other event.

Therefore, if events A and B are independent, $P(A \cap B) = P(A)\, P(B)$ (Figure 3.5).

Example 3.10

If $P(A) = 1/5$ and $P(B|A) = 1/3$, what is $P(A \cap B)$?

$P(A \cap B) = P(A)\, P(B|A) = 1/5 \times 1/3 = 1/15$.

Example 3.11

Card drawing

While laying a packet of cards, let A be the event of drawing a diamond and B be the event of drawing an ace. Then, $P(A) = 13/52 = 1/4$ and $P(B) = 4/52 = 1/13$.

Now, $A \cap B$ = getting a king card from hearts.

Then, $P(A \cap B) = 1/52$.

Now, $P(A|B) = P(A \cap B)/P(B) = (1/52)/(1/13) = 1/4 = P(A)$.

So, A and B are independent.

3.2.8 The Complement Rule

The complement rule says that if A is an event and A^c is its complement, the probability of A is equal to one minus the probability of A^c:

$$P\left(A^c\right) = 1 - P(A) \tag{3.10}$$

This will apply to all events and their complements.

Example 3.12

The chance of getting a white ball from a bag of four balls is 1/4. What is the chance of not drawing a white ball?

P (ball is not white) = 1 – 1/4 = 3/4.

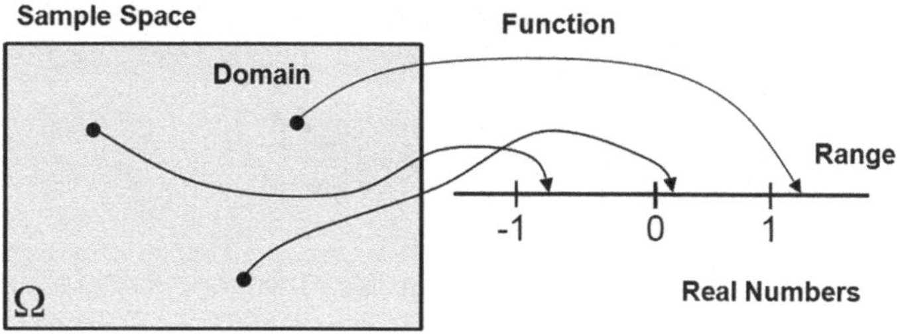

FIGURE 3.6 Random variable.

3.2.9 Random Variable

A random variable, X, is a variable whose feasible values are numerical outcomes of a random phenomenon from a statistical experiment. A random variable is a function from a sample space S into the real numbers (Figure 3.6).

There are two kinds of random variables: continuous and discrete. A discrete random variable is one that may take on only a countable number of definite values such as 0, 1, 2, 3, 4,…. Examples of discrete random variables are the number of typos in a page, the number of children in a family, the number of defective light bulbs in a box of ten, and the number of defective products per lot in manufacturing. On the other hand, a continuous random variable is one that takes on an infinite number of possible values in a real interval. It includes the departing time of airplanes, the time required to run a mile, height, and weight.

A random variable X is continuous if there is a function $f(x)$, such that for any $a \le b$, we have

$$P(a \le X \le b) = \int_a^b f(x)dx \tag{3.11}$$

The function $f(x)$ is called the probability density function (*pdf*).

The *pdf* always satisfies the following properties:

1. $f(x) \ge 0$ (f is nonnegative) (3.12)

2. $\int_{-\infty}^{\infty} f(x)dx = 1$ (This is equivalent to: $P(-\infty < X < \infty) = 1$) (3.13)

The probability density function is also called the probability distribution function or probability function. It is denoted by $f(x)$.

Example 3.13

Check whether the given probability density function is valid or not. The probability density function is

$$f(x) = 4x^3, 0 < x < 1$$

Since the function $4x^3$ is greater than 0, the condition $f(x) \geq 0$ is satisfied.
Consider

$$\int_{-\infty}^{\infty} f(x)dx = \int_{0}^{1} 4x^3 dx = 4\left[\frac{x^4}{4}\right]_0^1 = 1$$

Hence, the condition $\int_{-\infty}^{\infty} f(x)dx = 1$ is satisfied. Therefore, the given function $f(x) = 4x^3, 0 < x < 1$ is a valid probability density function.

3.2.10 Expectation Value and Variance

Assume X be a continuous random variable with range $[a, b]$ and probability density function $f(x)$. The expected value of X is defined as

$$E(X) = \int_{a}^{b} xf(x)dx \tag{3.14}$$

On the other hand, for a discrete random variable, the expected value of X is

$$E(X) = \sum_{i=1}^{n} x_i f(x_i) \tag{3.15}$$

The discrete formula takes a weighted summation of the values x_i of X, where the weights are the chances $f(x_i)$. $f(x)dx$ stands for the chance that X is in an infinitesimal range of width dx around x. Thus, we can interpret the formula for $E(X)$ as a weighted integral of the values x of X, where the weights are the chances $f(x)dx$. As mentioned before, the expected value is also defined as the average or mean.

Example 3.14

An automobile dealer who sells a car for a week obtains the following probability distribution. Let random variable X be the selling number of cars. Find out the expected number of car sales.

X_i	0	1	2	3	4	5	Sum
$f(x_i)$	0.1	0.1	0.2	0.3	0.2	0.1	1.0

$$E(X) = \sum_{i=1}^{n} x_i f(x_i)$$

$$= 0 \times (0.1) + 1 \times (0.1) + 2 \times (0.2) + 3 \times (0.3) + 4 \times (0.2) + 5 \times (0.1)$$

$$= 2.7$$

The properties of $E(X)$ for continuous random variables (or discrete ones) are

1. If X and Y are random variables on a sample space Ω,

$$E(X+Y)=E(X)+E(Y) \tag{3.16}$$

2. If a and b are constants,

$$E(aX+b)=aE(X)+b \tag{3.17}$$

Assuming that X is a random variable with mean μ, the variance of X is

$$Var(X)=E\left((X-\mu)^2\right)=\begin{cases}\sum_{i=1}^{n}(x_i-\mu)^2 f(x_i) \\ \int_{-\infty}^{\infty}(x-\mu)^2 f(x)\,dx\end{cases} \tag{3.18}$$

3.2.11 Properties of Variance

These are precisely identical to the case of discrete space.

1. If X and Y are independent, then $Var(X+Y)=Var(X)+Var(Y)$. (3.19)
2. For constants a and b, $Var(aX+b)=a^2Var(X)$. (3.20)
3. Theorem : $Var(X)=E\left(X^2\right)-E(X)^2 = E\left(X^2\right)-\mu^2$. (3.21)

Example 3.15

In the mid-term exam of statistics, the mean and the variance in a class are 60 and 100 points, respectively. To upgrade the test results, we are considering the following steps:

1. Add 20 points at each student.
2. Multiply 1.4 at each student.
3. Multiply 1.2 at each student and add 10 points.

Find the expected value and variance for three upgrading alternatives:

1. Because mean $E(X)$ is 60 and variance $Var(X)$ is 100 for random variable X = the grade of mid-term exam, the first step is to obtain the mean and the variance for a new random variable $X + 20$.
   $E(X + 20) = E(X) + 20 = 60 + 20 = 80$
   $Var\ (X + 20) = Var(X) = 100$
   For the first alternative, we know that the mean will only increase 20 because there is no change in the variance.

2. The second step is to obtain the mean and the variance for new random variable 1.4*X*.
   $E(1.4X) = 1.4E(X) = 1.4 \times 60 = 84$
   $Var(1.4X) = 1.4^2Var(X) = 1.96 \times 100 = 196$
   For the second alternative, we know that the mean increases 1.4 times and the variance does 1.4^2.
3. The third step is to obtain the mean and the variance for a new random variable 1.2*X* + 10. That is,
   $E(1.2X + 10) = 1.2E(X) + 10 = 1.2 \times 60 + 10 = 82$
   $Var(1.2X + 10) = 1.2^2Var(X) = 1.44 \times 100 = 144.$

3.3 PROBABILITY DISTRIBUTIONS – BINOMIAL, POISSON, EXPONENTIAL, AND WEIBULL

In real life, we can find that certain probability distributions occur regularly. A probability distribution is a mathematical expression that computes the chances of events with different possible outcomes in a statistical experiment. Probability distributions are a table or an equation that connects random variables (or outcomes of a statistical experiment) with their chances of events. There are two kinds of probability distributions: continuous and discrete.

A discrete probability distribution can be applied to the plots where the set of possible outcomes is discrete; for example, Poisson distribution and binomial distribution. On the other hand, a continuous probability distribution can be applied to the scenarios where the set of possible outcomes might take values in a continuous span, such as the temperature on a given day. The normal or Weibull distribution is a usually encountered continuous probability distribution. And there are general lifetime distributions that can model failure times arising from a wide span of products, such as Weibull distribution and exponential distribution.

3.3.1 Binomial Distribution

Binomial distribution occurs in daily life. The following are few general examples: (1) vote counts for two different applicants in an election, (2) the number of tails/heads in a sequence of coin flips, (3) the number of successful sales calls, (4) the number of female/male employees in a company, and (5) the number of problematic products in a manufacturing line.

There are several presumptions that explain a binomial distribution: (1) n fixed statistical experiments are performed, (2) each trial has one of the two outcomes – a success or a failure (Bernoulli trial), (3) the chance of 'success' p is the same for each outcome, (4) the outcomes of individual trials are independent, and (5) we are interested in the entire number of successes in these n trials.

Under the above presumptions, assuming that the random variable X is the total number of successes, the probability distribution of X is defined as the binomial distribution. Probability is expressed as follows:

$$P(X = x) = \binom{n}{x} p^x (1-p)^{n-x} = \frac{n!}{x!(n-x)!} p^x (1-p)^{n-x} \quad \text{for } x = 0,1,2,\ldots,n \quad (3.22)$$

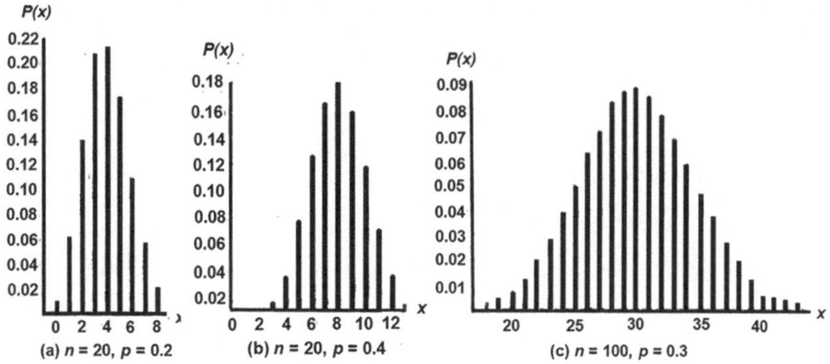

FIGURE 3.7 Shape of the binomial distribution according to n and p.

And binomial mean and variance can be expressed as follows:

$$\mu = E(X) = np, \quad Var(X) = np(1-p) \tag{3.23}$$

where the values of n and p are the parameters of the binomial distribution.

When $p = 0.5$, the binomial distribution is uniform – the median and the mean are equal. Even when $p < 0.5$ (or $p > 0.5$), the larger the value of N, the more uniform the form of the distribution. Because the binomial distribution can be cumbersome, there are estimations to the binomial that can be much easier to utilize when N is large (Figure 3.7).

Example 3.16

When a salesman in an insurance company meets customers, probability that they make contract is 20%. He or she will meet 10 persons today.

1. What will be the probability for three persons to make contract?
2. What will be the probability for more than two persons to make contract?
3. Find mean and standard variation.

For the first question, because binomial distribution has $n = 10$, $p = 0.2$, the probability that three persons make contract is

$$P(X=3) = \binom{10}{3} 0.2^3 (1-0.2)^7 = 0.2013$$

For the second question, it is better to use the concept of complement because there are more than two persons. That is,

$$P(X \geq 2) = 1 - \left(P(X=0) + P(X=1)\right)$$

$$= 1 - \left(\binom{10}{0} 0.2^0 (1-0.2)^{10} + \binom{10}{1} 0.2^1 (1-0.2)^9 \right)$$

$$= 1 - (0.1074 + 0.2684) = 0.6242$$

For the third question, the mean and the standard variation are

$$E(X) = np = 10 \times 0.2 = 2$$

$$Var(X) = np(1-p) = 10 \times 0.2 \times 0.8 = 1.6$$

3.3.2 Poisson Distribution

Poisson distribution arises based on how many times one should expect an event to occur in a fixed time period. The Poisson distribution is coined after Simeon Poisson (1781–1840), a French mathematician, and utilized in circumstances where large declines happen in a period with a particular mean rate, regardless of the elapsed time. More specifically, this distribution is utilized when the number of possible events is big, but the event probability over a particular period is small.

The probability distribution comes from a Poisson experiment: (1) the experiment outcomes can be categorized as successes or failures, (2) the mean number of successes (μ) that happen in a particular area is recognized, (3) the chance that a success will happen is proportional to the magnitude of the area, and (4) the chance that a success will happen in an extremely small region is almost zero.

The Poisson distribution occurs in (1) the number of hurricanes hitting Hawaii each year, (2) the hourly number of customers arriving at a bank, (3) the number of battery failures and replacements, (4) the number of hummingbirds seen while going to school each morning, (5) the daily number of accidents on a particular stretch of highway, (6) the everyday number of emergency calls in a city, (7) the monthly number of employees who had a non-attendance in a big company, (8) the number of typos in a book, (9) the hourly number of accesses to a specified web server.

Poisson distribution also has the following assumptions: (1) independence: events must be independent, (2) homogeneity: the average number of targets scored is presumed to be the same for all teams, and (3) time period (or space) must be fixed. If probability p is tiny and trial is far enough, Poisson probability function from Equation (3.22) means that binomial distribution can be estimated. That is,

$$
\begin{aligned}
{}_nC_x p^x (1-p)^{n-x} &= \frac{n(n-1)\cdots(n-x+1)}{x!}\left(\frac{m}{n}\right)^x \left(1-\frac{m}{n}\right)^n \Big/ \left(1-\frac{m}{n}\right)^x \\
&= \frac{m^x}{x!}\underbrace{1\left(1-\frac{1}{n}\right)\left(1-\frac{2}{n}\right)\cdots\left(1-\frac{x-1}{n}\right)}_{A}\underbrace{\left(1-\frac{m}{n}\right)^n}_{B} \Big/ \underbrace{\left(1-\frac{m}{n}\right)^x}_{C} \qquad (3.24)
\end{aligned}
$$

As n increases, A~C can be rearranged.

$$A = 1\left(1-\frac{1}{n}\right)\left(1-\frac{2}{n}\right)\cdots\left(1-\frac{x-1}{n}\right) \xrightarrow[n\to\infty]{} 1 \qquad (3.25a)$$

$$B = \left(1-\frac{m}{n}\right)^n \xrightarrow[n\to\infty]{} e^{-m} \qquad (3.25b)$$

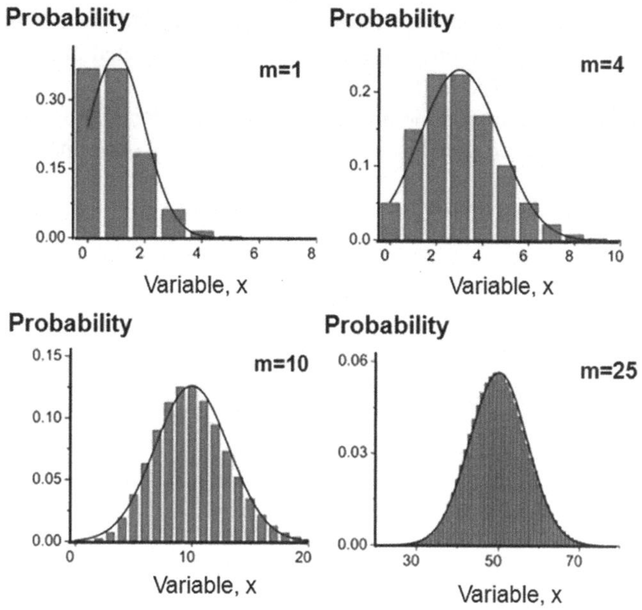

FIGURE 3.8 Shape of the Poisson distribution according to m.

$$C = \left(1 - \frac{\lambda}{n}\right)^x \xrightarrow[n\to\infty]{} 1 \tag{3.25c}$$

Therefore, we can summarize as follows (Figure 3.8):

$$P(X = x) = \frac{(m)^x e^{-m}}{x!} \tag{3.26}$$

3.3.3 Poisson Process

In probability theory, the Poisson process is one of the most main random processes. It is widely utilized to model random 'points' in space and time. Several critical probability distributions result normally from the Poisson process. In other words, the exponential distribution with a stable failure rate in bathtub curve might be expressed by a Poisson process. In the following, it is instructive to think that the Poisson process stands for discrete failure time.

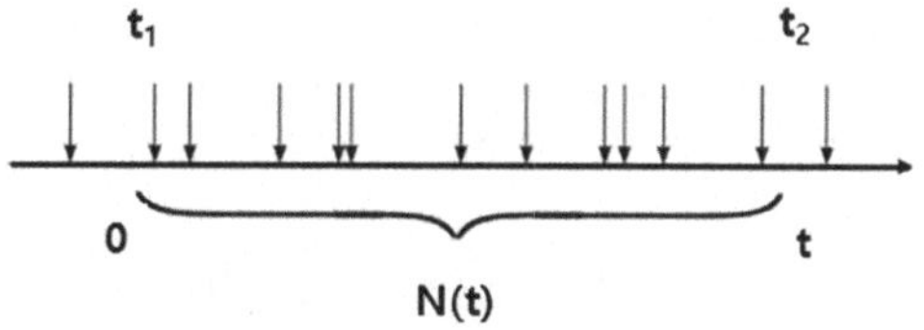

Mathematically, the process is expressed by the so-called counter process N_t or $N(t)$. The counter tells the number of failures that have happened in the interval (0, t) or, more usually, in the interval (t_1, t_2).

$N(t)$ = number of failures in the interval (0, t) (the stochastic process)

$N(t_1, t_2)$ = number of arrivals in the interval (t_1, t_2) (the increment process $N(t_2) - N(t_1)$).

The counting process $\{N(t), t \geq 0\}$ is a Poisson process with rate λ if all the following terms hold:

$$\text{i} \quad N(0) = 0, \tag{3.27a}$$

$$\text{ii} \quad N(t) \text{ has independent increments,} \tag{3.27b}$$

$$\text{iii} \quad N(t) - N(s) \sim \text{Poisson}(\lambda(t-s)) \text{for } s < t. \tag{3.27c}$$

Example 3.17

A fire insurance company has 1,860 customers. The probability of a customer to request for insurance money in a year is 1/600. Let random variable X be the frequency of insurance money requested. Find the probability for X = 0, 1, 2, 3.

Because n = 1,860 and p = 1/600, mean $m = np = 1860 \times 1/600 = 3.1$. So, if Poisson distribution with mean m = 3.1 is applied, the probability for X = 0, 1, 2, 3 is

$$P(X=0) = e^{-3.1}\frac{(3.1)^0}{0!} = 0.045$$

$$P(X=1) = e^{-3.1}\frac{(3.1)^1}{1!} = 0.140$$

$$P(X=2) = e^{-3.1}\frac{(3.1)^2}{2!} = 0.216$$

$$P(X=3) = e^{-3.1}\frac{(3.1)^3}{3!} = 0.223$$

Example 3.18

The number of typographical errors in a 'large' textbook is Poisson distributed with an average of 1.5 per 100 pages. Assume that 100 pages of the book are randomly chosen. What is the probability that there are no typos?

$$P(X=0) = e^{-m}\frac{m^x}{x!} = e^{-1.5}\frac{1.5^0}{0!} = 0.2231$$

Example 3.19

In a certain area, the average of traffic accidents occurs in one per two days. Find the chance of $x = 0, 1, 2$ accidents to happen in a given day.

Solution: one per two days means the mean of transport mishaps , $m = \lambda t = 0.5$,

$$X = 0,\ f(0) = \frac{(0.5)^0 e^{-0.5}}{0!} = 0.606$$

Accident days = 365 days × 0.606 = 221 days

$$X = 1,\ f(1) = \frac{(0.5)^1 e^{-0.5}}{1!} = 0.303$$

Accident days = 365 days × 0.303 = 110 days

$$X = 2,\ f(2) = \frac{(0.5)^2 e^{-0.5}}{2!} = 0.076$$

Accident days = 365 days × 0.076 = 27 days.

Example 3.20

Televisions sold in a certain area have an average failure rate of 1%/2000 hour. If a sample of 100 TV units are tested for 2,000 hours, find the chance that no accidents, $x = 0$, will happen.

Solution: $m = n \cdot \lambda \cdot t = 100 \times 0.01/2000 \times 2000 = 1.$
Because there is no mishap, the chance is

$$X = 0,\ f(0) = \frac{(1)^0 e^{-1}}{0!} = 0.36$$

We can estimate accidents for 100 TV units with confidence level of 63%. If there is no accident, $x = 0$, the confidence level is likely to increase to 90%, and then how many TV units will require?

$$X = 0,\ f(0) = \frac{(m)^0 e^{-m}}{0!} = 0.1$$

So if $m = 2.3$, the required sample size $n = 230$ will be obtained as

$$m = n \cdot \lambda \cdot t = n \times 0.01 / 2000 \times 2000 = 2.3$$

The mean and variance of Poisson distribution can be expressed as follows:

$$E(X) = m, \quad Var(X) = m \tag{3.28}$$

The reason why the variance of Poisson distribution has m can be understood intuitively. As you know, the variance of binomial distribution is npq. If mean m is fixed to np and p approaches 0, q will be 1. Therefore, we know that variance npq becomes m.

3.3.4 Exponential Distribution

In reliability engineering, the exponential distribution with only one unspecified parameter is the easiest of all life distribution models. Many engineering modules show a stable failure rate during the product lifetime if they suit the exponential distribution. Also, it is comparatively easy to manage in carrying out reliability analysis. The critical equations for the exponential distribution are expressed as follows:

$X(t)$ = the time it takes for one additional arrival, assuming that someone arrived at time t

By definition, the following conditions are equivalent:

$$(X(t) > x) \equiv (N(t) = N(t + x)) \tag{3.29}$$

The occurrence on the left-hand side of Equation (3.29) catches the possibility that no one has arrived in the time interval $[t, t + x]$, which indicates that the number of arrivals at time $t + x$ is similar to the number of arrivals at time t which is the possibility on the right side.

By the complement rule, we also have

$$P(X(t) \le x) = 1 - P(X(t) > x) \tag{3.30}$$

Using the equivalence of the two events that we expressed above, we can rewrite the above as follows:

$$P\big(X(t) \le x\big) = 1 - P\big(N(t + x) - N(t) = 0\big) \tag{3.31}$$

where $P(N(t + x) - N(t) = 0) = P(N(x) = 0)$.

Using the Poisson distribution Equation (3.26), where λ is the mean number of arrivals per time unit and x is the quantity of time units, right side in Equation (3.31) can be expressed as follows:

$$P\big(N(t + x) - N(t) = 0\big) = \frac{(\lambda x)^0}{0!} e^{-\lambda x} \tag{3.32}$$

Exchanging Equation (3.32) into Equation (3.30), we have

$$P\big(X(t) \le x\big) = 1 - e^{-\lambda x} \tag{3.33}$$

In other way, let X_1 be the time of the first failure. We can find reliability function $R(t)$ from Equation (3.22). That is,

$$R(t) = P(X_1 > t) = P\big(\text{no failure} \quad \text{in } (0,t]\big) = \frac{(m)^0 e^{-m}}{0!} = e^{-m} = e^{-\lambda t} \tag{3.34}$$

So, the cumulative distribution function as complement is also expressed as follows:

$$F(t) = 1 - e^{-\lambda t} \tag{3.35}$$

If the cumulative distribution function is differentiated, the probability density function is obtained as follows:

$$f(t) = \lambda e^{-\lambda t} \quad t \geq 0, \lambda > 0 \tag{3.36}$$

Failure rate $\lambda(t)$ is defined by

$$\lambda(t) = f(t)/R(t) = \lambda e^{-\lambda t}/e^{-\lambda t} = \lambda \tag{3.37}$$

Consider that the failure rate lessens the constant λ for any period. The exponential distribution is the only distribution to have a stable failure rate. Also, the other name for the exponential mean is the Mean Time To Fail or MTTF, and we have MTTF = $1/\lambda$. Generally, if a product suits the exponential distribution, its MTTF is 0.63 at $1/\lambda$ (Figure 3.9).

Some consideration points on exponential distribution model are as follows:

1. Due to its stable failure rate property, the exponential distribution is a good model for the 'useful life or design life' on the part of bathtub curve. Since most mechanical systems spend most of their lifetimes in this part of bathtub curve, this accounts for repeated usage of the exponential distribution (Figure 3.10).
2. Exactly as it is frequently functional to estimate a curve by piecewise straight line pieces, we can estimate any failure rate curve by week-by-week or month-by-month stable rates that are the mean of the real changing rate during the respective time periods.

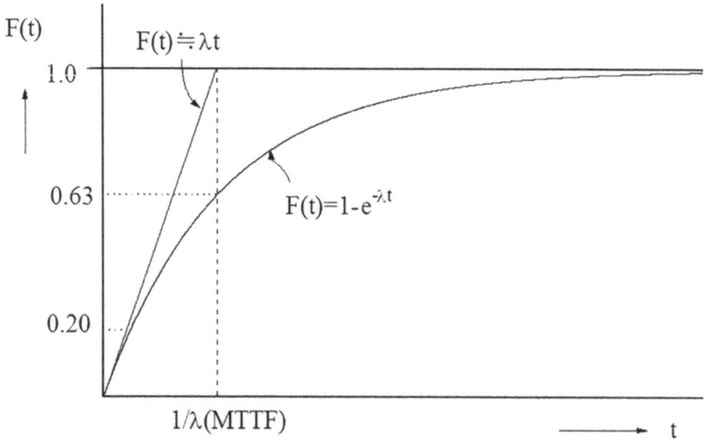

FIGURE 3.9 Cumulative distribution function $F(t)$ of exponential distribution.

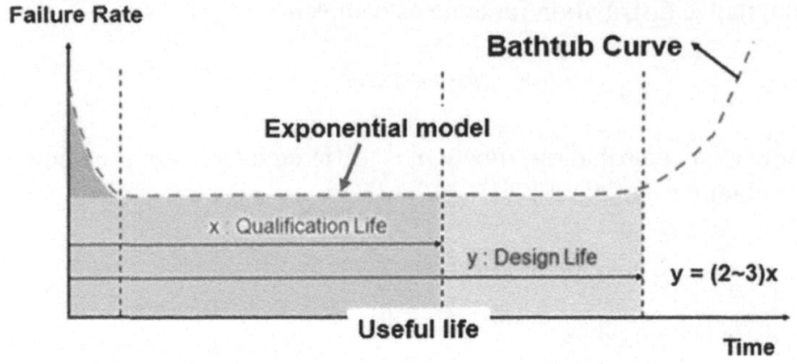

FIGURE 3.10 Design life versus qualification life.

3. An unspecified number of reasonable phenomena have a stable failure rate (or occurrence rate) property. The exponential model works well for inter-arrival times, while Poisson distribution reports the entire number of events in an assigned time period.
4. In field, the product recall happens rarely in its lifetime. Products follow the exponential distribution for this period in bathtub curve. This means that statistics only cannot explain the relationship between material (or design) and stress. So, we need a new statistical methodology such as parametric ALT that is combined with failure mechanics and design.

3.3.4.1 Expectation and Standard Variation of Exponential Distribution

It is convenient to use the unit step function expressed as

$$u(x) = 1 \quad \text{for } x \geq 0 \tag{3.38}$$

So, we can write the probability density function of an exponential random variable as

$$f_X(x) = \lambda e^{-\lambda x} u(x) \tag{3.39}$$

Let us find its cumulative density function, mean, and variance. For $x>0$, we have

$$F_X(x) = \int_0^x \lambda e^{-\lambda x}\, dt = 1 - e^{-\lambda x} \tag{3.40}$$

So, we can express the cumulative density function as

$$F_X(x) = \left(1 - e^{-\lambda x}\right) u(x) \tag{3.41}$$

Assuming X~Exponential(λ), we can obtain its expected value, utilizing integration by parts, as follows:

$$E(X) = \int_0^\infty x\lambda e^{-\lambda x}\, dx = \frac{1}{\lambda}\left[-e^{-\lambda x} - \lambda x e^{-\lambda x}\right]_0^\infty = \frac{1}{\lambda} \tag{3.42}$$

Now let's find *Var*(*X*). We have

$$E(X^2) = \int_0^\infty x^2 \lambda e^{-\lambda x}\,dx = \frac{1}{\lambda^2}\left[-2e^{-\lambda x} - 2\lambda x e^{-\lambda x} - (\lambda x)^2 e^{-\lambda x}\right]_0^\infty = \frac{2}{\lambda^2} \quad (3.43)$$

Thus, we obtain

$$V(X) = E\left(X^2\right) - (E(X))^2 = \frac{2}{\lambda^2} - \frac{1}{\lambda^2} = \frac{1}{\lambda^2} \quad (3.44)$$

If a random variable X is exponentially distributed with rate λ, the expectation and standard variation can be expressed as follows:

$$E(X) = \frac{1}{\lambda}, \quad V(X) = \left(\frac{1}{\lambda}\right)^2 \quad (3.45)$$

3.3.5 Weibull Distribution

In distinguishing the failure periods of definite parts, one usually utilizes the Weibull distribution. As it was suggested by Weibull in the early 1950s, this distribution might be used to stand for many different failure behaviors. Many other extensions of the Weibull distribution have been suggested to extend its capability to be the right shape of diverse lifetime data since 1970s.

If a random variable X is characterized as the individuals of a population, the distribution function of X, designated $F(x)$, may be expressed as the number of all individuals having an $X \le x$, divided by the entire number of individuals. This function also gives the probability P of selecting at random an individual having a value of X equal to or less than x, and thus we have

$$P(X \le x) = F(x) \quad (3.46)$$

Any distribution function may be expressed as follows:

$$F(x) = 1 - e^{-\phi(x)} \quad (3.47)$$

Though this seems to be a complication, this formal change relies on the relation

$$(1 - P)^n = e^{-n\phi(x)} \quad (3.48)$$

The advantages of Equation (3.48) can be explained on a product failure. Presume that we have a product composing several parts. If we have found, by testing, the chance of failure P at any load x applied to a 'single' part (or link), and if we want to find the chance of failure P_n of a product (or chain) composing n parts (or links), we have to base our deductions upon the proposition that the whole chain fails, if any one of its parts fails. Therefore, the probability of non-failure of the chain $(1 - P_n)$ is equal to the probability of the simultaneous non-failure of all the links.

Thus, we have $(1 - P_n) = (1 - P)^n$. If the distribution function of a single link has the form of Equation (3.47), we get

$$P_n = 1 - e^{-n\phi(x)} \tag{3.49}$$

Equation (3.49) gives the proper mathematical expression for the principle of the weakest chain, or, more usually, for the size effect on failures in solids. The same way of reasoning may be applicable to a big group of problems, where the occurrence of an event in any part of an object may be said to have happened in the object as a whole, e.g., static or dynamic strengths, electrical insulation breakdowns, the phenomena of yield limits, life of electric bulbs, or even death of a man, as the chance of not having died from many different causes.

Now we have to specify the function $\phi(x)$. The only requisite common condition this function has to assure is to be a positive, non-decreasing function. The simplest function satisfying this condition is

$$F(x) = \int_0^\infty f(x)dx = 1 - e^{-\left(\frac{x}{\eta}\right)^\beta} \tag{3.50}$$

where η and β are characteristic life and shape parameters, respectively.

3.3.6 Normal Distribution

Normal distribution was suggested by French mathematician Abraham de Moivre (1667–1754). After that, German mathematician and physicist Johann Carl Friedrich Gauss (1777–1855) made major contributions to a lot of areas such as astronomy and physics. The normal distribution function or Gaussian distribution function is expressed as follows:

$$f(x) = \frac{1}{\sqrt{2\pi}\sigma}\exp\left\{-\frac{1}{2}\left(\frac{x-\mu}{\sigma}\right)^2\right\} \quad \text{for} -\infty < x < \infty \tag{3.51}$$

Therefore, the Gaussian distribution curve has two parameters, namely, mean μ and standard deviation σ, which is symmetric around the mean μ and has a bell shape. If a random variable X is normally distributed with mean μ and variance σ^2, it will be expressed as $X \sim N(\mu, \sigma^2)$. Its characteristics can be summarized as follows:

- Continuous for all values of X between $-\infty$ and ∞ and bell shaped.
- Symmetric around the average μ. Chance for the left and the right of mean are 0.5, respectively.
- Relies on parameters – μ and σ, and there are infinite normal distributions.
- The probability for interval $[\mu - \sigma \leq X \leq \mu + \sigma]$ is 0.6826. The probability for interval $[\mu - 2\sigma \leq X \leq \mu + 2\sigma]$ is 0. 9544. And the probability for interval $[\mu - 3\sigma \leq X \leq \mu + 3\sigma]$ is 0.997. In other words, most of the data in normal distribution locate around mean, and there is very little data at more than three times of standard deviation.

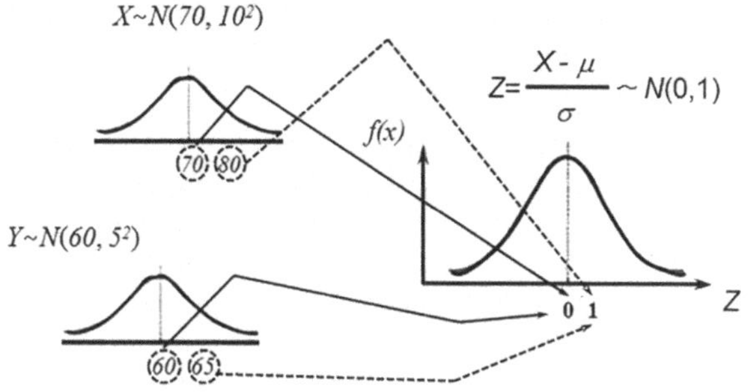

FIGURE 3.11 Standardization normal distribution.

In statistics, the normal distribution is the most critical distribution, since it results normally in numerous areas. The major reason is that big sums of (small) random variables usually prove to be normally distributed. When random variable X suits $N(\mu, \sigma^2)$, the probability for interval $[a, b]$ shall be the region of $f(x)$ that is surrounded by a and b on x axis. The mathematical region is expressed as follows:

$$P(a \le X \le b) = \int_a^b \frac{1}{\sqrt{2\pi}\sigma} \exp\left\{-\frac{1}{2}\left(\frac{x-\mu}{\sigma}\right)^2\right\} dx \tag{3.52}$$

However, Equation (3.52) is very hard to figure out. Luckily, in case of a normal random variable X with arbitrary parameters μ and σ, we can transform it into a standardized normal random variable $Z = \dfrac{X-\mu}{\sigma}$ with parameters 0 and 1 (Figure 3.11).

For a standard normal distribution function, we can also refer to its table that represents probability, namely, area from the end of left to real value z. Table 3.1 is a part of standard normal distribution.

It is better to memorize several probabilities of standard normal distribution because of their frequent usage. Figure 3.12 shows probability in case of excluding both ends equally. That is, the probabilities are 90% ($P(-1.645 \le z \le 1.645)$), 95% ($P(-1.96 \le z \le 1.96)$), and 99% ($P(-2.575 \le z \le 2.575)$).

Using this table of standard normal distributions, we can find the chance of normal distribution. If a random variable X has mean μ and standard deviation σ^2, the probability for interval $[a, b]$ is given by

$$P(a \le X \le b) = P\left(\frac{a-\mu}{\sigma} \le Z \le \frac{b-\mu}{\sigma}\right) \tag{3.53}$$

Example 3.21

If random variable X suits normal distribution that has a mean of 70 and a standard deviation of 10, we can get the following probability:

1. $P(X<94.3)$, (2) $P(X>57.7)$, (3) $P(57.7<X<94.3)$

TABLE 3.1
Standard Normal Distribution Table

Z	0.000	0.01	0.02	0.03	0.04	0.05	0.06	0.07	0.08	0.09
0.0	0.5000	0.5040	0.5080	0.5120	0.5160	0.5199	0.5239	0.5279	0.5319	0.5359
0.1	0.5398	0.5438	0.5478	0.5517	0.5557	0.5596	0.5636	0.5675	0.5714	0.5753
0.2	0.5793	0.5832	0.5871	0.5910	0.5948	0.5987	0.6026	0.6064	0.6103	0.6141
0.3	0.6179	0.6217	0.6255	0.6293	0.6331	0.6368	0.6406	0.6443	0.6480	0.6517
0.4	0.6554	0.6591	0.6628	0.6664	0.6700	0.6736	0.6772	0.6808	0.6844	0.6879
0.5	0.6915	0.6950	0.6985	0.7019	0.7054	0.7088	0.7123	0.7157	0.7190	0.7224
0.6	0.7257	0.7291	0.7324	0.7357	0.7389	0.7422	0.7454	0.7486	0.7517	0.7549
0.7	0.7580	0.7611	0.7642	0.7673	0.7704	0.7734	0.7764	0.7794	0.7823	0.7852
0.8	0.7881	0.7910	0.7939	0.7967	0.7995	0.8023	0.8051	0.8078	0.8106	0.8133
0.9	0.8159	0.8186	0.8212	0.8238	0.8264	0.8289	0.8315	0.8340	0.8365	0.8389
1.0	0.8413	0.8438	0.8461	0.8485	0.8508	0.8531	0.8554	0.8577	0.8599	0.8621
1.1	0.8643	0.8665	0.8686	0.8708	0.8729	0.8749	0.8770	0.8790	0.8810	0.8830
1.2	0.8849	0.8869	0.8888	0.8907	0.8925	0.8944	0.8962	0.8980	0.8997	0.9015
1.3	0.9032	0.9049	0.9066	0.9082	0.9099	0.9115	0.9131	0.9147	0.9162	0.9177
1.4	0.9192	0.9207	0.9222	0.9236	0.9251	0.9265	0.9279	0.9292	0.9306	0.9319
1.5	0.9332	0.9345	0.9357	0.9370	0.9382	0.9394	0.9406	0.9418	0.9429	0.9441
1.6	0.9452	0.9463	0.9474	0.9484	0.9495	0.9505	0.9515	0.9525	0.9535	0.9545
1.7	0.9554	0.9564	0.9573	0.9582	0.9591	0.9599	0.9608	0.9616	0.9625	0.9633
1.8	0.9641	0.9649	0.9656	0.9664	0.9671	0.9678	0.9686	0.9693	0.9699	0.9706
1.9	0.9713	0.9719	0.9726	0.9732	0.9738	0.9744	0.9750	0.9756	0.9761	0.9767

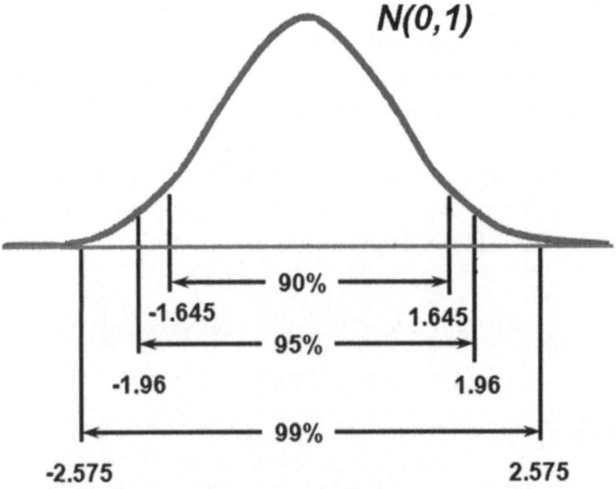

FIGURE 3.12 Probabilities in the case of excluding both ends equally are 90%, 95%, and 99%.

Solution

1. $P(X < 94.3) = P\left(\frac{X-70}{10} < \frac{94.3-70}{10}\right) = P(Z < 2.43) = 0.9925$

2. $P(X > 57.7) = P\left(\frac{X-70}{10} > \frac{57.7-70}{10}\right) = P(Z > -1.23) = 0.8907$

3. $P(57.7 < X < 94.3) = P\left(\frac{57.7-70}{10} < \frac{X-70}{10} < \frac{94.3-70}{10}\right)$

$= P(-1.23 < Z < 2.43) = 0.8832$

Example 3.22

If the battery lifetime in a company follows a normal distribution that has a mean of 110 hours and a standard deviation of 10, we can find the following probability:

1. Probability that the battery lifetime is less than 90 hours: $P(X \le 90)$.
2. Probability that the battery lifetime is greater than 100 hours and less than 115 hours: $P(100 \le X \le 115)$.
3. Probability that the battery lifetime is greater than 120 hours: $P(X \ge 120)$.

Solution

1. $P(X \le 90) = P\left(\frac{X-110}{10} \le \frac{90-110}{10}\right) = P(Z \le -2) = 0.0228$

2. $P(100 \le X \le 115) = P\left(\frac{100-110}{10} \le \frac{X-110}{10} \le \frac{115-110}{10}\right)$

$= P(-1 \le Z \le 0.5) = 0.6915 - 0.1587 = 0.5328$

3. $P(X \ge 120) = P\left(\frac{X-110}{10} \frac{120-110}{10}\right) = P(Z \ge 1)$

$= 1 - P(Z < 1) = 1 - 0.8413 = 0.1587$

3.4 SAMPLE DISTRIBUTIONS

We have studied to examine a sample on the premise that the characteristics of a population are investigated. However, in actual life or academic research, we have frequently experienced the situations where we have to find out its characteristics if populations are unknown (Figure 3.13).

If we would like to be aware of the average capacity of a battery manufactured in a company, we have to look over its entire products. However, it is very difficult to accomplish this task because it consumes a lot of time and cost. If we select a proper sample from the entire products and calculate its statistics, we can estimate the parameters of all products such as mean and standard deviations in certain confidence level. In the same way, choosing a sample from a population and thus drawing conclusion on the entire population is called statistical inference.

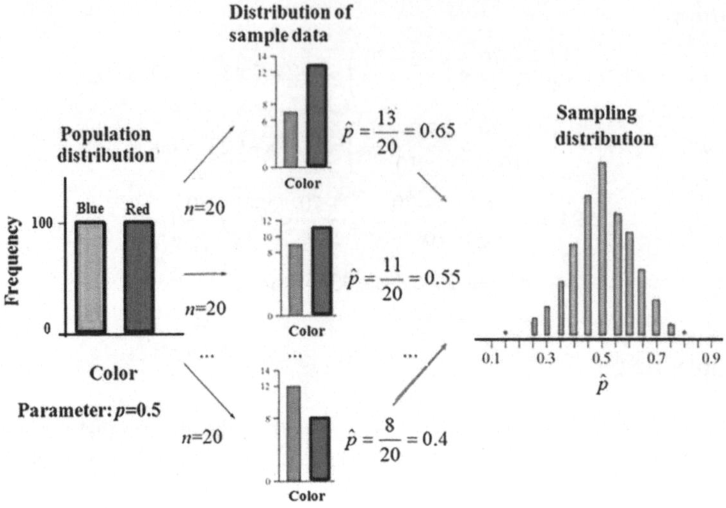

FIGURE 3.13 Sample distributions in the case of a known population.

If we choose a random sample, the numerical descriptive standards are called statistics. These statistics change for each different random sample we choose. In other words, they are random variables. If the sampling is carried out randomly, the value of a statistic will be random. Since statistics are random variables, they have the sampling distribution. It supplies the following information: (1) what values of the statistic can happen? and (2) what is the chance of each value to happen?

For each random sample, the sample mean $\overline{x}$ is different, which is called the sample variability. From possible random sampling obtained from students in certain class, we can generate the sampling distribution. That is,

1. The value of $\overline{x}$ differs from one random sample to another (sample variability).
2. Some samples produced $\overline{x}$ values larger than μ, whereas others produce $\overline{x}$ smaller than μ.
3. They can be fairly close to the mean μ, or also quite far off the population mean μ.

The sampling distribution of $\overline{x}$ provides important information about the behavior of the statistic $\overline{x}$ and how it relates to the population mean μ. If the population increases, it is impossible to find the distribution of sample mean by checking out all possible samples. However, the facts observed in the above example can be true for a greater population or other distributions.

3.4.1 The Distribution of Sample Mean

Sample mean $\overline{x}$ that is calculated from a big sample tends to be closer to μ than does $\overline{x}$ based on a small n. Assume that $X_1, \ldots, , X_n$ are random variables with the same distribution with mean μ and population standard deviation σ. Now look at the random variable $\bar{X}$.

$$\bar{X} = \frac{1}{n}(X_1 + X_2 + \cdots + X_n) \tag{3.54}$$

$$E(\bar{X}) = \frac{1}{n}(E(X_1) + E(X_2) + \cdots E(X_n)) = \frac{1}{n} \cdot n\mu = \mu \tag{3.55}$$

$$V(\bar{X}) = \frac{1}{n^2}(V(X_1) + V(X_2) + \cdots + V(X_n)) = \frac{1}{n^2} \cdot n\sigma^2 = \frac{\sigma^2}{n} \tag{3.56}$$

The expectation value is $E(\bar{X}) = \mu$, and the variation is $V(\bar{X}) = \sigma^2/n$. That is,

1. The population mean of $\bar{X}$, indicated as $\mu_{\bar{X}}$, is equal to μ.
2. The population standard deviation of $\bar{X}$, indicated as $\sigma_{\bar{X}}$, is equal to $\sigma_{\bar{X}} = \dfrac{\sigma}{\sqrt{n}}$.

It signifies that the sampling distribution of $\bar{x}$ is always pivoted at μ. As n increases, the second statement means that the spread of the sampling distribution (sampling variability) will decrease. The standard deviation of a statistic is called the standard error of the statistic. The standard error gives the precision of the statistic for estimating a population parameter. The smaller the standard error, the higher the precision. The standard error of the mean $\bar{X}$ is $SE(\bar{X}) = \sigma/\sqrt{n}$.

We learned about the average and the standard deviation of the sampling distribution for the sample mean, we might ask, if there is anything, we can tell about the shape of the density curve of this distribution.

If the population is infinite and sample size n is big enough, we know that the distribution of sample mean is approximately normally distributed regardless of the population characteristics. Under rather general conditions, the central limit theorem states that the means of random samples drawn from one population tend to have an approximately normal distribution. We find that it does not matter which kind of distribution we find in the population. It can even be discrete or extremely skewed (Figure 3.14).

3.4.2 Central Limit Theorem

From any population with finite mean μ and standard deviation σ, as n is large, if random samples of n observations are chosen, the sampling distribution of the mean $\bar{X}$ is approximately normally distributed, with mean μ and standard deviation $\sigma/\sqrt{n}$. That is,

$$\bar{X} \sim N\left(\mu, \frac{\sigma^2}{n}\right) \tag{3.57}$$

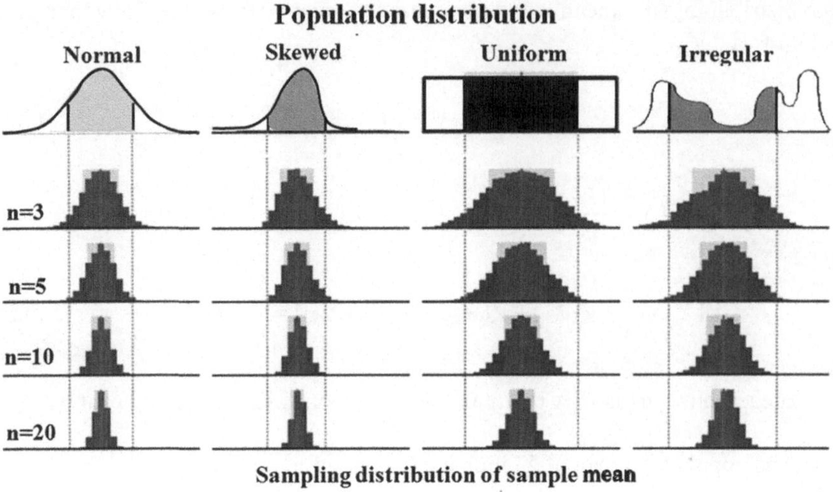

FIGURE 3.14 Central limit theorem.

The central limit theorem becomes very substantial one in the modern statistics. Look at a binomially distributed random variable *X*. With probability *p* (Bernoulli trial), we can carry out the experiment of *n* trials. According to this theorem, as *n* increases infinitely, the random variable *X* will be distributed normally.

The random variable *X* is binomially distributed $B(n, p)$ with mean pn and standard deviation $\sqrt{p(1-p)}$. Since $\hat{p}$ is simply the value of *X* expressed as a proportion, the sampling distribution of $\hat{p}$ is identical to the probability distribution of *X*. Then, it is $\hat{p} = \sum X_i/n = \bar{X}$. According to the central limit theorem, as *n* increases infinitely, it will be normally distributed $N(np, np(1-p))$, where the mean is np and the variation is $np(1-p)$. As *n* increases infinitely, we know that the random variable *X* is normally distributed:

$$\frac{X - np}{\sqrt{np(1-p)}} \sim N(0,1) \tag{3.58}$$

Example 3.23

If the defective proportion of a product manufactured in a factory is 5% and a total of 100 products in some day are sampled, find out (1) the probability that is less than 2 pieces and (2) the probability that is from 3 to 7 pieces.

Solution: Let *X* be the number of defective products. *X*, which is binomially distributed, is $n = 100$ and $p = 0.05$. As *n* increases infinitely, we can calculate the probability using the normal distribution. The mean of binomial distribution is $np = 100 \times 0.05 = 5$, and its variation is $np(1 - p) = 100 \times 0.05 \times (1 - 0.05) = 4.75$. So, if normal distribution $N(5, 4.75)$ is used, the probability is

1. $P(X \le 2) = P\left(Z \le \frac{2-5}{\sqrt{4.75}}\right) = P(Z \le -1.375) = 0.0845$

2. $P(3 \le X \le 7) = P\left(\frac{3-5}{\sqrt{4.75}} \le Z \le \frac{7-5}{\sqrt{4.75}}\right) = P(-0.918 \le Z \le 0.918) = 0.642$

Distribution of Sample Variation: If n sample is chosen randomly in a population with variation σ^2 that is normally distributed, $(n-1)S^2/\sigma^2$ follows chi-squared distribution with $(n - 1)$ degrees of freedom. That is,

$$(n-1)S^2/\sigma^2 \sim \chi^2(n-1) \tag{3.59}$$

where $S^2 = \frac{\sum_{i=1}^{n}(X_i - \bar{X})^2}{n-1}$.

4 Descriptive Statistics
Lifetime Analysis of Products

4.1 INTRODUCTION

The reliability of an item is to perform a required (intended) function under given environmental and operational conditions for a specified time period (ISO8402) [1]. Most theories in reliability engineering had been derived from the main ideas of probability and statistics, which were mainly utilized as tools by nineteenth-century maritime and life insurance companies to compute profitable rates to charge their customers.

Invented in 1904 by John Ambrose Fleming, vacuum tubes were a fundamental part of electronic products throughout the first half of the twentieth century. They were also used in the mechanical products, such as tanks and airplanes, as semi-conductors that could connect two systems. However, vacuum tubes became problematic parts due to the problems arose in their usage in the combat missions of World War II. Many engineers tried to solve these problems. One of such efforts was to establish the organization of Advisory Group on the Reliability of Electronic Equipment (AGREE). AGREE suggested the basic reliability concepts such as bathtub and initiated the use of modern concepts in reliability engineering. Ironically, as the problems still remained thereafter, vacuum tubes were replaced by the use of silicon as a semiconductor in the modern electronics.

The reliability of mechanical products in the R&D phase is achieved by the design robustness of the products' (intended) functions under customer usage or environmental conditions. As a mechanical system is subjected to repeated stress (or loads) during its lifetime, mechanical structures are designed to have proper strength and stiffness to endure it. Strength, the resistance against irreversible deformation, is always required to be high. Requirements on stiffness, the resistance against reversible deformation, may depend on their applications.

A product's failure in its lifetime seldom happens for a small sample in the field so that the characteristics, i.e., failure rate versus time, could be described in the bathtub that composes three elements [2]. First, there is less failure rate in the initial stage of the lifetime of the products ($\beta < 1$). Then, there is a stable failure rate ($\beta = 1$). Finally, there is an increasing failure rate toward the end of the products' life ($\beta > 1$). On bathtub, products would follow an exponential distribution for useful lifetime. If there are design faults in the products, the permanent deformation (i.e. failure) due to repeated loads and design faults may become product failure such as fatigue and even lead to loss of product functionality. The analysis of product lifetime, therefore, should be replaced with the Weibull distribution that was derived from the model of the weakest chain [3].

Based on testing or field data, an engineer can estimate the product lifetime such as MTTF or BX lifetime using the Weibull analysis. To assess its lifetime through the analysis of testing data, the engineer should also recognize new concepts for combining the modern definitions of reliability engineering with design concepts of mechanical system. Eventually, we can derive a general lifetime failure model and a sample size equation based on that. So, reliability testing such as parametric accelerated life testing (ALT) will reveal the design faults of a product and modify them. Finally, it assures whether the lifetime target of the final designs is achieved. This chapter, therefore, will discuss the fundamental concepts of reliability engineering – reliability (or unreliability), bathtub curve, Weibull distribution, Weibayes method, Maximum Likelihood Estimate (MLE), MTTF, BX life, etc.

4.2 RELIABILITY AND BATHTUB CURVE

4.2.1 Product Reliability

Let the survival time, T, for individuals in a population have a density function $f(t)$. The equivalent distribution function is the fraction of the population failing by the time t. That is,

$$F(t) = \int_{-\infty}^{t} f(s)ds \tag{4.1}$$

The complementary function $1 - F(t)$, often defined as the reliability function, is the fraction still surviving at time t. The failure rate function $\lambda(t)$ measures the instantaneous risk, in which $\lambda(t)\delta t$ is the probability of failing in the next small interval δt given survival to time t. From the relation,

$$\text{pr(survival to } t + dt) = \text{pr(survival to } t) \text{ pr(survival to } \delta t \mid \text{survival to } t) \tag{4.2}$$

where pr is the probability.

We have the following equations:

$$1 - F(t + \delta t) = \{1 - F(t)\}\{1 - \lambda(t)\delta(t)\} \tag{4.3}$$

$$\delta t F'(t) = \{1 - F(t)\}\lambda(t)\delta(t) \tag{4.4}$$

so that the failure rate function is given by

$$\lambda(t) = \delta(t)F'(t)/\{1 - F(t)\}\delta(t) = f(t)/\{1 - F(t)\} \tag{4.5}$$

A distribution for survival times should have a failure rate function with proper properties. Thus, for a large t, the failure rate function will not decrease, because at certain point in time, the probability of breakdown does not normally decrease. For a small t, various forms can be justified, including the one that initially declines with t, for such distribution could describe the behavior of a mechanical part with a settling-in period, where reliability increases once the initial period is over.

The simplest failure rate function implies an exponential distribution of survival times with a Poisson process. For if T has the density

$$f(t) = \lambda e^{-\lambda t}; \quad t \geq 0, \tag{4.6}$$

then

$$F(t) = 1 - e^{-\lambda t}, \tag{4.7}$$

and from Equation (4.5), we can find the hazard rate function:

$$h(t) = \lambda(t) = f(t)/\{1 - F(t)\} = \lambda e^{-\lambda t}/e^{-\lambda t} = \lambda \tag{4.8}$$

4.2.2 Bathtub Curve

When a product is initially designed and manufactured, the life features of the problematic product, such as a vacuum tube, can be described as the bathtub curve because it has high failure rate and short lifetime. It was produced by functioning the rate of initial "infant mortality" failures ($\beta < 1$), the rate of random failures with a stable failure rate during its "useful life" ($\beta = 1$), and eventually the rate of "wear-out" failures ($\beta > 1$), which pursue the Weibull distribution.

Early potential origins of initial failure such as handling, storage, and installation error are dominated in bathtub curve. Aging test can immediately eliminate them. In the mid-life of bathtub, the failure rate of a product is stable because it follows an exponential distribution. In this period, the product may endure random failures due to, for example, overuse by customers and exposure to overstress. The product will experience wear failure due to long usage in the latter part of bathtub. However, if any design problems exist in the system, the failure rate of the product might increase suddenly and catastrophically in useful life. So, this area will be analyzed by the Weibull distribution (Figure 4.1).

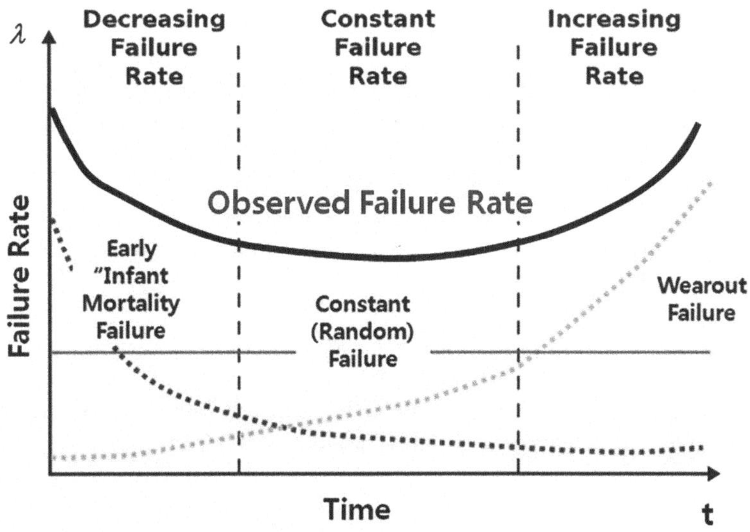

FIGURE 4.1 Bathtub curve.

If a product pursuits the bathtub curve, it may face difficulties to be successful in the marketplace. Because of the higher failure rates and short lifetimes due to the inherited design faults, companies may incur financial losses throughout the whole product life cycle. Therefore, they should emphatically improve the design of a product by planning reliability goals for new products to (1) reduce initial failures, (2) lessen random failures during the product operating time, and (3) increase product lifetime.

To do this, firstly, the life of parts might be lengthier than the anticipated life of the set product. Secondly, the failure rate of the product is the sum of the failure rates of all parts and the related connections between them. The lifetime of a mechanical product depends on its problematic modules (or units), which will be discussed in Chapter 8. Thirdly, early failure of all parts should be prevented.

In other words,

- Early Failures: Because these require a short test time, they can easily be removed by aging test prior to shipment.
- Random Failures: These require specified environmental tests related to usage/shipping/disposal which include the following:
  - Shipping Testing: transportation by vehicle/ship/rail, storage
  - Usage Testing: high voltage, shock, temperature & humidity, lightning, EMC (EMI)
  - Disposal (pass/fail testing, low frequency).
- Catastrophic Disasters: These usually occur within 1–2 years after manufacture and are caused by design failures. These can be fixed through identification and modification by parametric ALT described in Chapters 8 and 9 (Figure 4.2).

4.2.3 Cumulative Distribution Function *F(t)*

Reliability is the chance that a mechanical system will satisfactorily work for a design life under the operating/environmental conditions. If T is the random variable indicating the time to failure, the reliability function at time t might be defined as

$$R(t) = P(T > t) \tag{4.9}$$

As the complement of $R(t)$, the cumulative distribution function might also be defined as follows:

$$F(t) = 1 - R(t) \tag{4.10}$$

As seen in Figure 4.3, if the time to failure, T, has a probability density function $f(t)$, Equations (4.9) and (4.10) can be redefined as

$$R(t) = 1 - F(t) = 1 - \int_0^t f(\xi)\, d\xi \tag{4.11}$$

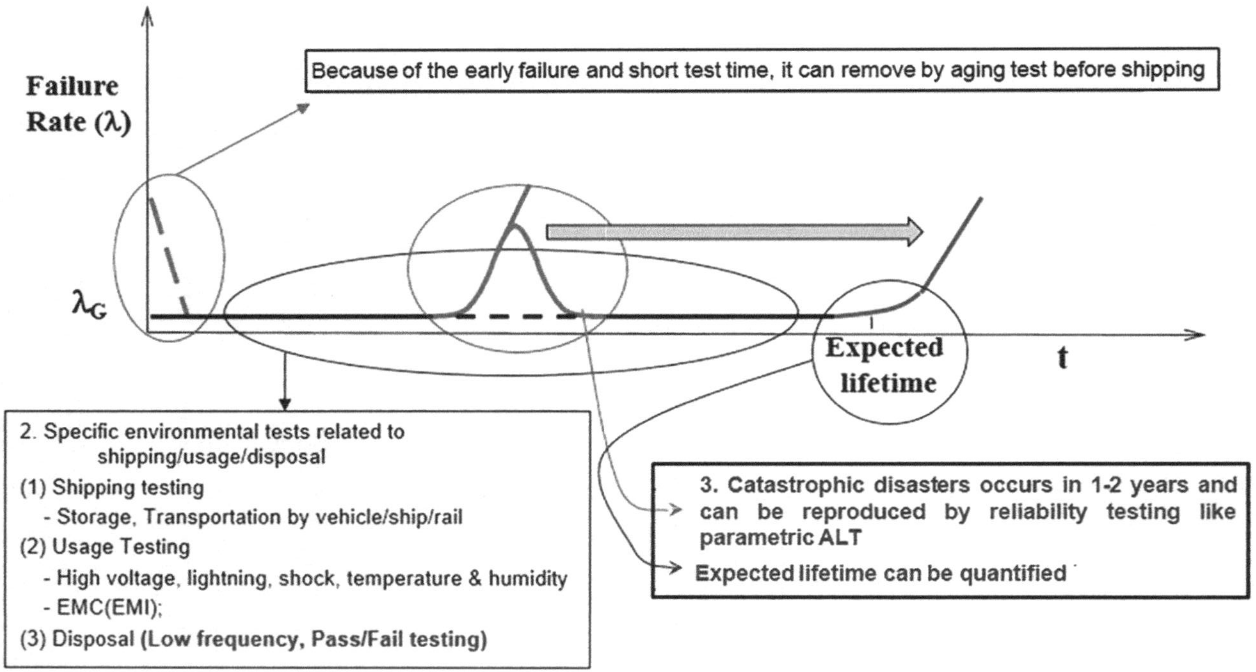

FIGURE 4.2 Three patterns of product tests associated with the failure rate.

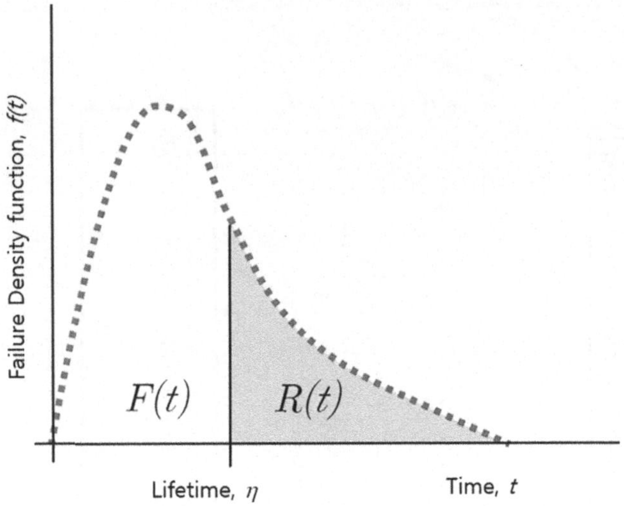

FIGURE 4.3 Reliability function $R(t)$ and cumulative distribution function $F(t)$.

The failure rate in a time interval $[t_1, t_2]$ might be defined as the chance that a failure rate unit time happens in the interval given that no failure has happened earlier to t_1, the beginning of the interval. Thus, the failure rate is defined as follows:

$$\frac{R(t_1)-R(t_2)}{(t_2-t_1)R(t_1)} \tag{4.12}$$

If we take the place of t_1 by t and t_2 by $t+\Delta t$, we might redefine as follows:

$$\frac{R(t)-R(t+\Delta t)}{\Delta t R(t)} \tag{4.13}$$

As Δt comes near to zero, the instantaneous failure rate can be redefined as follows:

$$\lambda(t)=\lim_{\Delta t\to 0}\frac{R(t)-R(t+\Delta t)}{\Delta t R(t)}=\frac{1}{R(t)}\left[-\frac{d}{dt}R(t)\right]=\frac{f(t)}{R(t)} \tag{4.14}$$

A hazard function is to examine the anticipated duration of time until one or more events occur, such as failure in mechanical systems. Cumulative hazard rate function $\Lambda(t)$ is expressed as follows:

$$\Lambda(t)=\int_0^t \lambda(x)\,dx \tag{4.15}$$

Assume that the failure rate $\lambda(t)$ is recognized. Then, it can get $f(t)$, $F(t)$, and $R(t)$:

$$f(t)=\frac{dF(t)}{dt}=-\frac{dR(t)}{dt}\Rightarrow\lambda(t)=-\frac{dR/dt}{R} \tag{4.16}$$

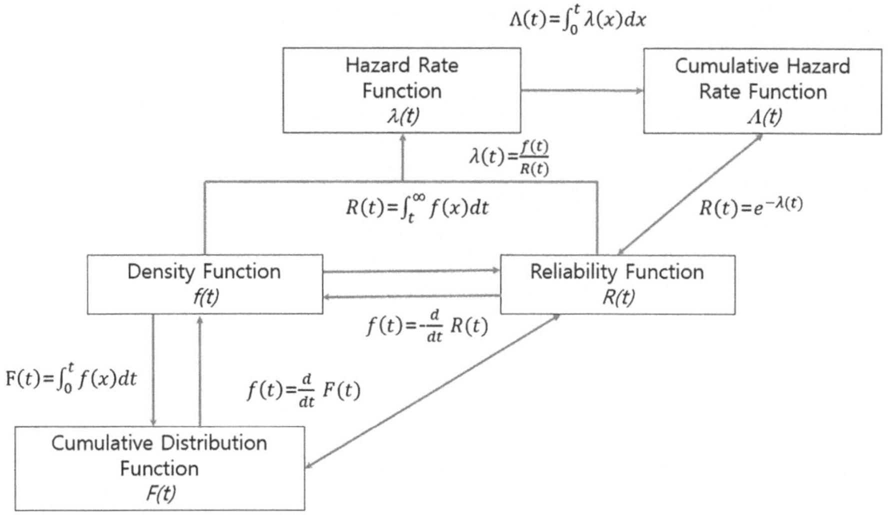

FIGURE 4.4 Relationship between reliability function $R(t)$ and cumulative distribution function $F(t)$ in reliability engineering.

If Equation (4.16) is integrated, reliability function can be defined as follows:

$$R(t) = \exp\left[-\int_0^t \lambda(\tau)d\tau\right] \tag{4.17}$$

Therefore, cumulative distribution function and density function are redefined as follows:

$$f(t) = \lambda(t)\exp\left[-\int_0^t \lambda(\tau)d\tau\right] \tag{4.18}$$

$$F(t) = 1 - \exp\left[-\int_0^t \lambda(\tau)d\tau\right] \tag{4.19}$$

Relationship between cumulative distribution function $F(t)$ and reliability function $R(t)$ in reliability engineering can be abridged as Figure 4.4.

4.3 LIFETIME METRICS

When a customer utilizes a product, failure reveals in the weakest part of the product due to a design fault. As time passes, the trend of product reliability can be expressed as failure rate in bathtub curve. For useful lifetime, the product pursues the exponential distribution with a constant failure rate. Engineers usually explain a design's lifetime to stand for an engineer's specifications of product usage under which the reliability must be verified. Generally, the design's lifetime (or useful lifetime) is 2–3 times longer than BX life (qualification lifetime). In other words, to make sure

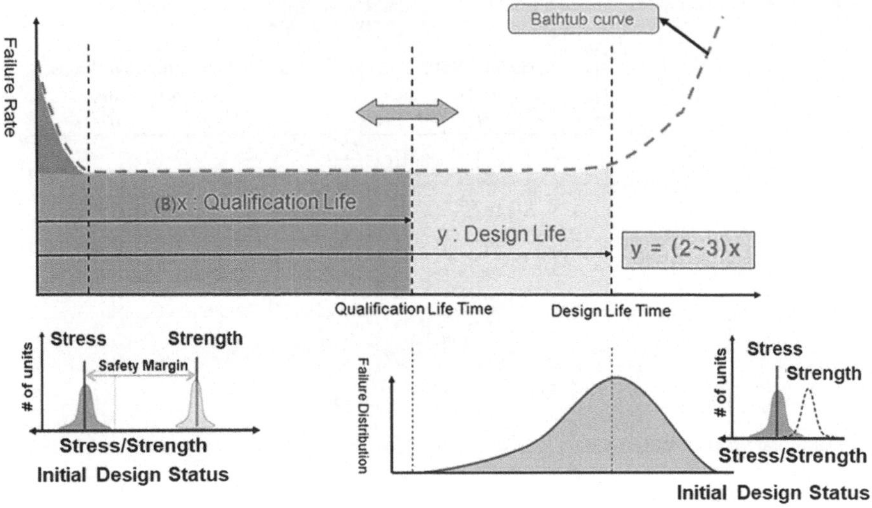

FIGURE 4.5 BX life (qualification life) and design life.

that product quality does not result in customer's discontent due to the ending of the useful life, it is critical idea and to go into the combination of this specification (see Figure 4.5).

The main goal of reliability designers is to evaluate lifetime from product failures in field or testing data. The design life of a product is the time period in which the product functions rightly in its lifetime. The following are metrics of the design lifetime:

- MTTF (Mean Time To Failure) – time to reach the accumulate failure rate of 60%
- MTBF (Mean Time Between Failures)
- *BX* life – time to reach the accumulate failure rate of *X*%

Qualification of products can be done by conducting the tests that are performed on prototypes or on the last product to make sure that it will satisfactorily operate across all environmental/operational conditions and will follow all government regulations. MTTF, MTBF, and *BX* are utilized as qualification standards to specify design lifetime. Fans, aluminum electrolytic capacitors, and batteries will be unsuccessful in achieving their goal due to overuse before they could achieve MTBF. *BX* life is another metric that can be used for design lifetime.

4.3.1 Mean Time To Failure (MTTF)

MTTF is the span of time that a product is anticipated to end in operation. MTTF is a fundamental lifetime standard of reliability to state the lifetime of non-repairable products such as "one-shot" tools like light bulbs. Non-repairable system is one for which single items that fail are eliminated forever from the population. It is the average time until a piece of equipment fails at first statistically. MTTF is the average over a long period of time with a big unit:

$$\text{MTTF} = \frac{t_1 + t_2 + \cdots + t_n}{n} \tag{4.20}$$

Example 4.1

Find the MTTF of an automobile from Figure 4.6.

So, we know that MTTF in Figure 4.6 is calculated from

$$\text{MTTF} = \frac{15{,}000 + 20{,}000 + 28{,}000 + 30{,}000}{4} \approx 23{,}000 \text{ km}.$$

MTTF might also be expressed with other mathematical terms:

$$\text{MTTF} = E(T) = \int_0^\infty t \cdot f(t)\,dt = -\int_0^\infty t \frac{dR(t)}{dt}\,dt = \int_0^\infty R(t)\,dt \tag{4.21}$$

where $f(t) = \frac{d}{dt}F(t) = -\frac{d}{dt}R(t)$.

Example 4.2

Consider a system with reliability function:

$$R(t) = \frac{1}{(0.2t+1)^2}, \quad \text{for } t > 0$$

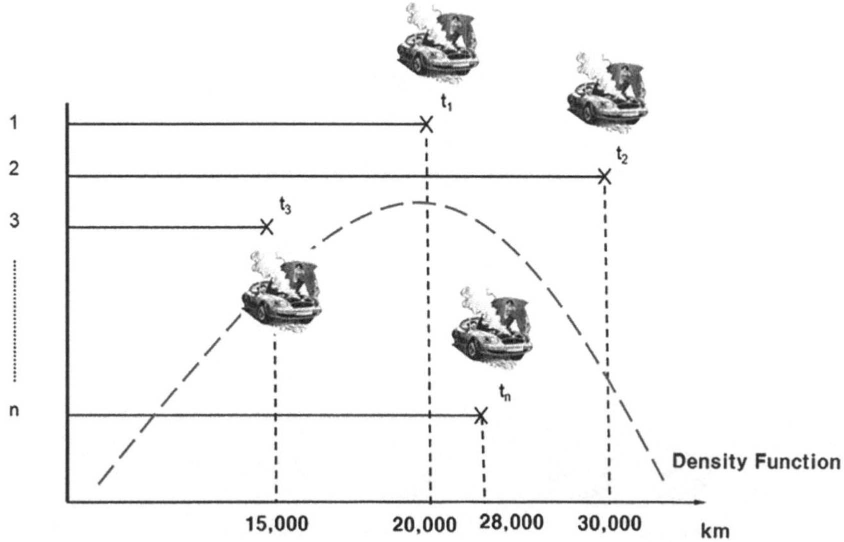

FIGURE 4.6 Concept of MTTF.

Get the probability density function, failure rate, and MTTF:

Probability density $f(t) = -\frac{d}{dt}R(t) = \frac{0.4}{(0.2t+1)^3}$

Failure rate $\lambda(t) = \frac{f(t)}{R(t)} = \frac{0.4}{(0.2t+1)}$

On the other hand, Mean Time To Failure is $\text{MTTF} = \int_0^\infty R(t)dt = 5\ \text{months}$.

4.3.2 Mean Time Between Failures (MTBF)

MTBF is the forecasted elapsed time between the failures of a mechanical system during normal system operation. MTBF can be calculated as the average time between failures of a system. MTBF is a reliability standard utilized to report the average lifetime of repairable components such as airplanes, automobiles, construction machines, and refrigerators. A repairable system is the one that can be replaced to adequate operation by any action, including component replacements or changes to adaptable settings.

MTBF is a fundamental standard of system reliability for most systems, though it still is changed. MTBF is more critical for integrators and industries than for customers (Figure 4.7).

$$\text{MTBF} = \frac{T}{n} \tag{4.22}$$

MTBF value is corresponding to the anticipated number for operating hours (service life) before a product fails. There are several variables that can affect failures. Aside from part failures, customer use/installation can also result in a failure. MTBF is often computed, based on a calculation process in which factors in all of a product's parts are summed to find the product's life cycle in hours. MTBF is considered a system failure. It is still regarded as a useful tool when considering the purchase and installation of a product. For repairable complex systems, failures are considered to be those out of design conditions which place the system out of service and into a state for repair. Technically, MTBF is used only in reference to a repairable item, while MTTF is used for non-repairable items like electric components.

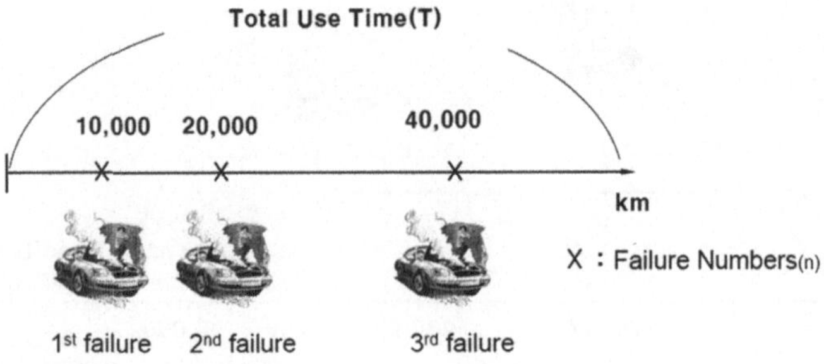

FIGURE 4.7 Concept of MTBF.

4.3.3 *BX* Life

A proper measure of product lifetime, such as *BX* life, needs to be chosen for performing parametric ALT. BX life is expressed as the time at which *X% of the items in a population will fail.* For example, if the lifetime of a product has B20 life 10 years, then 20% of the population will fail during 10 years of operation time. 'BX life Y years' helps to satisfactorily decide the cumulative failure rate of a system and its lifetime with respect to market requirements. The MTTF as the inverse of the failure rate cannot be used for the lifetime, because it equates to a B60 life which is too big. BX life reflects a more adequate estimate of lifetime, compared to MTTF.

The design life of products, BX, might disagree with the lifetime metrics – MTTF or MTBF. The use of BX life metric started from the ball and roller bearing industry, but it has grown today as a product lifetime standard utilized across a variety of industries. It's especially useful in defining warranty time for a system. The BX % life or "Bearing Life" is the lifetime standard which fails *X%* of the sample units in a population. For instance, if an item has a B10 life of 1,000 km, 10% of the population will fail by 1,000 km of operation.

On the other hand, the B10% life has the 90% reliability of a population at a specific end in product lifetime. The B10 life metric became favorable among product industries due to the industry's severe requirement. Now B1, B10, and B50 lifetime values serve as measurements for the reliability of a product. For instance, if the MTTF of a product is 100,000 km, then its B10 life will be 10,000 km (Figure 4.8).

The particular design lifetime supplies a usage or time frame for reliability testing such as parametric ALT. Some organizations might straightforwardly choose to design a product to be reliable over a stated warranty (qualification) period. Because it is commensurate with how long the product is expected to be used in the field, an improved organization might choose a design life. Depending on the designer's perspective, the life specification might be based on any of the design lifetime – BX life.

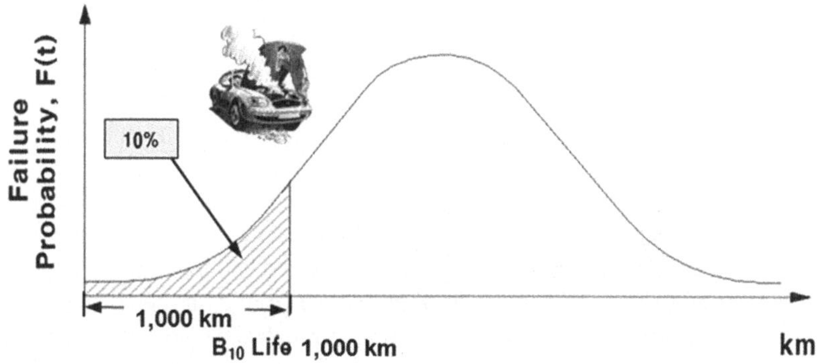

FIGURE 4.8 Concept of BX% life.

4.3.4 The Insufficiency of the MTTF (or MTBF) and the Another Standard – BX Life

There are two typical standards of product reliability – failure rate and lifetime. The failure rate is sufficient for comprehending circumstances that comprise unit periods, such as annual failure rate. However, the lifetime is often indexed using the MTTF. The MTTF are misunderstood. For example, presume that the MTTF of a printed circuit unit in a TV is 40,000 hours. The total usage reaches 40,000 hours divided by 2,000 hours per year and thus becomes 20 years, which is regarded as the lifetime of the unit. The mean lifetime of the television printed circuit assembly (PCA) is presumed to be 20 years. But because we know from actual experience that the lifetime of a television is about 10 years, this can lead to misjudgments or overdesign that wastes material.

MTTF is usually presumed to be the same as lifetime because customers comprehend the MTTF as, simply, the mean lifetime. As customers understand the mean lifetime of their appliances, they suppose that products will operate well until they reach the MTTF. In reality, this does not occur. By definition, the MTTF is an arithmetic mean; specifically, it is identical to the period from the begin of usage to the time that the 63rd item fails among 100 sets of one production lot when laid out in the sequence of failure times.

Under this definition, the number of failed televisions before the reach of MTTF would be so high that customers would never receive the MTTF as a lifetime index in the current competitive market situation. The products of first-class companies have fewer failures in a lifetime than would occur at the MTTF. In the case of home appliances, customers expect no failure for the first 10 years. The failure of a TV is accepted from the customer's view in the later period. Customers would expect the failure of all televisions once the anticipated usage time is exceeded – 12 years in the case of a television set – but they will not accept major problems within the first 10 years.

As the lifetime index of a product design, the MTTF is not proper. Alternatively, it is logical to explain the lifetime as the time period when the accumulated failure rate reaches *X*%. This is called the BX life. The value *X* may depend on products, but for home appliances, the time period to achieve a 10%–30% cumulative failure rate, B20–30 life, surpasses 10 years. Thus, a mean annual failure rate equals to 1%–3%.

Now we can start to compute the B10 life from the MTTF of 40,000 hours. Since the yearly usage is 2,000 hours, the B10 life is 2 years, which means that the yearly failure rate would be 5%. The reliability level of a television would not be satisfactory in light of the current annual failure rate of 1%–3%. The wrong interpretation of reliability using an MTTF of 20 years would lead to higher service costs if the product were released into the marketplace without further improvement. The lifetime of a television is 12–14 years, not 20 years. Since random failures cannot explain the sharply increasing failure rate, the MTTF developed from a random failure or on an exponential distribution is clearly not the same as the design lifetime of the product (See Table 4.1).

TABLE 4.1
Results of 1987 Army SINCGARS Study[a]

Vendor	*MIL – HDBK-217* MTBF(hour)	*Actual Test* MTBF(hour)
A	2,840	1,160
B	1,269	74
C	2,000	624
D	1,845	2,174
E	2,000	51
F	2,304	6,903
G	3,080	3,612
H	811	98
I	2,450	472

[a] The transition from statistical field failure–based models to physics of failure–based models for reliability assessment of electronic packages, EEP-Vol. 10–2, 619–625, *Advances in Electronic Packaging ASME* 1995, – T.J. Stadterman et al.

4.4 WEIBULL DISTRIBUTIONS AND ITS APPLICATIONS

4.4.1 Introduction

If a random variable T is represented as the lifetime of a product, the Weibull cumulative distribution function of T, denoted as $F(t)$, might be defined as all individual samples having an $T \leq t$, divided by the total number of individuals. This function also gives the probability P of selecting at random an individual sample having a value of T equal to or less than t, and thus we have

$$P(T \leq t) = F(t) \tag{4.23}$$

If a module in a product follows the Weibull distribution, the accumulated failure rate, $F(t)$, is expressed as follows:

$$F(t) = 1 - e^{-\left(\frac{t}{\eta}\right)^{\beta}} \tag{4.24}$$

Equation (4.24) gives the proper mathematical expression for the principle of the weakest chain in the product, where the occurrence of a failure in any part of the product may be said to have happened in the product as a whole, e.g., the phenomena of yield limits, static or dynamic strengths of a product, life of electric bulbs, and even death of a man.

As seen in Figure 4.9, the sum of $F(t)$ and $R(t)$ is one all the time as follows:

$$F(t) + R(t) = 1, \quad \text{for } t \geq 0. \tag{4.25}$$

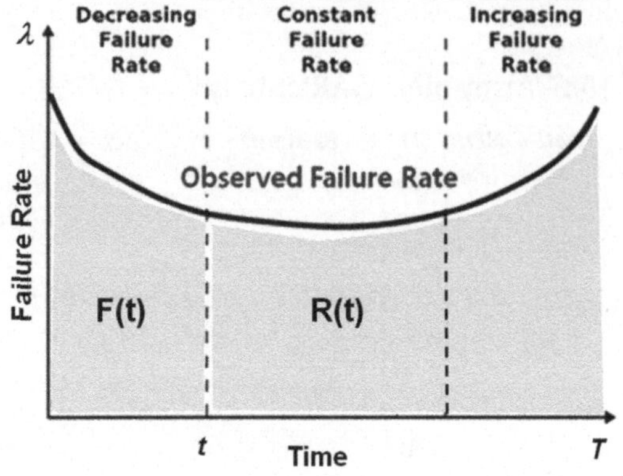

FIGURE 4.9 Relationship between $F(t)$ and $R(t)$ at time t that might be described on bathtub.

So if the Weibull distribution is followed, the product reliability is expressed as follows:

$$R(t) = 1 - F(t) = e^{-\left(\frac{t}{\eta}\right)^{\beta}} \tag{4.26}$$

where t is the time, η is the characteristic life, and β is the shape parameter (the intensity of wear-out failure).

This is mostly due to the product's weakest chain properties, but the other reasons of its increasing failure rate are, for example, age of its parts and variety of distribution shapes. The increasing failure rate explains to some extent for fatigue failures. The density function depends on the shape parameter β. For low β values ($\beta < 1$), the failure behavior can be similar to the exponential distribution. For $\beta > 1$, the density function starts at $f(t) = 0$, increases to a maximum as the lifetime passes, and decreases slowly again. A two-parameter fit is more common in reliability testing and more efficient with the same sample size. If the shape parameter is presumed to be well known, it can reduce fit to one parameter (Figure 4.10).

For $\beta = 1$ and 2, the exponential and Rayleigh distributions are mainly called in the Weibull distribution, respectively. Abernathy (1996) advised Weibayes as the best application for all small samples, 20 failures or less, if a reasonable point estimate of β is available. Using the Weibayes method, $\hat{\beta}$ can be presumed from some prior knowledge or historical data, leaving $\hat{\eta}$ as the single parameter, which can be estimated using the maximum likelihood as follows:

$$\hat{\eta} = \left[\sum_{i=1}^{N} \left(\frac{t_i^{\beta}}{r} \right)^{\frac{1}{\beta}} \right] \tag{4.27}$$

where t is the testing time, r is the number of failed sample, N is the total number of failures.

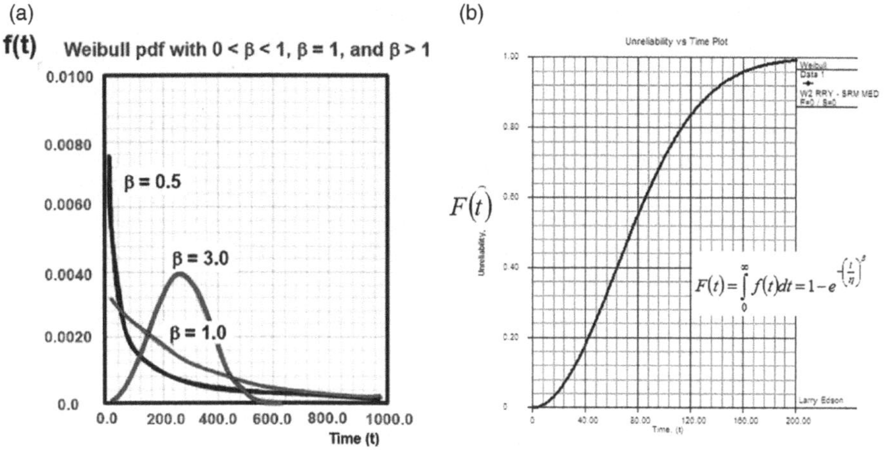

FIGURE 4.10 (a) Probability density and (b) cumulative distribution function on the Weibull distributions.

The various failure rates of the Weibull distribution stated in bathtub curve can be split into three regions:

- $\beta < 1.0$: Failure rates decrease with increasing lifetime (initial failure).
- $\beta = 1.0$: Failure rates are constant.
- $\beta > 1.0$: Failure rates increase with increasing lifetime.

For an exponential distribution, the characteristic lifetime η is allocated to the cumulative distribution function $F(t) = 63.2\%$. A shape parameter approximated from the data influences the shape of a Weibull distribution, but it does not condition the location or scale of its distribution. The spread of the shape parameters stands for the confidence intervals and a dependency of the stress level. A summary of the resolute shape parameters is roughly described as follows:

- High temperature, high pressure, high stress: $2.5 < \beta < 10$
    - Low cycle fatigue: depend on cycle times
      e.g., disk, shaft, turbine
- Low temperature, low pressure, low stress: $0.7 < \beta < 2$
    - Degradation: depend on use time
      e.g., electrical appliance, pump, fuel control value

Shape parameter β of a certain part might be unvarying, but its characteristic life η changes according to operation condition and material status. Thus, shape parameter (β) might be estimated and then will be affirmed after a test. The hazard rate function and density function for the Weibull distribution vary from shape parameters $\beta \approx 1.0$–5.0.

4.4.2 Weibull Parameter Estimation

There are two generally utilized general methods that might approximate life distribution parameters from a particular data set: (1) graphical estimation in the Weibull plotting, (2) median rank regression (MRR) and maximum likelihood estimation (MLE). The Weibull plotting is a graphical method for informally checking on the assumption of the Weibull distribution model and also for approximating the two Weibull parameters – characteristic life and shape parameter.

For a Weibull probability plot, sketch a parallel line from the y axis to the fitted line at the 62.3 percentile point. That estimate line bisects the line through the points at a time that is the approximate of the characteristic life parameter η. In order to approximate the gradient of the fitted line (or the shape parameter β), select any two points on the fitted line and divide the change in the y variable by the change in x variable.

There are several different methods for approximating the Weibull parameters such as MRR and MLE. Olteanu and Freeman [1] have studied the performance of MRR and MLE methods, and finished that the MRR method is the finest combination of accuracy and ease of interpretation when the sample size and the number of suspensions are compact. This method is widespread in the industry because fitting can be simply imagined.

The median ranks method is used to get an approximate of the unreliability for each failure.

First, we will analyze the fitting of two-parameter Weibull using the MRR method. MRR determines the best-fit straight line by least squares regression curve fitting. This method begins as follows:

1. Classify the times-to-failure in accordance with increasing order $t_1 < t_2 \ldots < t_n$

i	1	2	3	………	$r-1$	r
t_i	t_1	t_2	t_3		t_{r-1}	t_r

2. The cumulative distribution function $F(t)$ has an S-like curve (Figure 4.10a). If the function $F(t)$ is marked in a Weibull probability paper, it is useful to assess the lifetime of a mechanical system in reliability testing.

   After taking an inverse number and logarithmic transformation from reliability Equation (4.26), it might be expressed as follows:

$$\ln\left(1-F(t)\right)^{-1}=\left(\frac{t}{\eta}\right)^{\beta} \tag{4.28}$$

   After one more time taking logarithmic transformation, it might be expressed as follows:

$$\ln\left(\ln\frac{1}{1-F(t)}\right)=\beta\cdot\ln t-\beta\ln\eta \quad (4.29)$$

If F is sufficiently small, then Equation (4.29) can be restated as follows:

$$\ln\left(\ln\frac{1}{1-F(t)}\right)\cong\ln F(t)=\beta\cdot\ln t-\beta\ln\eta \quad (4.30)$$

Equation (4.30) corresponds to a linear equation. In other words,

$$y=ax+b \quad (4.31)$$

with the variables

$$a=\beta \quad \text{(slope)} \quad (4.32a)$$

$$b=-\beta\ln\eta \quad \text{(axis intersection)} \quad (4.32b)$$

$$x=\ln t \quad \text{(abscissa scaling)} \quad (4.32c)$$

$$y=\ln(-\ln(1-F(t))) \quad \text{(ordinate scaling)} \quad (4.32d)$$

Two-parametric Weibull distribution might be stated as a straight line on the Weibull probability paper. The gradient of its straight line is the shape parameter β (see Figure 4.11b).

3. Compute median ranks: Classify failure times in an increasing order. Mean ranks are less correct for the skewed Weibull distribution; therefore, median ranks are better. Median ranks might be computed as follows:

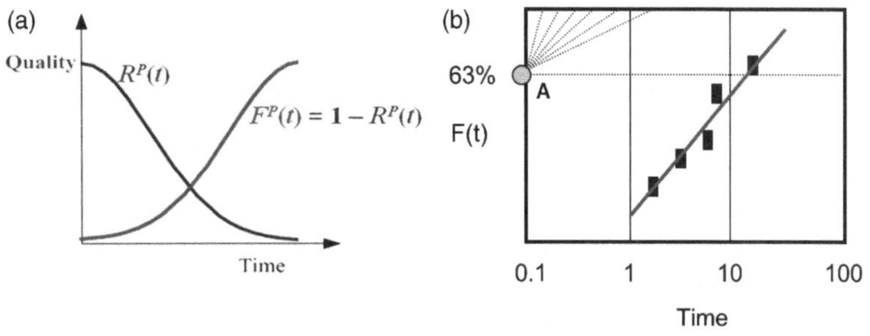

FIGURE 4.11 A plotting of the Weibull probability paper: (a) $R(t)=1-F(t)=e^{-\left(\frac{t}{\eta}\right)^{\beta}}$ and (b) $\ln\left(\ln\frac{1}{1-F(t)}\right)\cong\ln F(t)=\beta\cdot\ln t-\beta\ln\eta$.

$$\sum_{k=i}^{N}\binom{N}{k}(MR)^k(1-MR)^{N-k}=0.5=50\% \tag{4.33}$$

Bernard utilized an estimation of it as follows:

$$F(t_i)\approx\frac{i-0.3}{n+0.4}\times 100 \tag{4.34}$$

where i is the failure-order number and N is the total sample size.

4.5 RELIABILITY TESTING

4.5.1 Introduction

Based on the AGREE, reliability is the 'probability of working without failure a (intended) function under stated conditions for a certain period of time'. This meaning might be particularly analyzed as follows:

- Working signifies that the item does not fail.
- The stated conditions include the customer usage conditions and total physical environmental conditions, i.e., regional climate, shipping, electrical, mechanical, and thermal conditions.
- The stated time interval means that the product lifetime can be very long (20 years, for telecommunication equipment), long (a few years), or short (a few hours or weeks, for space research equipment). The time can also be replaced by other parameters such as kilometer (of an airplane or an automobile) or the number of cycles (i.e., for a relay unit or a capacity).

Many manufactures use reliability testing to decide whether the last designs satisfy customers' request for high-quality and reliable products. In a reliability test, the following are basic questions: (1) How many units have to be performed on the test? (2) How long should you test? (3) If acceleration modeling is part of the experimental plan? (4) What collaboration of a pattern of stresses (or loads) and how many experimental cells? (5) How many units go in each cell?

The answers to these questions relies on: (1) What kind of life models are you assuming? (2) How to perform the accelerated life testing? (3) What conclusions or decisions do you want to make after running the test and analyzing the data? (4) What uncertainty are you willing to take of making conclusions or wrong decisions?

If product failures utilize the strength for load concepts under serious test conditions, reliability testing will be utilized to determine whether the product is sufficient to satisfy the reliability requirements under accelerated testing. If the accelerated factor and sample size are evolved, products can also approximate the lifetime under actual lifetime specifications. Therefore, reliability testing during design and development is obligatory to show whether the product lifetime is adequate for customer needs.

In analyzing the data obtained from reliability testing, engineers can estimate the product lifetime by fitting a Weibull probability distribution to the testing data from a typical sample of units. Data analysis requires to (1) gather the life data of the product from reliability testing or field, (2) choose a lifetime distribution that will fit the data and model the life of the product, (3) estimate the parameters that will fit the distribution to the data, and (4) generate plots and results that estimate the life characteristics of the product.

4.5.2 Censoring

The reliability data obtained from operation/environment conditions incorporate the values reported as "below detection limit" along with the specified detection limit. A sample contains censored observations if only information about some of the observations is that they are below or above a specified value. In the environmental literature, such values are often called 'non-detects'. Although this result has some loss of information, we can still do graphical and statistical analyses of data with non-detects.

Engineers determine that reliability test criteria can be stated in units of time (Type I censoring) or failure number (Type II censoring). In Type I censoring, the items will be tested for a certain amount of time, whereas in Type II censoring, the items will be tested until a certain number of failures. Because it is not uncommon in time-terminated tests to have many survived units, engineers prefer to utilize the failure number (Type II censoring).

A sample is Type I censored when the censoring levels are known in advance. The number of censored observations is a random outcome, even if the total sample size, N, is fixed. Environmental data are almost always Type I censored. A sample is Type II censored if the sample size N and the number of censored observations are fixed in advance. The censoring level(s) are random outcomes. Type II censored samples most commonly arise in time-to-event studies that are planned to end after a specified number of failures (Figure 4.12).

The time and effort in testing might be notably reduced by censored tests, and they can approximate the product lifetime. If a test trial is interrupted before all n test units have failed, a censored test may produce. If an interruption happens after a given time, one is dealing with censoring of Type I. On the other hand, if a trial is interrupted after a given amount of test unit r has failed, one is dealing with censoring of Type II. The trials stop after four failures. The point in time at which the failure r happens is a random variable. Thus, leaving the entire trial time length opens until the end of the trial.

The fact that $n-r$ test units have not failed is taken into account by substituting r for n in the denominator of the approximation equation. With Type I or II censoring, it is necessary to estimate the characteristic lifetime η in the Weibull chart by extrapolating the best fit line beyond the final failure time. This is generally problematic as long as further failure mechanisms cannot be neglected. A statistical statement affecting the failure behavior can be found on the observed lifetime. The procedures and methods for the assessment of complete data or censored data can be found in Table 4.2.

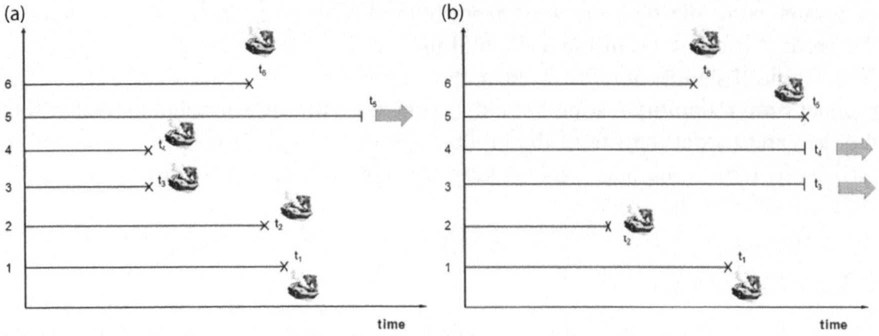

FIGURE 4.12 Schematic of (a) type I and (b) type II censoring.

TABLE 4.2
Overview of Procedures for Evaluation of Censored Data

Data Type	Type of Censor	Description	Procedure
Complete Data $r = n$	No censoring	All samples have failed	Median Procedure $F(t_i) \approx \frac{i-0.3}{n+0.4}$ For $j = 1,2, \ldots, n$
Censored Data $r<n$	Censoring Type I or Type II	Lifetime characteristics of all entire units are bigger than the lifetime characteristics of the units r which failed	Median Procedure $F(t_i) \approx \frac{i-0.3}{n+0.4}$ For $i = 1,2, \ldots, r$

4.5.3 Lifetime Estimation – Maximum Likelihood Estimation (MLE)

To forecast the product lifetime, MLE is a favored method based on the observed data of reliability testing. MLE is a statistical method for approximating the parameters of a statistical model – some unknown mean and variance that are given to a data set. It chooses a set of values of the model parameters that make as large as possible the likelihood function.

For a fixed set of data and an underlying statistical model, the method of maximum likelihood approximates a set of values of the model parameters that maximize the likelihood function. Assume that there is a sample $t_1, t_2, \ldots, t_n$ of n independent and identically distributed failure times, coming from a distribution with an unknown probability density function f_0 (·). On the other hand, suppose that the function f_0 belongs to a certain family of distributions $\{f(\cdot|\theta), \theta \in \Theta\}$ (where θ is a vector of parameters), called the parametric model, so that $f_0 = f(\cdot|\theta_0)$. The unknown value θ_0 is expected to be a true value of the parameter vector. An estimator $\hat{\theta}$ would be fairly close to the true value θ_0. The observed variables t_i and the parameter θ are vector parts:

$$L(\Theta;t_1,t_2,\ldots,t_n) = f(t_1,t_2,\ldots,t_n \mid \Theta) = \prod_{i=1}^{n} f(t_i \mid \Theta) \tag{4.35}$$

This function is defined as the likelihood. The plan of this procedure is to discover a function *f*, for which the product *L* is maximized. Here, the function should have high values of the density function *f* in the corresponding region with some failure times t_i. Simultaneously, only low values of *f* in regions with few failures are found. Thus, the real failure behavior is correctly represented. If determined in this manner, the function *f* offers the best chance to explain the test samples.

It is frequently more successful to utilize the log-likelihood function. Therefore, the product equation becomes an addition equation, which greatly clarifies the differentiation. Since the natural log is a monotonic function, this footstep is mathematically logical:

$$\ln L(\Theta; t_1, t_2, \ldots, t_n) = \sum_{i=1}^{n} \ln f(t_i \mid \Theta) \tag{4.36}$$

Differentiating Equation (4.36), the maximum of the log-likelihood function and thus the statistically optimal parameters Θ_l can be found as follows:

$$\frac{\partial \ln(L)}{\partial \theta_l} = \sum_{i=1}^{n} \frac{1}{f(t_i;\Theta)} \cdot \frac{\partial f(t_i;\Theta)}{\partial \theta_l} = 0 \tag{4.37}$$

Because these equations can be nonlinear, it is usually useful to apply the parameter estimates in Weibull distribution by numerical procedures such as Newtion-Raphson method. By the likelihood function value, the opportunity is given to approximate the quality of the adaptation of a distribution to the failure data. The greater the likelihood function value, the better the conclusive distribution function, which represents the actual failure behavior. However, based on MLE, the characteristic life η_{MLE} from the reliability testing (or lifetime testing) can be approximated on the Weibull chart.

Example 4.3

Normal distribution – maximum likelihood estimation

The likelihood function of normal distribution can be defined as

$$L(\Theta) = \prod_i f_{\mu,\sigma}(x_i) = \prod_i \frac{1}{\sqrt{2\pi}\sigma} \exp\left(\frac{-(x_i - \mu)^2}{2\sigma^2} \right) \tag{4.38}$$

If differentiated for mean μ and the variance σ^2, the maximum value is obtained as follows:

$$\frac{\partial}{\partial \theta_l} L^*(\Theta) = \frac{1}{\sigma^2} \sum_i (x_i - \mu) = \frac{1}{\sigma^2} \left(\sum_i x_i - n\mu \right) = 0 \tag{4.39a}$$

$$\frac{\partial}{\partial \mu} L^*(\Theta) = -\frac{n}{\sigma} + \frac{1}{\sigma^2} \sum_i (x_i - \mu)^2 = 0 \tag{4.39b}$$

Lastly, we can obtain the maximum likelihood approximation for normal distribution at

$$\mu = \left(\sum_i x_i \right) \Big/ n \tag{4.39c}$$

$$\sigma^2 = \sum_i (x_i - \mu)^2 \Big/ n \tag{4.39d}$$

Example 4.4

Obtain the shape parameter and characteristic life by utilizing the maximum likelihood method and the censored data listed below.

Cycles	Status	Cycles	Status
1,500	Failure	4,300	Failure
1,750	Suspension	5,000	Suspension
2,250	Failure	7,000	Failure
4,000	Failure		

Approximate the shape parameter and characteristic life. The maximum likelihood estimate, β, is the root of the equation

$$\frac{\sum_{i=1}^{7} t_i^{\beta} \ln x_i}{\sum_{i=1}^{7} t_i^{\beta}} - \frac{1}{5}\sum_{i=1}^{5} \ln t_i - \frac{1}{\beta} = 0$$

The Weibull plot estimate, 1.8, was utilized as the first value of β. The estimates of β are listed below with the corresponding values of $G(\beta)$. By utilizing a modified Newton–Raphson procedure, the value of β giving $G(\beta) = 0$) is obtained.

1.800	−0.1754 (1st iteration)
1.802	−0.1746 (2nd iteration)
2.179	−0.0255 (3rd iteration)
2.182	−0.0248 (4th iteration)
2.253	−0.0007 (5th iteration)
2.256	−0.0005 (6th iteration)
2.257	−0.0000 (maximum likelihood estimate of beta)

The maximum likelihood estimate of η is

$$\hat{\eta} = \left(\frac{\sum_{i=1}^{7} t_i^{2.257}}{5} \right)^{1/2.257} = 4900.1$$

4.5.4 Time-to-Failure Models

Since time is a usual standard of life, life data points are defined as 'times to failure'. The failure time T of a product is a random variable. Time can take on different meanings depending on operational time, the distance driven by a vehicle, and the number of cycles for a periodically operated system. A time-to-failure model usually supplies all the tools for reliability testing, especially data analysis of accelerated life testing. It is designed to use with complete (time-to-failure), right-censored (suspended), interval, or left-censored data.

The following are different kinds of time-to-failure model:

- Arrhenius: a single stress model generally utilized when temperature is the accelerated stress.
- Inverse power law (IPL): a single stress model generally utilized with a non-thermal stress, such as vibration, voltage, or temperature cycling.
- Eyring: a single stress model generally used when temperature or humidity is the accelerated stress.

Based on the conventional time-to-failure models, the most important issue for a reliability test is how early the potential failure mode can be found. To do this, we should formulate a failure model and determine the related coefficients. First of all, we can configure the life-stress (LS) model, which includes stresses and reaction parameters. This equation can explain various failures such as fatigue in the mechanical structure. As electronic technology advances, engineers also have understood product failures from micro-void coalescence (MVC), which is observed in some engineering plastics or majority of metallic alloys. Because system failure originates in a micro-depletion (void) regardless of whether it is mechanical or electronic, the LS model should be applicable to electronic parts.

Fatigue failure occurs not due to theoretical stresses in an ideal part, but rather due to pre-existing defect or the presence of a small crack on the surface of a part that the stress and strain become plastic. To better understand it, engineers recognize how a small crack or pre-exiting material defects in material generate. That is, because a system failure starts from the presence of a material defect formed at a microscopic level when repeatedly subjected to a variable tensile and compression load in a random way, we might define the LS model from such a standpoint. It combines the design principles of a mechanical system by using general LS model and accelerated life testing based on load analysis. It will be discussed in Chapters 5 and 8.

REFERENCES

1. IEEE Std 610.12-1990, 1990, *IEEE standard glossary of software engineering terminology*, New York: Standards Coordinating Committee of the Computer Society of IEEE.
2. Klutke, G., Kiessler, P.C., Wortman, M.A., 2015, A critical look at the bathtub curve, *IEEE Transactions on Reliability*, 52 (1), 125–129.
3. Lei, W., 2018, A generalized weakest-link model for size effect on strength of quasi-brittle materials, *Journal of Materials Science*, 53, 1227–1245.

Part II

Parametric Accelerated Life Testing

5 Mechanical Structure Including Mechanisms and Load Analysis

5.1 INTRODUCTION

The mechanical systems convert power into forces (effort) and velocity (flow) that eventually provides mechanical advantages to accomplish a task by adapting product mechanisms. A product mechanism is a kind of kinematic chain that makes motions such as rotary motion, oscillation, and linear motion. For example, an airplane moves forward by the thrust from a jet engine, which is known as the Brayton thermodynamic cycle [1]. An automobile is also a wheeled motor vehicle utilized for transportation. It is comprised of engine, transmission, drive, electric, and body systems. Under the Carnot cycle, an engine goes through four stages: (1) induction (fuel enters), (2) compression, (3) ignition, and (4) emission (exhaust out). The engine is designed to convert fuel into mechanical energy through the slider-crank mechanism. The automobile obtains the power to move forward through transmission and drive systems (Figure 5.1).

In the power transmission, a product will be subjected to a variety of stress due to repeated loads. If there is a design fault in a product that causes an inadequacy in the product's stiffness (or strength) when subjected to loads, it may cause a structural failure of parts in a mechanical system, i.e., the (low/high) cycle fatigue that is characterized by repeated plastic deformation. While designing a mechanical system, an engineer should recognize what the structural loading of the mechanical system is.

There are two kinds of loadings: statics and dynamics. Static loads – tension, compression, shear, or twist – cause displacement, strain, and stress. A product will experience a failure such as permanent deformation if repeatedly subjected to static loads in its lifetime. The mechanical system is designed to provide sufficient strength to a product in its lifetime to bear such repeated (static) loads:

- Structural load and deflection versus material stress and strain
- Tension and compression loads
- Torsion and bending loads.

A dynamic load, also called as probabilistic (or random) load, is a force applied by a moving body on a resisting member, generally in a relatively short time period. Because such loads are usually changeable, we can define the dynamic load. Dynamic loads cause motions like vibration, cumulative damage, fatigue, and fracture.

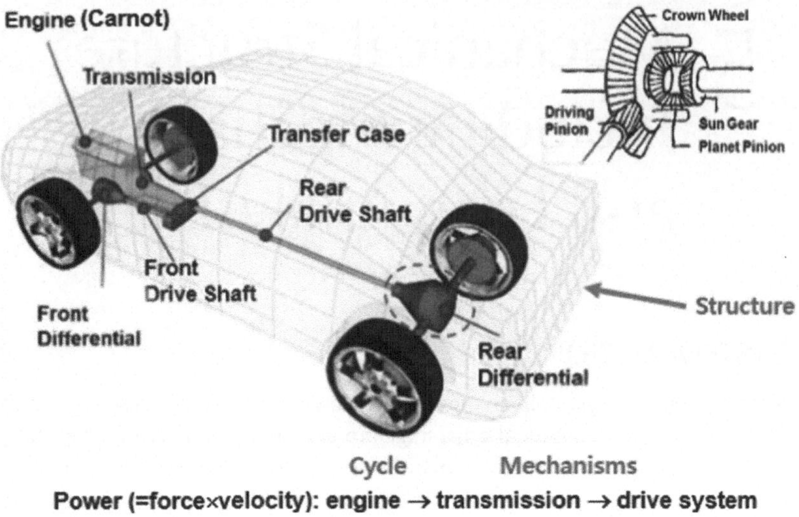

FIGURE 5.1 Automobile: cycle, structure, and mechanisms (example).

Therefore, a product should be designed to have sufficient stiffness in its structure during its lifetime to endure repeated (dynamic) loads such as the following:

- Impact, vibration, and shock loads
- Unbalanced inertia loads.

To be prepared for any failure that can occur in a product's design in its lifetime, we should recognize the design process of the mechanical product, i.e., (1) design specifications, (2) the implemented structure including its mechanism and load-bearing capacity, and (3) the final robust design. In this chapter, first of all, we will discuss the design principles of a mechanism that will implement the product's (intended) functions for its specifications. Secondly, we discuss how to traditionally model the product system by Newtonian method, D'Alembert's principle, or Lagrangian method. Finally, we will learn how to deal with a new modeling of bond graph – especially, power transmission by a mechanism in a mechanical system.

5.2 MECHANICAL STRUCTURE INCLUDING MECHANISMS

5.2.1 Introduction

Structures such as building, bridge, and mechanical product – vehicle, airplane, refrigerator, etc. – are a core part of modern life, which are assembly of rigid bodies connected by joints with no mobility and involving no mechanism. Early architects and engineers were designing the structures for their own purposes of facility (or product) with the capacity to withstand loads and avoid mechanical failures such as fatigue. Many structures, supervised by the Pharaohs of ancient Egypt and the Caesars of Rome, are still standing in the historical places. In Europe, a lot of buildings and bridges built in the Renaissance Period are still functioning. These structures are

FIGURE 5.2 Golden gate bridge. (Wikipedia.)

built in thousands of shapes and different sizes in nature which provide them inherent robustness (Figure 5.2).

Mounted on the mechanical structure of a product, various mechanisms convert power to work. Mechanical systems from internal combustion engines to helicopters and machine tools also mount many mechanisms. Many tools such as wrenches, screwdrivers, hammers, and jacks also have their own mechanisms. Besides, the hands and feet, legs, arms, and jaws play an important role (mechanisms) in human bodies as do the paws and legs, wings, tails, and flippers in animal bodies.

A mechanism is an assembly of moving parts that performs some function, which generates some motions such as slider-crank. Mechanical mechanism and structure are joined together in a product as an arrangement of parts and also provide enough strength and stiffness to bear loadings. For example, the bicycle is driven by a big driver gear wheel with pedals attached. When the back wheel turns, the bicycle moves forward. We know that a mechanism such as the chain and sprocket in a bicycle is distinctly constructed around a frame structure. Now we discuss the basic mechanical mechanisms (Figure 5.3).

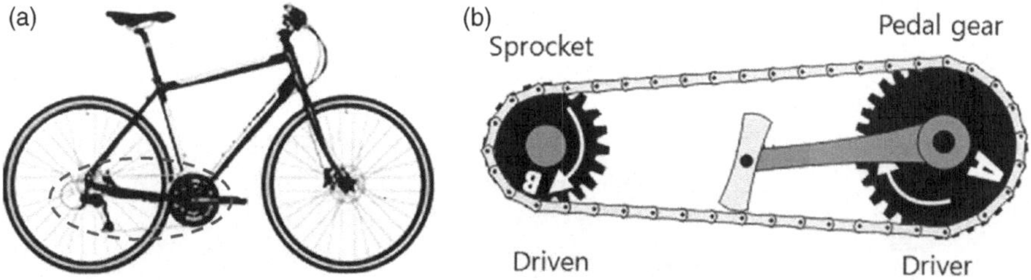

FIGURE 5.3 Bicycle and its mechanism: (a) bicycle (structure + mechanism) and (b) chain and sprocket mechanism.

5.2.2 Mechanical Mechanisms

A variety of mechanisms in automobiles, refrigerators, bicycles, etc. are a mechanical portion that can transfer motion and forces from a power source to the other. Most mechanisms are designed to provide a mechanical advantage. They also form a kinematic chain in which at least one link is attached to a frame of reference (ground) and makes four different motions, namely, rotary motion, oscillation, linear motion, and reciprocating motion. Rotary motion is the starting point for many mechanisms, mostly provided by an engine or dc motors. Oscillation is a back and forth motion about a pivot point. Linear motion is a one-dimensional motion along a straight line. Reciprocating motion is again a back and forth motion.

5.2.2.1 (Lever) Mechanisms and Their History

From the old ages, the lever is the most basic mechanism with a rigid beam rotating on a fulcrum used for thousands of years. "If you had a lever long enough and a suitable fulcrum, you could move the earth" [2]. A small effort from one end of the beam will lift a heavy load at the other end. That is, by moving the fulcrum of a long beam nearer to the load, the lever can move a big load with minimal effort.

As seen in Figure 5.4, there are three types of levers. Class 1 lever is positioned along the length of the lever where the effort is put in. For example, the crowbar is a single lever, but the pliers make use of two levers joined together at the fulcrum. Class 2 lever is positioned along the length of the lever where the load is put in. Class 3 lever is positioned along the length of the lever where the fulcrum or pivot point is located.

Leonard Euler (1707–1783) was one of the earliest experts in mathematics to investigate the mathematics of linkage design (synthesis). Most linkages

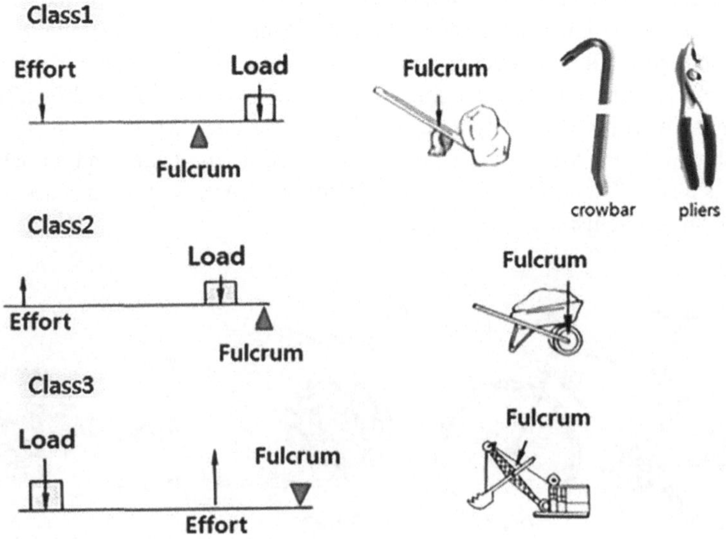

FIGURE 5.4 Three classes of levers.

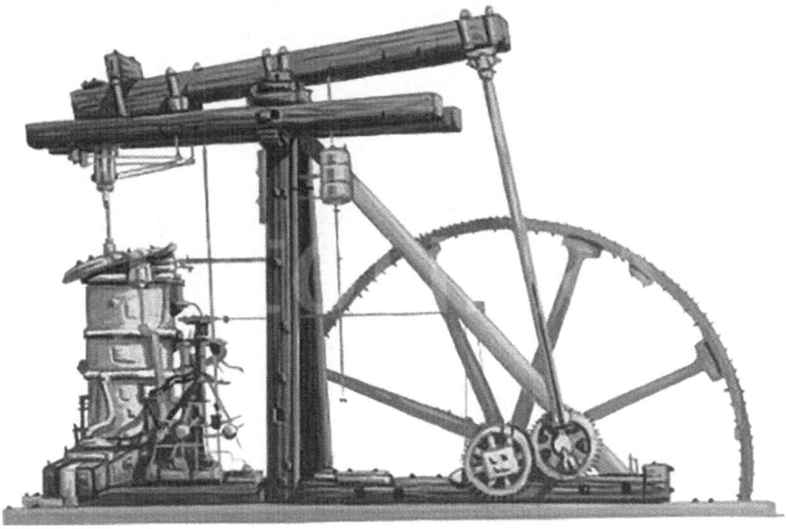

FIGURE 5.5 James Watt's industrial steam engine.

are planar; their movement is enclosed to a plane. The general study of linkage motions, planar and spatial, is defined as screw theory. Sir Robert Stawell Ball (1840–1913) is deemed the father of screw theory. Near industrial revolution, lots of the weaving of cloth suggested to the need for more complicated machines to convert waterwheels' rotary motion into complex motions. The design of the steam engine generated a huge necessity for newly designed mechanisms and machines. Lengthy linear motion travel was required to robustly use steam power. Especially, though he did not understand it, James Watt (1736–1819) applied thermodynamics and rotary joints and long links to generate structured straight line motion (Figure 5.5).

5.2.2.2 Mechanical Advantage

When a man uses a stake to lift a rock by a lever mechanism, the man applies a small quantity of effort at the end of the lever to move the rock (class 1). Levers are used to change the direction, distance, or velocity of movement or to decrease the effort required to lift a load. This reduction of effort is known as mechanical advantage. With its fulcrum at *X*, the lever is a bar of length *AB*, dividing the length of the bar into components: L_1 and L_2. To raise a load *W* through a height of *h*, a force *F* should be employed downward through a distance *s*. If ignoring friction, the triangles *AXC* and *BXD* are alike and proportionate:

In the example, the effort of 10 kg_f is required to lift a load of 50 kg_f. So, we can calculate the mechanical advantage. That is, mechanical advantage (*MA*) = load/effort = $s/h = L_1/L_2 = 50\ kg_f/10\ kg_f$. Therefore, this system has an *MA* of 5. There are no units because it is a ratio of the same units (Figure 5.6).

If a weight *W* of 300 kg is to vertically be lifted through a height *BC* of 1 m, a force *F* of 300 kg might be acted. However, for a sloped plane, the weight is moved

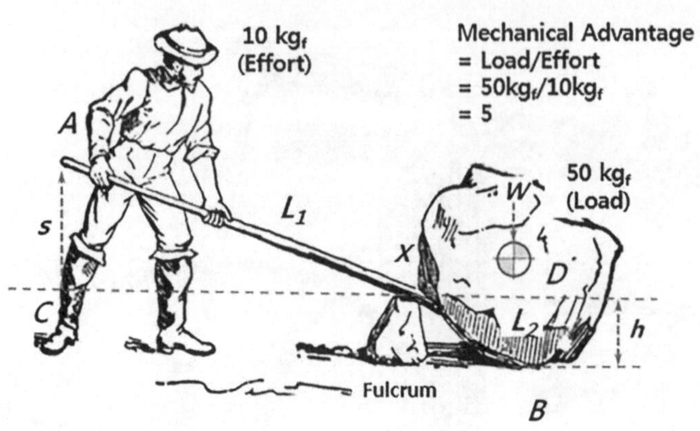

FIGURE 5.6 Mechanical advantage.

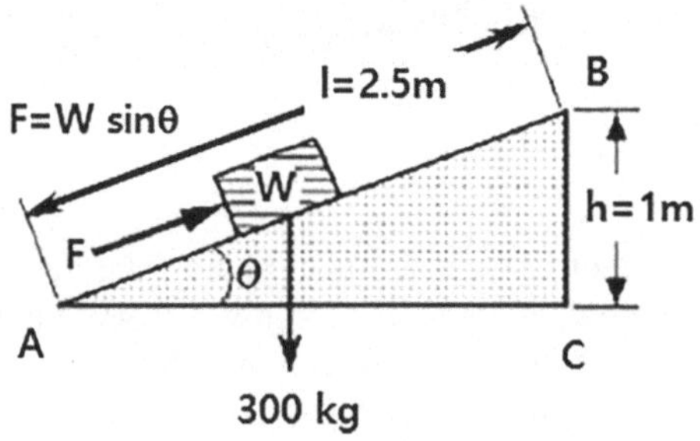

FIGURE 5.7 Diagram for computing mechanical advantage of a sloped plane.

over the lengthy distance of 2.5 m, but a force F of 120 kg (only 2/5 of 300 kg) would be required because the weight is moved through a longer distance. So, we can decide the mechanical advantage of the sloped plane. That is, MA = load/effort = W/F = 300/120 = 2.5 (Figure 5.7).

5.2.2.3 Efficiency of Machines

On the basis of power and mechanical advantage, we can evaluate the efficiency of simple machines. If expressed as energy or power, the efficiency of a machine is the ratio of energy output to energy input. As getting a bigger force from a machine than the force applied upon it, this refers only to force and not energy. According to the conservation law of energy, more work cannot be obtained from a machine than the energy applied to it. In other words, machine cannot have 100% efficiency because some energy in all moving machinery is dissipated into heat due to friction.

5.3 DESIGN OF MECHANISMS

Now we discuss the concepts of mechanism design. As previously mentioned, the mechanical system converts power to forces and movement for accomplishing its own tasks by adapting its mechanism. Most mechanisms, therefore, are designed to supply a mechanical advantage. To fulfil a set of performance requirements for the machine, a mechanism design is required to prescribe the sizes, shapes, material compositions, and arrangements of parts by a proper kinematic analysis and synthesis. The resulting machine will perform the prescribed tasks.

5.3.1 Classification of Mechanisms

There are three applications of mechanisms: (1) path generation mechanisms, (2) function generation mechanisms, (3) motion generation mechanisms. So, kinematic synthesis might include three distinct subcategories. That is, mechanisms for path generation are to control a specified point whether it pursues a specified path in 2D space. Mechanisms for function generation require whether it has a particular output for each specified input, like a black box with input/output. For example, a sprinkler with a four-bar lawn sprinkling mechanism should oscillate within a specific range to avoid over sprinkling or under sprinkling in the lawn area (see Figure 5.9b). Mechanisms for motion generation are to control a rigid body whether its location and orientation related to an established coordinate system are achieved in advance. Mechanism dimensions are computed to achieve or approximate predefined rigid-body path points (path generation)/link displacement angles (function generation)/rigid-body position (motion generation) (Figures 5.8 and 5.9).

Degree of freedom (DOF) of space in the mechanism can be classified as follows: (1) planar mechanisms, (2) spherical mechanisms, and (3) space mechanisms. Mechanism types can also be classified as follows: (1) screw mechanisms, (2) cam mechanisms, (3) wheel mechanisms (gear mechanisms or roller mechanisms), (4) belt mechanisms, (5) crank mechanisms (sometimes also called 'link mechanisms'), and (6) ratchet and lock mechanisms (including Geneva drives). Though still designing a simple linkage, design is generally an exercise of trade-offs between what you can obtain and what you want. Design requirements include geometric, kinematic, dynamic, and any combination of these (Figure 5.10).

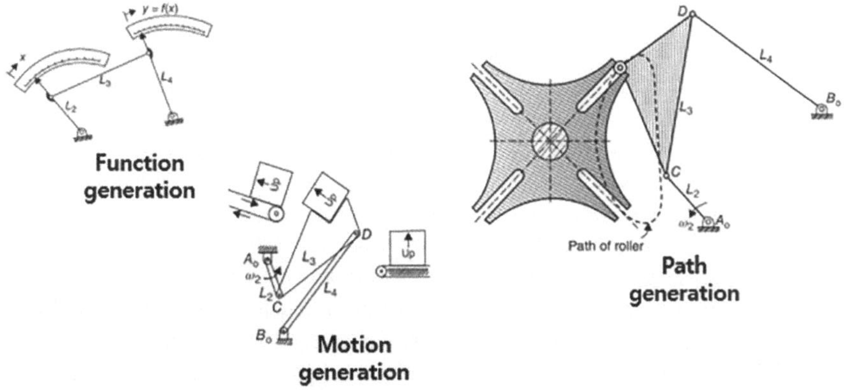

FIGURE 5.8 Classifications of mechanisms based on their applications.

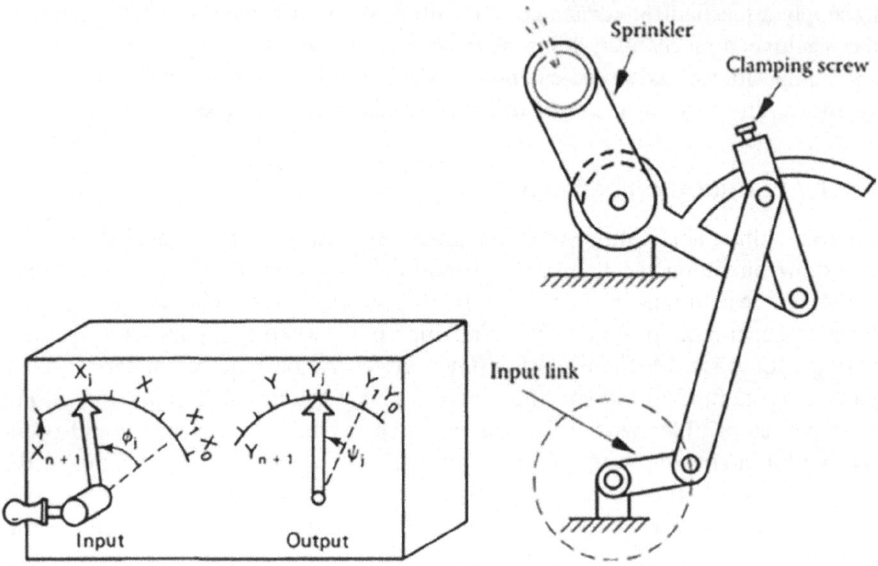

(a) (b)

FIGURE 5.9 Function generation (example): (a) black box and (b) sprinkler.

Analysis is a technique that permits a designer to critically inspect an already existing, or proposed, design to decide its suitability for the assignment. The analysis of mechanisms and machines aims, therefore, at understanding the relationships between the motions of the machine parts (kinematics) and the forces that produce the motions (kinetics). In a mechanism design, kinematic analysis is required to determine displacement, position, rotation, velocity, speed, and acceleration of the mechanism. On the other hand, synthesis is a process of contriving a scheme or a method of accomplishing a given purpose. In other words, it is a process of evolving a mechanism to fulfil a set of performance needs for the machine. Especially, in mechanisms, synthesis is the design of a linkage to generate a desired output motion for a given input motion (Figure 5.11).

There are two types of synthesis in mechanism design: (1) type synthesis and (2) dimensional synthesis. Type synthesis is the type of mechanism that is suitable to satisfy a requirement. It refers to the choice of the kind of mechanism to be used; it might be a linkage, a geared system, a cam system, or even belts and pulleys. This beginning stage of the total design process usually involves the consideration of design factors, such as manufacturing processes, materials, space, safety, and economics. On the other hand, dimensional synthesis pursues to decide the notable dimensions and the beginning position of a mechanism of a predefined type for a specified task and a prescribed performance. In dimensional synthesis, the objective is to calculate the mechanism dimensions that are required to approximate prescribed mechanism output parameters. For example, in a linkage mechanism or a computed mechanism, the dimensions are link positions, link lengths, and joint coordinates.

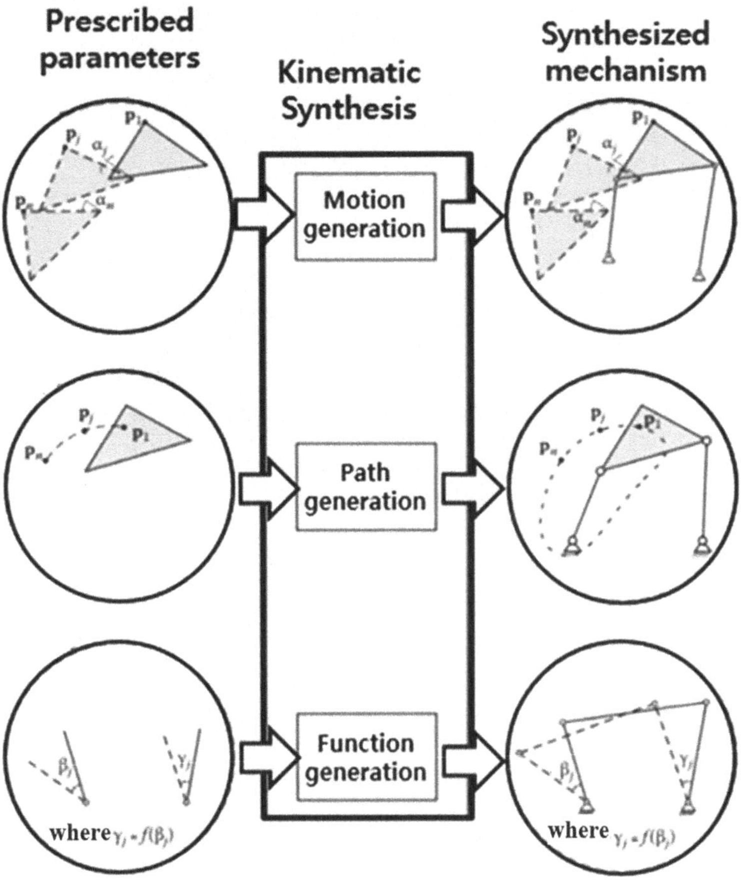

FIGURE 5.10 Design of mechanisms through the kinematic synthesis.

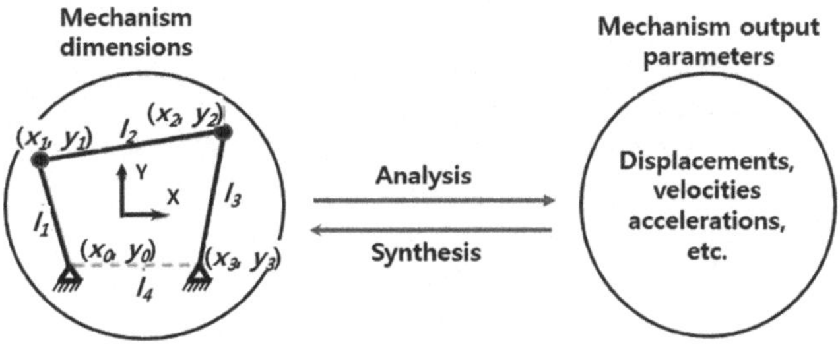

FIGURE 5.11 Kinematic synthesis versus analysis.

5.3.2 Terminologies

Kinematic chain or linkage is an assembly of links and joints that supplies a desired output motion in response to a stated input motion. Links are individual parts of the

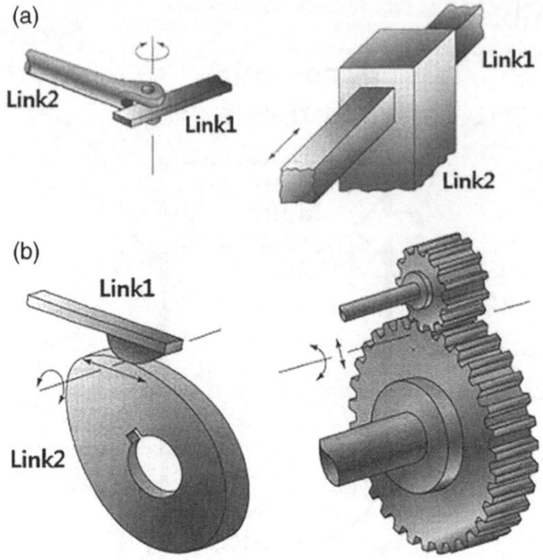

FIGURE 5.12 Joints: (a) primary joints: pin and sliding and (b) higher-order joints: cam joint and gear joint.

mechanism that possesses at least two nodes. They are considered rigid bodies and are connected with other links to transfer forces and motion.

A joint is a movable attachment between links and permits relative motions between the links. Revolute and sliding joints are primary joints. The revolute joint is defined as a hinge joint or pin. It permits pure rotation between the two links that it joins. The sliding joint is also called a piston or prism joint. It permits linear sliding between the links that it connects. A cam joint allows for both rotation and sliding of the two links that it connects. A gear connection permits rotation and sliding of the teeth as they mesh (Figure 5.12).

5.3.3 Mobility

The DOF is the number of independent inputs needed to exactly position all links of a mechanism with respect to the ground. It can also be defined as the number of actuators required to work the mechanism. The number of DOFs of a mechanism is also called the mobility, and it is given by the symbol M:

$$M = \text{degree of freedom} = 3(n-1) - 2jp - jh \tag{5.1}$$

where n is the total number of links, jp is the number of primary joints, and jh is the number of higher-order joints.

5.3.4 Kinematic Model/Diagram

Kinematic model is a representation of a mechanism that only manifests the dimensions that effect the motion. To draw the kinematic model from a mechanism, remove any superfluous detail, and simplify the drawing of the mechanism for further analysis.

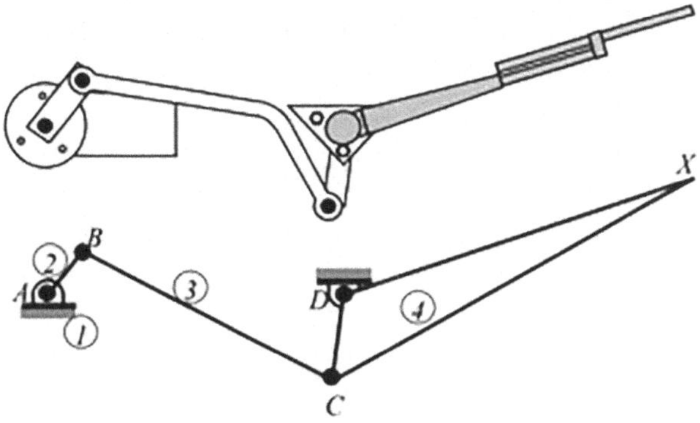

FIGURE 5.13 Rifle and its kinematic mechanism.

Then, we can carry out the displacement, velocity, and acceleration analysis of a mechanism in machines (Figure 5.13).

Example 5.1

Find the DOF from the following figures:

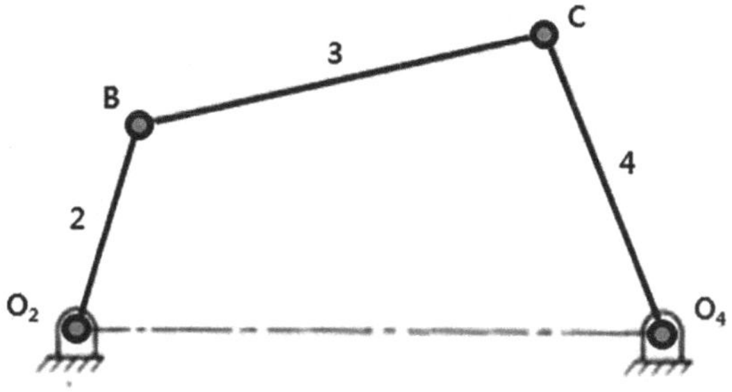

$$M = 3(n-1) - 2jp - jh$$

Here, $n = 4$, $jp = 4$ and $jh = 0$

$$M = 3(4-1) - 2(4) = 1$$

i.e. one input to any one link will result in a definite motion of all the links.

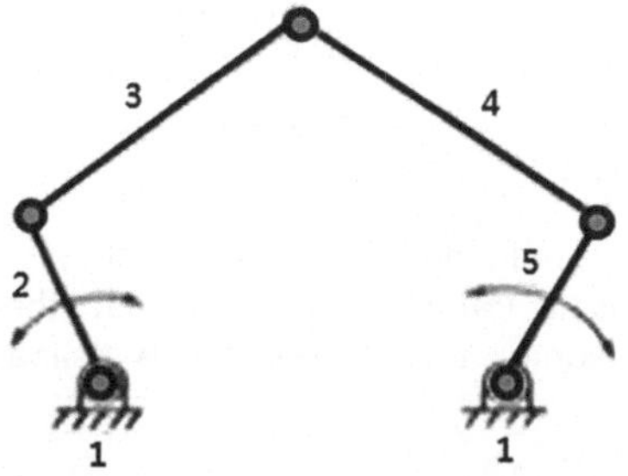

$$M = 3(n-1) - 2jp - jh$$

Here, $n = 5, jp = 5$ and $jh = 0$

$$M = 3(5-1) - 2(5) = 2$$

i.e. two inputs to any two links are required to yield a definite motion of all the links.

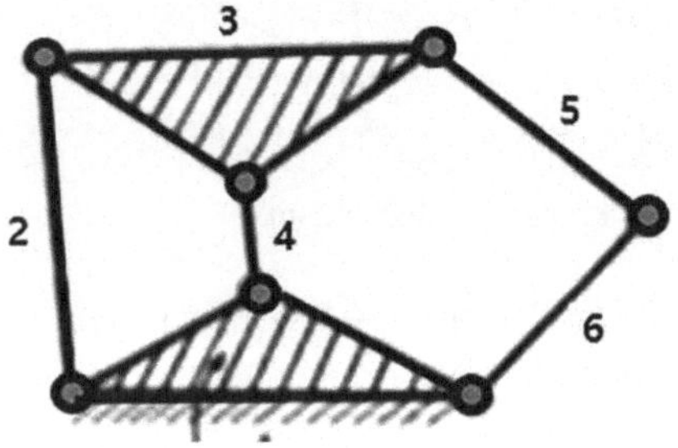

$$M = 3(n-1) - 2jp - jh$$

Here, $n = 6, jp = 7$ and $jh = 0$

$$M = 3(6-1) - 2(5) = 2$$

i.e. one input to any one link will result in a definite motion of all the links.

Example 5.2

Draw a kinematic diagram, a shear press shown that is used to cut and trim electronic circuit board laminates.

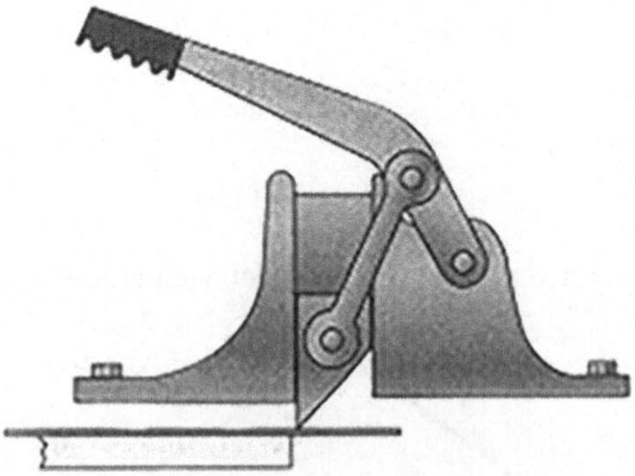

Solution: The first step in building a kinematic diagram is to determine the part that will be appointed as the frame. The motion of all other links will be decided relative to the frame. In some cases, its selection is clear as the frame is securely fastened to the ground. In this problem, the big base that is bolted to the table is designated as the frame. The motion of all other links is decided relative to the base.

The base is numbered as link 1. Cautious watching shows three other moving parts: (2) handle, (3) cutting blade, and (4) bar that attaches the cutter with the handle. Pin joints are used to attach these three different parts. Pin joints are labeled *A* through *C*. In addition, the cutter glides up and down, along the base. This gliding joint is labeled *D*. Finally, the motion of the ending of the handle is desired. This is appointed as the point of interest *X*. The kinematic diagram is given in the following figure.

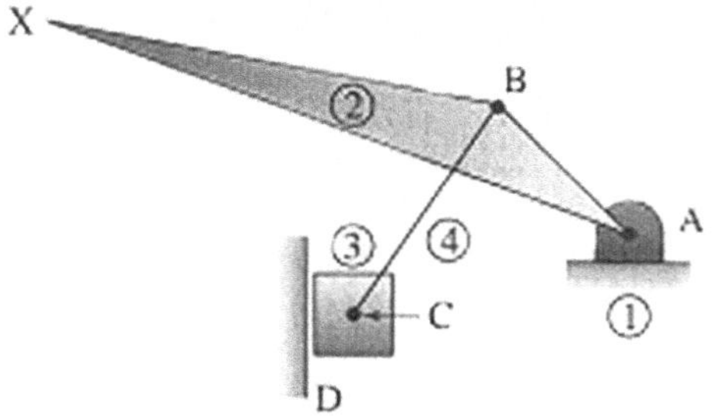

Example 5.3

The following figure shows a pair of vice grips. Draw a kinematic diagram.

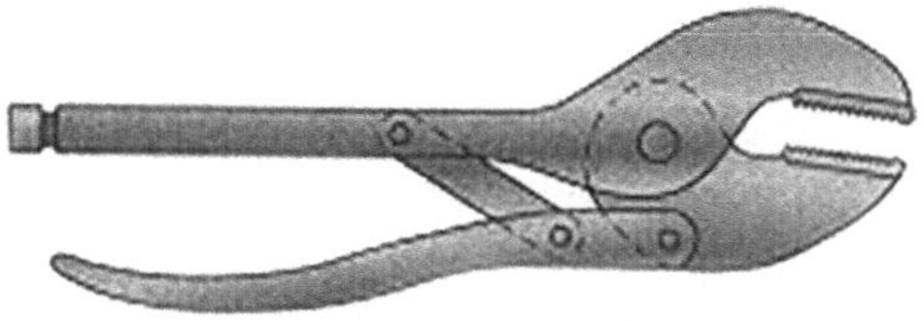

Solution: The first step is to determine the part that will be appointed as the frame. In this problem, no parts are attached to the ground. Therefore, the choice of the frame is rather random. The top handle is appointed as the frame. The motion of all other links is decided relative to the top handle.

The top handle is numbered as link 1. Cautious observation shows three other moving parts: (2) bottom handle, (3) bottom jaw, and (4) bar that attaches the top and bottom handles. Four-pin joints are used to attach these different parts. These joints are labeled A through D. In addition, the motion of the ending of the bottom jaw is desired. This is appointed as the point of interest X. Finally, the motion of the ending of the lower handle is also desired. This is designated as the point of interest Y. The kinematic diagram is given in the following figure.

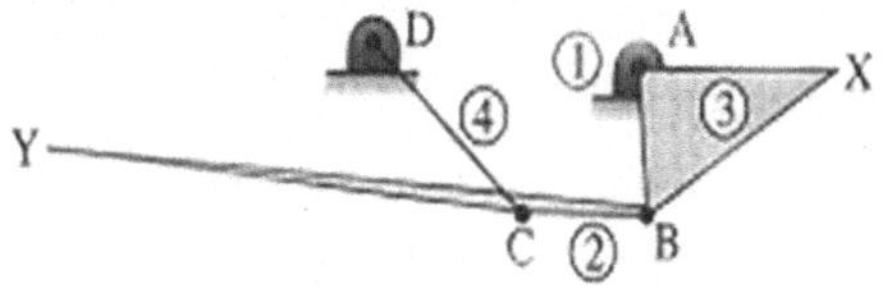

Example 5.4

The following figure displays a toggle clamp. Draw a kinematic diagram utilizing the clamping surface and the handle as points of interest. Also calculate the DOFs for the clamp.

Four-pin joints are used to attach these different parts. These joints are labeled A through D. In addition, the motion of the clamping surface is desired. This is designated as the point of interest X. Finally, the motion of the ending of the handle is also desired. This is designated as the point of interest Y. The kinematic diagram is shown in the following figure.

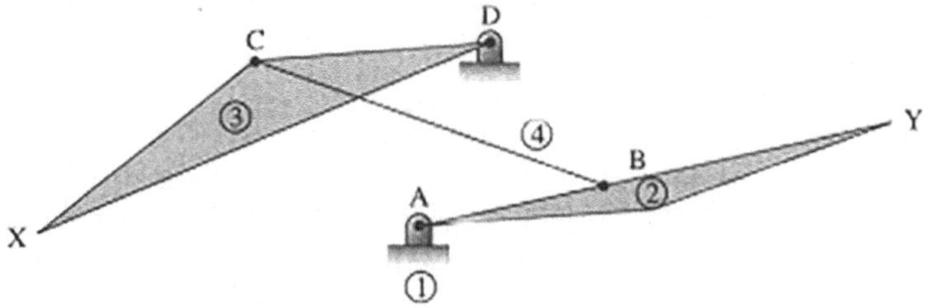

The figure above displays a toggle clamp. Draw a kinematic diagram using the clamping surface and the handle as points of interest. Also calculate the DOFs for the clamp.

Calculating the mobility of the mechanism, it is seen that there are four links. There are also four-pin joints. Therefore,

$$n = 4, jp = 4\text{pins}, jh = 0 \text{ and:}$$

$$M = 3(n-1) - 2jp - jh = 3(4-1) - 2(4) - 0 = 1$$

With one DOF, the clamp mechanism is restricted. Moving only one link, the handle exactly positions all other links in the clamp.

5.3.5 Position Analysis of a Mechanism

If a, b, c, d, the ground position, and angle θ_2 are given, find the angles θ_3 and θ_4.

5.3.5.1 Graphical Solution

First, draw an arc of radius b centered at A. And then draw an arc of radius c centered at O_4. We can find that the intersections are the two possible positions for the linkage, open and crossed (Figure 5.14).

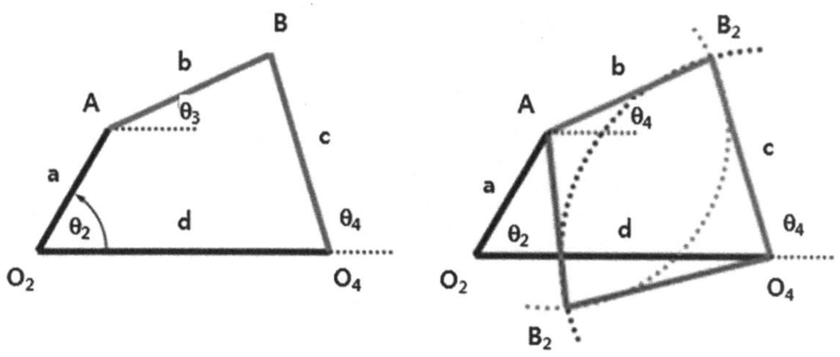

FIGURE 5.14 Graphical solution (example).

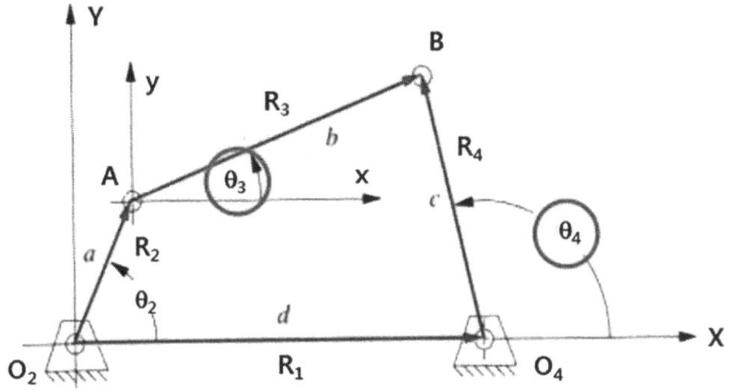

FIGURE 5.15 Analytical solution for four-bar mechanism (example).

5.3.5.2 Analytical Solution

- Obtain coordinates of point A in Figure 5.15:

$$A_x = a\cos\theta_2 \tag{5.2a}$$

$$A_y = a\sin\theta_2 \tag{5.2b}$$

- Obtain coordinates of point B:

$$b^2 = (B_x - A_x)^2 + (B_y - A_y)^2 \tag{5.3a}$$

$$c^2 = (B_x - d)^2 + B_y^2 \tag{5.3b}$$

- We can find the positions B_x and B_y if you solve two equations.

5.3.6 Velocity and Acceleration Analysis of Mechanisms

To complete the kinematics analysis of mechanisms, we will discuss the most common and currently practiced methods that describe the motion through velocity and

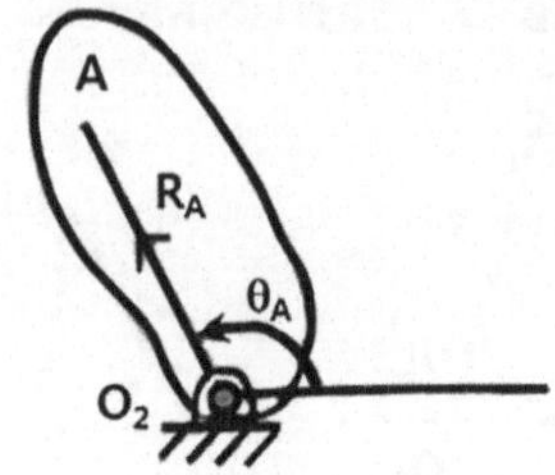

FIGURE 5.16 Analytical solution for simple bar mechanism (example).

acceleration. Firstly, position and velocity analysis must be completed before acceleration analysis. All acceleration components should be expressed as one coordinate system – the inertial frame of reference of the fixed link of the mechanism. For example, we simply consider that a link at center O_2 rotates with radius R_A. Find velocity and acceleration at point A in Figure 5.16.

If this link is represented as polar coordinate, we can describe the displacement of point A:

$$R_A = r_A e^{i\theta_A} \tag{5.4}$$

If Equation (5.4) is differentiated, we can find velocity of link:

$$V_A = \frac{dR_A}{dt} = i\omega_2 r_A e^{i\theta_A} \tag{5.5}$$

If Equation (5.5) is differentiated, we can find the acceleration of link:

$$\begin{aligned} a_A &= \frac{dV_A}{dt} = i\frac{d\omega_2}{dt} r_A e^{i\theta_A} + i\omega_2 r_A \frac{d\left(e^{i\theta_A}\right)}{dt} \\ &= i\alpha_2 r_A e^{i\theta_A} + i\omega_2 r_A i\omega_2 e^{i\theta_A} \\ &= \underbrace{i\alpha_2 R_A}_{a_A^t} - \underbrace{\omega_2^2 R_A}_{a_A^n} = \left(-\omega_2^2 + i\alpha_2\right) R_A \end{aligned} \tag{5.6}$$

5.4 MODELING OF MECHANICAL SYSTEM (POWER SYSTEM)

5.4.1 INTRODUCTION

Modeling of a mechanical system including the structure and mechanism is a mathematical description of the dynamic structural systems to understand its attribute under the work of forces. General modeling methods utilized in dynamic system are Newton, Lagrange, D'Alembert's law, and Hamiltonian mechanics. As an output, models might report the system behavior that can be depicted as random variables in state space. The state space is shown as vectors (displacement, velocity, and acceleration) in time. It also supplies a convenient and dense way to examine systems with multiple inputs and outputs.

When detected in most mechanical parts, field loads suit a random curve that load amplitudes fluctuate in time. For instance, automobiles have totally random stochastic load curves due to the road roughness, vehicle velocity, and environmental conditions. In an airplane, a mean load change like drag due to wind repeatedly happens on the wings when it takes off, taxi or lands. On the other hand, though the load sequence is still variable, the load of the gas turbine blade in an airplane is to a big extent deterministic that there is no randomness in the system conditions.

With uncomplicated algorithms and fast processors, an on-line load measurement for components can be straightly calculated during operation. However, a measurement during operation is completely time-consuming and literally unfeasible to figure out the entire transmitted loads in product lifetime. To efficiently understand the load history of a product in time, engineers rely on the mathematical modeling such as Newtonian and its response analysis (Figure 5.17).

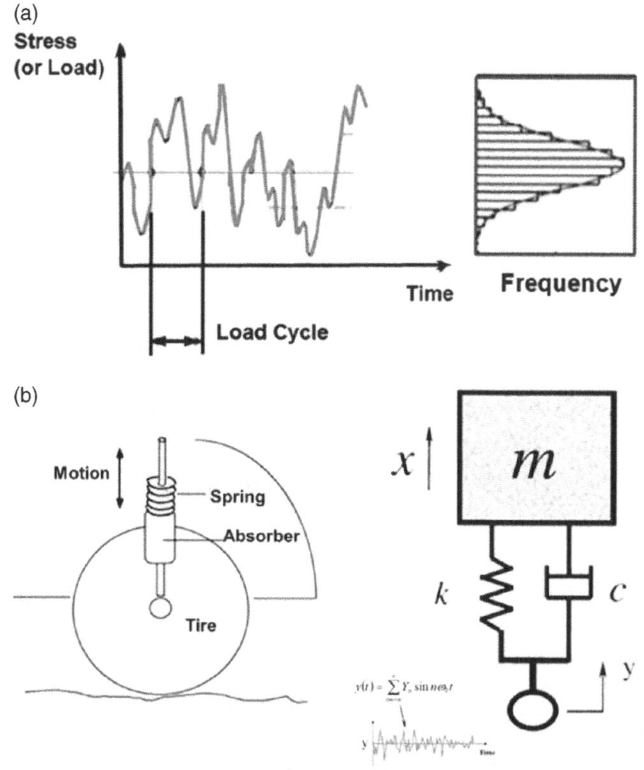

FIGURE 5.17 Random loads and modeling of automobiles by Newtonian modeling: (a) base random vibrations and (b) a simplified modeling of automobiles.

5.4.2 Newton's Mechanics

Mechanics is a physical science that discusses the condition of rest or movement of bodies under the action of forces. No other subjects play a more crucial role in engineering analysis than does mechanics. Therefore, the start of engineering originated from the research of mechanics. Modern research and development in the field of system reliability might depend on the fundamental principles of mechanics including probability and statistical methods.

Mechanics is the oldest of the physical sciences. The earliest study was that of Archimedes (287–212 B.C.) which has to do with the principle of the buoyancy and lever. The earliest investigation of a dynamic problem was ascribed to Galileo (1564–1642) in connection with his experiments with dropping stones. The precise formation of the law of gravitation, as well as the laws of motion, was established by Newton (1642–1727). Basic contributions to the development of mechanics were also achieved by D'Alembert, Laplace, Lagrange, etc.

The concepts of mechanics as a science are dependent on the strict mathematics. On the other hand, the aim of engineering mechanics is the implementation of its principles to the product design. The fundamental principle of mechanics is comparatively few in number, but they have broad fields such as vibrations, automatic control, engine performance, fluid flow stability and strength of structures and machines, and rocket and spacecraft design.

For example, only utilizing a few equations, scientists can outline the motion of a projectile flying through the air and the pull of a magnet, and forecast eclipses of the moon. The mathematical study of the motion is defined as Newtonian mechanics because nearly the whole study constructs on the work of Isaac Newton. Some mathematical principles and laws at the core of Newtonian mechanics include the following:

- Newton's First Law of Motion: In an inertial reference frame, an object is either at rest or continuously moves at a constant velocity, unless acted upon by a force.
- Newton's Second Law of Motion: The acceleration of an object is directly related to the net force and inversely related to its mass.

$$F = \frac{dp}{dt} = \frac{d(mv)}{dt} \tag{5.7}$$

- Newton's Third Law of Motion: For every action, there is an equal and opposite reaction. That is, when one body applies a force on a second body, the second body at the same time applies a force equal in magnitude and opposite in the direction of the first body.

5.4.3 D'Alembert's Principle for Mechanical System

D'Alembert's principle is a statement of the basic classical laws of motion. The principle of virtual work states that the sum of the incremental virtual works performed by all external forces F_i acting in conjunction with virtual displacements δs_i of the point on which the associated force is acting is zero:

$$\delta W = \sum_i F_i \cdot \delta r_i = 0 \tag{5.8}$$

Utilizing D'Alembert's principle and free body diagram, an engineer can model a mechanical system. For example, if there is an automobile in transportation, we can model a simple dynamic system with a mass that is separated from a wall by a spring and a dashpot. The mass could represent an automobile, with the spring and dashpot representing the automobile's bumper. If only horizontal motion and forces like wind are considered, it can be represented as in Figure 5.18.

The free body diagram is a graphical method displaying all external forces applying on a body. There is only one position in this system defined by the variable "x" that is positive to the right. It is presumed that $x=0$ when the spring is in its loosened state. As seen in Figure 5.19, there are four forces to evolve a model from the free

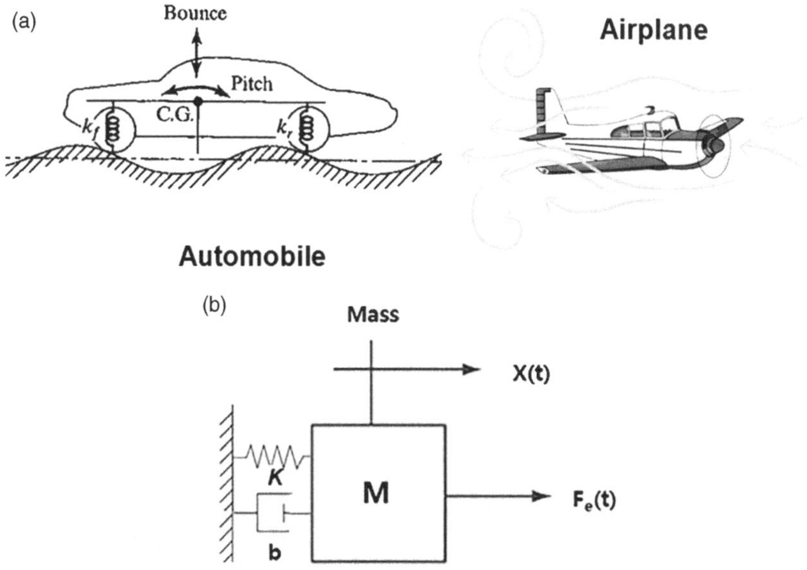

FIGURE 5.18 Typical mechanical automobile modeling: (a) typical automobile and airplane subjected to an external force and (b) mass-spring-dashpot system.

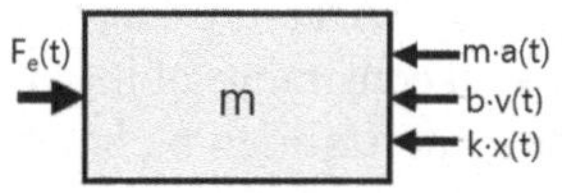

FIGURE 5.19 Completed free body diagram for automobile modeling.

body diagram: (1) an external force (F_e) such as air-resistance force and friction force; (2) a spring force that will be a force from the spring, $k \cdot x$, to the left; (3) a dashpot force that will be a force from the dashpot, $b \cdot v$, to the left; and (4) finally, there is an inertial force which is defined to be opposed to the defined direction of motion. This is represented by $m \cdot a$ to the left.

Newton's second law explains that an object accelerates in the direction of an acted force, and that this acceleration is oppositely proportional to the force, or

$$\sum_{\text{all lexternal}} F = m \cdot a \tag{5.9}$$

Taking away the right-hand side results in D'Alembert's principle:

$$\sum_{\text{all lexternal}} F - m \cdot a = 0 \tag{5.10}$$

If we think about the $m \cdot a$ term to be an inertia force (or D'Alembert's force), D'Alembert's law will be left

$$\sum_{\text{all}} F \cdot \delta r = 0 \tag{5.11}$$

To imagine this, think about pushing against a mass (in the absence of friction) with your hand in the positive direction. Your hand experiences a force in the direction opposite to that of the direction of the force (this is the $-ma$ term). The inertial force is always in the direction opposite to the prescribed positive direction. We sum all of these forces to zero and get

$$F_e(t) - ma(t) - bv(t) - k \cdot x(t) = 0 \tag{5.12}$$

That is, we can change

$$m\frac{d^2x}{dt^2} + b\frac{dx}{dt} + kx(t) = F_e(t) \tag{5.13}$$

5.4.4 Derivation of Lagrange's Equations from D'Alembert's Principle

The movement of particles in space often is confined by constraints. An engineer would work only with independent DOFs (coordinates) without constraints (Figure 5.20).

Movements of N particles, $n=3N$ DOFs, are subject to k equations relating coordinates. In classical mechanics, if all constraints of the system are holonomic, a system may be called holonomic. For a constraint to be holonomic, it should be expressed as follows:

$$f_j(x_1,\ldots,x_n,t) = c_j \quad j = 1,2,\ldots,k \tag{5.14}$$

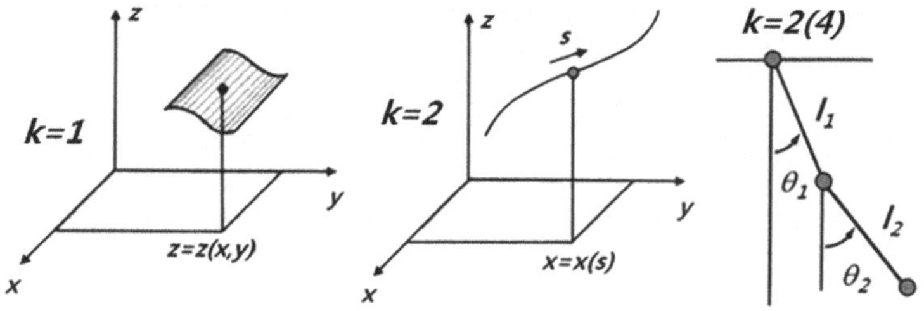

FIGURE 5.20 The motion of particles restricted by constraints.

That is, a holonomic constraint relies only on the coordinates x_j and time t. The generalized coordinates describe the motion of the system relative to Cartesian coordinates subject to k constraints:

$$\begin{aligned} x_1 &= x_1\left(q_1, q_2, \ldots, q_{n-k}, t\right) \\ &\cdots \\ x_n &= x_n\left(q_1, q_2, \ldots, q_{n-k}, t\right) \end{aligned} \tag{5.15}$$

where any set of $n - \mathrm{k} = 3N - \mathrm{k}$ independent coordinates that can totally explain the system.

The change of a Cartesian coordinate can be induced from changes in generalized coordinates in dt:

$$dx_1 = \frac{\partial x_1}{\partial q_1} dq_1 + \ldots + \frac{\partial x_1}{\partial q_{n-k}} dq_{n-k} + \frac{\partial x_1}{\partial t} dt \tag{5.16}$$

In a compact way, we can describe as follows:

$$dx_i = \sum_{\sigma=1}^{n-k} \frac{\partial x_i}{\partial q_\sigma} dq_\sigma + \frac{\partial x_1}{\partial t} dt \quad i = 1, \ldots, n \tag{5.17}$$

So, the virtual displacements are called infinitesimal, instantaneous displacements of the coordinates consistent with the constraints. That is,

$$\delta x_i = \sum_{\sigma=1}^{n-k} \frac{\partial x_i}{\partial q_\sigma} \delta q_\sigma \quad i = 1, \ldots, n \tag{5.18}$$

Because forces of constraint are perpendicular to the direction of motion, the forces of constraint do not work under a virtual displacement (Figure 5.21).

$$\delta W = \tau \cdot \delta s_1 - \tau \cdot \delta s_2 = -\tau \cdot \delta s_r = 0 \tag{5.19}$$

We can rewrite Newton's second law as follows:

$$\dot{p}_i = F_i^{(a)} + R_i \quad i = 1, \ldots, n \text{ or } \sum_i \left(F_i^{(a)} + R_i - \dot{p}_i\right)\delta x_i = 0 \tag{5.20}$$

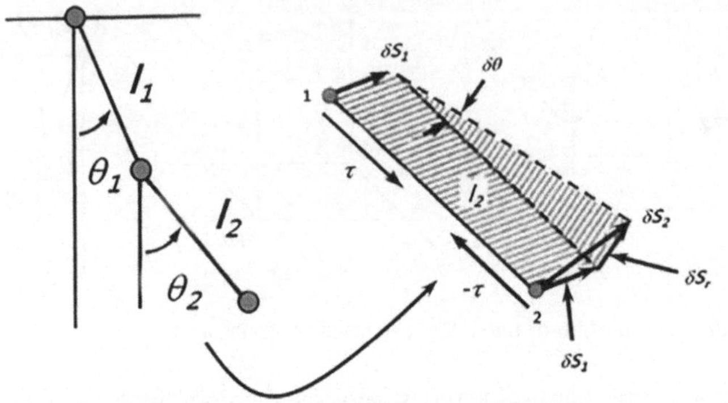

FIGURE 5.21 Forces of constraint perpendicular to the direction of motion.

Because $\sum_i R_i \delta x_i = 0$, the virtual work by D'Alembert's principle can be refined as follows:

$$\sum_i \left(F_i^{(a)} - \dot{p}_i\right)\delta x_i = 0 \tag{5.21}$$

We can also rewrite D'Alembert's principle in terms of the generalized coordinates. The virtual work done by applied forces under virtual displacement is defined as follows:

$$\delta W\left(= \sum_i \left(F_i^{(a)} - \dot{p}_i\right)\delta x_i\right) = \sum_i F_i \delta x_i = \sum_{\sigma=1}^{n-k}\left(\sum_{i=1}^{n} F_i \frac{\partial x_i}{\partial q_\sigma}\right) dq_\sigma = \sum_{\sigma=1}^{n-k} Q_\sigma \delta q_\sigma \tag{5.22}$$

where $\delta x_i = \sum_{\sigma=1}^{n-k}\left(\frac{\partial x_i}{\partial q_\sigma}\right) dq_\sigma$ and generalized forces $Q_\sigma = \sum_{i=1}^{n} F_i \frac{\partial x_i}{\partial q_\sigma}$.

Let Cartesian coordinates be subjected to k constraints $x_i = x_i\left(q_1,\ldots,q_{n-k},t\right)$. The change of a Cartesian coordinate induced from changes in generalized coordinates in dt is defined as follows:

$$dx_i = \sum_{\sigma=1}^{n-k}\left(\frac{\partial x_i}{\partial q_\sigma}\right) dq_\sigma + \frac{\partial x_i}{\partial t} dt \quad i = 1,\ldots,n \tag{5.23}$$

Taking with time derivatives from Equation (5.23):

$$\frac{dx_i}{dt} \equiv \dot{x}_i = \sum_{\sigma=1}^{n-k}\left(\frac{\partial x_i}{\partial q_\sigma}\right)\dot{q}_\sigma + \left(\frac{\partial x_i}{\partial t}\right) \quad i = 1,\ldots,n \tag{5.24}$$

where $\dot{x}_i = x_i\left(q_1,\ldots,q_{n-k};\dot{q}_1,\ldots,\dot{q}_{n-k};t\right)$.

If other variables are kept constant in taking partial derivatives,

$$\frac{\partial \dot{x}_i}{\partial \dot{q}_\sigma} = \frac{\partial x_i}{\partial q_\sigma} \frac{d}{dt}\left(\frac{\partial x_i}{\partial q_\sigma}\right) = \frac{\partial}{\partial q_\sigma}\left(\frac{dx_i}{dt}\right) \tag{5.25a}$$

$$\text{lhs} = \sum_\lambda \left(\frac{\partial}{\partial q_\lambda}\frac{\partial x_i}{\partial q_\sigma}\right)\dot{q}_\lambda + \frac{\partial}{\partial t}\frac{\partial x_i}{\partial q_\sigma} \tag{5.25b}$$

$$\text{rhs} = \sum_\lambda \left(\frac{\partial}{\partial q_\sigma}\frac{\partial x_i}{\partial q_\lambda}\right)\dot{q}_\lambda + \frac{\partial}{\partial q_\sigma}\frac{\partial x_i}{\partial t} \tag{5.25c}$$

Here, the order of partial derivatives can be interchanged. The momentum equations might be described as follows:

$$\sum_i \dot{p}_i \delta x_i = \sum_i m_i \ddot{x}_i \delta x_i = \sum_\sigma \left(\sum_i m_i \ddot{x}_i \frac{\partial x_i}{\partial q_\sigma}\right)\delta q_\sigma \tag{5.26}$$

Because $\dot{p}_i = m_i \ddot{x}_i$ and $\delta x_i = \sum_{\sigma=1}^{n-k} \frac{\partial x_i}{\partial q_\sigma}\delta q_\sigma$, Equation (5.26) might be described as follows:

$$\sum_i m_i \ddot{x}_i \frac{\partial x_i}{\partial q_\sigma} \cong \sum_i m_i \frac{d\dot{x}_i}{dt}\frac{\partial x_i}{\partial q_\sigma} = \sum_i m_i \left[\frac{d}{dt}\left(\dot{x}_i \frac{\partial x_i}{\partial q_\sigma}\right) - \dot{x}_i \frac{d}{dt}\frac{\partial x_i}{\partial q_\sigma}\right] \tag{5.27}$$

Because $\frac{\partial \dot{x}_i}{\partial \dot{q}_\sigma} = \frac{\partial x_i}{\partial q_\sigma}$ and $\frac{d}{dt}\left(\frac{\partial x_i}{\partial q_\sigma}\right) = \frac{\partial}{\partial q_\sigma}\left(\frac{dx_i}{dt}\right)$, Equation (5.27) can be expressed as follows:

$$\sum_i m_i \frac{d}{dt}\left(\dot{x}_i \frac{\partial \dot{x}_i}{\partial \dot{q}_\sigma}\right) = \frac{d}{dt}\left[\frac{\partial}{\partial \dot{q}_\sigma}\left(\frac{1}{2}\sum_i m_i \dot{x}_i^2\right)\right] \tag{5.28a}$$

$$\sum_i m_i \dot{x}_i \frac{\partial}{\partial q_\sigma}\frac{dx_i}{dt} = \frac{\partial}{\partial q_\sigma}\left(\frac{1}{2}\sum_i m_i \dot{x}_i^2\right) \tag{5.28b}$$

From Equation (5.26), the momentum equations can be expressed as follows:

$$\sum_i \dot{p}_i \delta x_i = \sum_\sigma \left(\frac{d}{dt}\frac{\partial T}{\partial \dot{q}_\sigma} - \frac{\partial T}{\partial q_\sigma}\right)\delta q_\sigma \tag{5.29}$$

where kinetic energy $T = \frac{1}{2}\sum_i m_i \dot{x}_i^2 = T\left(q_1,\ldots,q_{n-k};\dot{q}_1,\ldots,\dot{q}_{n-k};t\right)$.

From Equations (5. 21) and (5. 22), we can rewrite D'Alembert's principle in terms of the generalized coordinates as follows:

$$\sum_{\sigma}\left(\frac{d}{dt}\frac{\partial T}{\partial \dot{q}_\sigma}-\frac{\partial T}{\partial q_\sigma}-Q_\sigma\right)\delta q_\sigma=0 \tag{5.30}$$

If all q are independent and arbitrary, Lagrange's equations are expressed as follows:

$$\frac{d}{dt}\frac{\partial T}{\partial \dot{q}_\sigma}-\frac{\partial T}{\partial q_\sigma}=Q_\sigma \quad \sigma=1,\ldots,n-k \tag{5.31}$$

$n-k$ equations for $n-k$ independent generalized coordinates are equivalent to Newton's second law. For conservative forces, potential energy depends only on the position:

$$V(x_1,\ldots,x_n)=V(q_1,\ldots,q_{n-k},t) \qquad \frac{\partial V}{\partial \dot{q}_\alpha}=0 \tag{5.32}$$

Generalized forces are given by negative gradients with respect to corresponding generalized coordinate:

$$Q_\sigma \equiv \sum_i F_i\frac{\partial x_i}{\partial q_\sigma}=-\sum_i\left[\frac{\partial}{\partial x_i}V(x_i,\ldots,x_n)\right]\frac{\partial x_i}{\partial q_\sigma}=-\frac{\partial}{\partial q_\sigma}V(q_1,\ldots,q_{n-k},t) \tag{5.33}$$

Lagrange's equations for conservative forces:

$$\frac{d}{dt}\frac{\partial L}{\partial \dot{q}_\sigma}-\frac{\partial L}{\partial q_\sigma}=0 \quad \sigma=1,\ldots,n-k \tag{5.34}$$

5.5 BOND-GRAPH MODELING FOR LOAD ANALYSIS

5.5.1 Introduction

A mechanical system such as automobile or airplane generates power from engine and transfers it to transmission via drive system through mechanism. When different physical domains – electric, magnetic, mechanical, hydraulic, etc. – are involved in a mechanical system, one of the powerful tools that can model engineering systems is bond graph. It is a direct graphical tool for modeling dynamic systems involving several academic disciplines, that is, different engineering areas such as mechanical, electrical, thermal, and hydraulic system. When designing a newly designed dynamic system, it is a capable method to utilize a graphical representation for communicating other engineers on its dynamic modeling.

In 1959, Professor Henry Payner and his former students at MIT suggested the bond graph method, which gave the thoroughgoing concepts for portraying multiport systems in terms of power bonds, connecting the elements of the physical system to the so-called junction structures which were demonstrations of the constraints. This power exchange portray of a system is defined as bond graph. It is suitable to

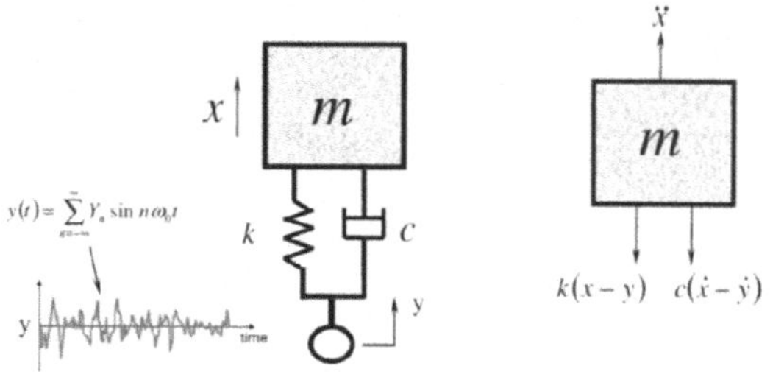

FIGURE 5.22 Typical modeling of automobiles subjected to repetitive random vibrations.

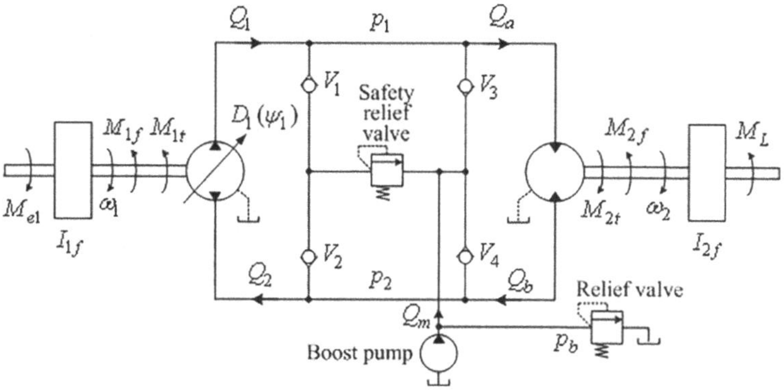

FIGURE 5.23 General hydrostatic transmission modeling.

model a mechanical product because a multi-port system is composed and power generated from the engine is transmitted to other systems – transmission, drive system, etc.

In 1961, Professor Henry Paynter published a book entitled *Analysis and Simulation of Simulation of Multiport Systems*. Three authors in 2006 had issued the fourth edition entitled *System Dynamics – Modeling and Simulation of Mechatronic Systems*. Now several disciplines of bond graph have been widely accepted in the world as a modeling methodology (Figures 5.22 and 5.23).

5.5.2 Basic Elements, Energy Relations, and Causality of Bond Graph

A bond graph is a graphical display of a physical dynamic multi-port system. Behavior with respect to energy is domain independent, regardless of electric or mechanical system. That is, we can conclude that it is the same in all engineering disciplines when contrasting the RLC circuit for electrical system with the damped mass spring in a mechanical system. Based on mutual energy exchange between different systems, the symbols in bond graph mean a bi-directional exchange of physical energy. Therefore, bond graph might be applicable in multi-energy domain – mechanical, electrical, hydraulic system, etc.

The analysis of dynamic systems might be comparatively straightforward when it has few DOFs or a steady-state behavior. However, it is complex to model a multi-port

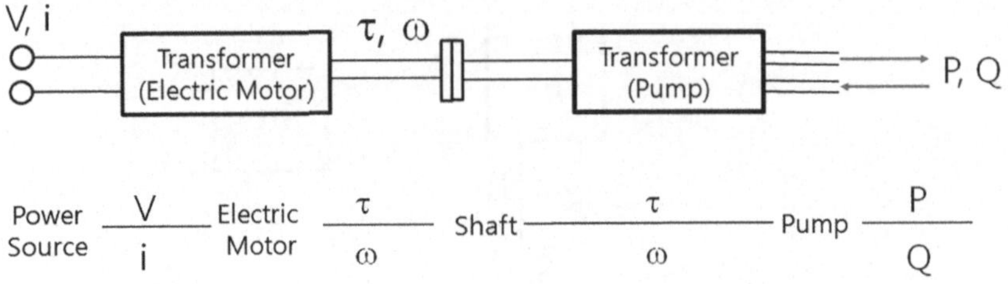

FIGURE 5.24 Power flow in bond graph for electric-hydraulic system.

system and derive its governing equation. In most of the cases, the principal matter of engineers is to demonstrate a mathematical model that represents the dynamic behavior of the system and how the different parameters affect the system behavior, because the system dynamic equations are normally partial differential equations, whose solutions require extensive mathematical understanding.

As the essential knowledge of the bond graph theory, energy flow is a fundamental component in a system. It runs in from one or more sources, is momentarily stored in system components or partly dissipated in resistances as heat, and lastly reaches at "loads" where it creates some desired effects. Power, therefore, is the rate of energy flow without direction.

Bond graph represents the power flow between two systems. For example, think about an electric-hydraulic system. Generated electric power works motor which operates pump. This energy flow is represented through an arrow (bond) in Figure 5.24. Each bond represents the on-the-spot energy flow or power. The flow in each bond is designated by a pair of variables called 'power-conjugated variables' whose product is the instantaneous power of the bond. Because (generated) power in mechanical system is difficult to directly compute, engineers can use two transient variables such as flow and effort. Every system domain has a pair of effort and flow variables. For example, in a mechanical system, flow represents the "velocity" and effort the "force", while in an electrical system, flow represents the "current" and effort the "voltage". The product of both transient variables, that is, power is represented as follows:

$$P = e(t) \cdot f(t) \tag{5.35}$$

The method enables to simulate multiple physical domains such as mechanical, thermal, electrical, and hydraulic. Flows and efforts should be recognized with a specific variable for each particular physical domain that is working. Table 5.1 also displays the physical definitions of the variables in different domains.

The bond graph is composed of the "bonds" which join together "1-port", "2-port", and "3-port" elements. Whether or not power in bond graph is continuous, every element is represented by a multi-port. Ports are attached by bonds. The fundamental blocs of the common bond graph theory are itemized in Table 5.2.

For 1-ports, there are effort sources, flow sources, *I*-type elements, *C*-type elements, and *R*-type elements that can discontinuously connect power. For 2-ports,

TABLE 5.1
Energy Flow in the Multi-Port Physical System

Modules	Effort, e(t)	Flow, f(t)
Mechanical translation	Force, $F(t)$	Velocity, $V(t)$
Mechanical rotation	Torque, $\tau(t)$	Angular velocity, $\omega(t)$
Compressor, pump	Pressure difference, $\Delta P(t)$	Volume flow rate, $Q(t)$
Electric	Voltage, $V(t)$	Current, $i(t)$
Thermal	Temperature, T	Entropy change rate, ds/dt
Chemical	Chemical potential, μ	Mole flow rate, dN/dt
Magnetic	Magneto-motive force, e_m	Magnetic flux, φ

TABLE 5.2
Basic Elements of Bond Graph

Elements		Symbol	Relation Equations
1-Port elements (sources)	Effort	S_e ——	$S_e = e(t)$
	Flow	S_f ——	$S_f = f(t)$
1-Port elements	C-type elements	C ——	$\frac{de(t)}{dt} = \frac{1}{C} f(t)$
	I-type elements	I ——	$\frac{df(t)}{dt} = \frac{1}{I} e(t)$
	R-type elements	R ——	$e(t) = R \cdot f(t)$
2-Port elements	Transformer	$\overset{1}{——}\ TF\ \overset{2}{——}$	$e_2(t) = TF \cdot e_1(t)$ $f_2(t) = \frac{1}{TF} \cdot f_1(t)$
	Gyrator	$\overset{1}{——}\ GY\ \overset{2}{——}$	$e_2(t) = GY \cdot f_1(t)$ $f_2(t) = \frac{1}{GY} \cdot e_1(t)$
3-Port junction elements	0-Junction	$\overset{1}{——}\ 0\ \overset{2}{——}$	$e_2(t) = e_1(t)$
	1-Junction	$\overset{1}{——}\ 1\ \overset{2}{——}$	$f_2(t) = f_1(t)$

there are gyrator and transformer that can continuously connect power. For 3-ports, there are 1-junction and 0-junction that can make up the network.

Power bonds may join at one of the two categories of junctions: a "0" junction and a "1" junction. In these connections, no energy can be generated or dissipated. This is defined as power continuity. In a "0" junction, the flow and the efforts fulfill Equations (5.36) and (5.37):

$$\sum \text{flow}_{\text{input}} = \sum \text{flow}_{\text{output}} \tag{5.36}$$

$$\text{effort}_1 = \text{effort}_2 = \ldots = \text{effort}_n \tag{5.37}$$

This matches to a node in an electrical circuit (where Kirchhoff's current law applies). On the other hand, in a "1" junction, the flow and the efforts fulfil Equations (5.38) and (5.39):

$$\sum \text{effort}_{\text{input}} = \sum \text{effort}_{\text{output}} \tag{5.38}$$

$$\text{flow}_1 = \text{flow}_2 = \ldots = \text{flow}_n \tag{5.39}$$

This matches to force balance at a mass in a system. An instance of a "1" junction is a resistor in series. In junction, the assumption of energy conservation is presumed, but no loss is allowed. There are two extra variables significant in the account of dynamic systems.

For any element with a bond with power variables such as *effort* and *flow*, the energy variation from t_0 to t might be expressed as follows:

$$H(t) - H(t_0) = \int_{t_0}^{t} e(\tau) f(\tau) d\tau \tag{5.40}$$

For C-type storage elements, e (effort) is a function of q (displacement) as capacitor or spring (Figure 5.25).

If displacement is differentiated, flow is expressed as

$$q(t) = \int f(t) dt \Rightarrow \dot{q} = \frac{dq}{dt} = f(t) \tag{5.41}$$

which is defined as a balance equation and forms a part of the constitutive equations of the storage element. If Equation (5.41) is changing variables from t to q, the linear case can be expressed as follows:

$$H(q) - H(q_0) = \frac{1}{2C}\left(q^2 - q_0^2\right) \tag{5.42}$$

For I-type storage elements, f (flow) is a function of p (momentum) as inductor or mass (Figure 5.26).

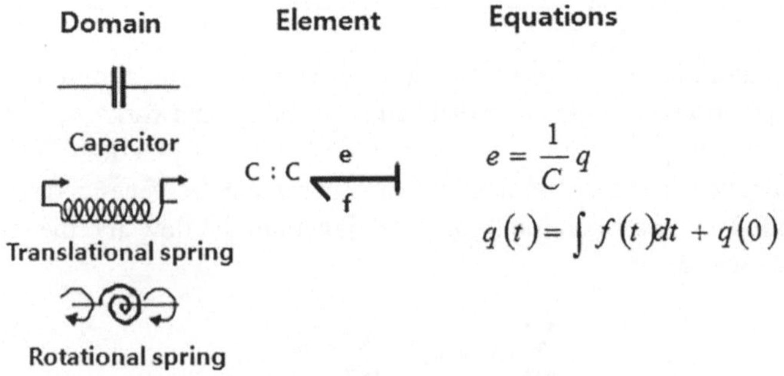

FIGURE 5.25 Examples of C elements.

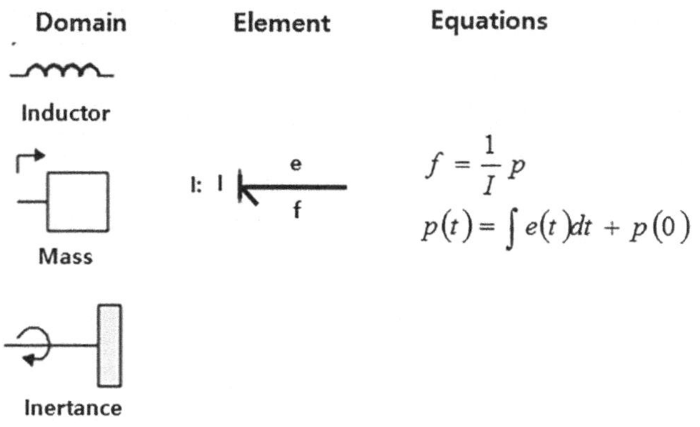

FIGURE 5.26 Examples of *I* elements.

If momentum is differentiated, effort is expressed as

$$p(t) = \int e(t) dt \Rightarrow \dot{p} = \frac{dp}{dt} = e(t) \tag{5.43}$$

which is called the balance equation. If Equation (5.43) is changing variables from t to p, the linear case can be expressed as follows:

$$H(p) - H(p_0) = \frac{1}{2I}\left(p^2 - p_0^2\right) \tag{5.44}$$

Resistor elements represent situations where energy dissipates such as mechanical damper, electrical resistor, and coulomb frictions. In these kinds of elements, there is a relationship between effort and flow. The value of "R" might be stable or function of any system parameter including time:

$$e(t) = R \cdot f(t) \tag{5.45}$$

Compliance elements represent the circumstances where energy stores – mechanical springs, electrical capacitors, etc. In these kinds of elements, there is a relationship between effort and displacement variables. The value of "K" might be stable or function of any system parameter including time:

$$e(t) = K \cdot q(t) \tag{5.46}$$

Inertia elements represent the relationship between the "flow" and momentum (mass, moment of inertia, electrical coil, etc.) as Equation (5.47) shows. The value of "I" tends to be stable

$$p(t) = I \cdot f(t) \tag{5.47}$$

A transformer adds on no power but changes it, such as a lever or an electrical transformer. Transformers show those physical phenomena that are difference of the gains of output flow and effort on the gains of input flow and effort. If the transformation ratio is denoted by the "TF" value, then the relationship between input and output can be expressed as in Equations (5.48) and (5.49):

$$e_{\text{output}}(t) = TF \cdot e_{\text{input}}(t) \tag{5.48}$$

$$f_{\text{output}}(t) = \frac{1}{TF} \cdot f_{\text{input}}(t) \tag{5.49}$$

The process of determining the computational direction of the bond variables is defined as causal analysis. One is the "half-arrow" sign convention. This defines the presumed direction of positive energy flow. As with electrical circuit diagrams and free-body diagrams, the selection of positive direction is random, with the warning that the analyst must be consistent throughout with the chosen definition. The other attribute is the "causal stroke". This is a vertical bar placed on only one end of the bond. It is not random (Figure 5.27).

On each bond, one of the two variables should be the cause and the other one the effect. This can be concluded by the relationship indicated by the arrow direction. Effort and flow causalities always act in opposite directions in a bond. The causality allocation procedure selects what sets for each bond. Causality allocation is a requisite to change the bond graph into computable coded.

Any port (single, double, or multi) attached to the bond will state either "effort" or "flow" by its causal stroke, but not both. The port attached to the end of the bond with the "causal stroke" states the "flow" of the bond. And the bond imposes "effort" upon that port. Similarly, the port on the end without the "causal stroke" imposes "effort" to the bond, while the bond imposes "flow" to that port.

Once the system is represented in the formation of bond graph, the state equations that govern its behavior can be obtained straight as a first-order differential equations in terms of generalized variables defined above, utilizing straightforward and standardized procedures, regardless of the physical domain to which it associates, even when connected to one another across domains.

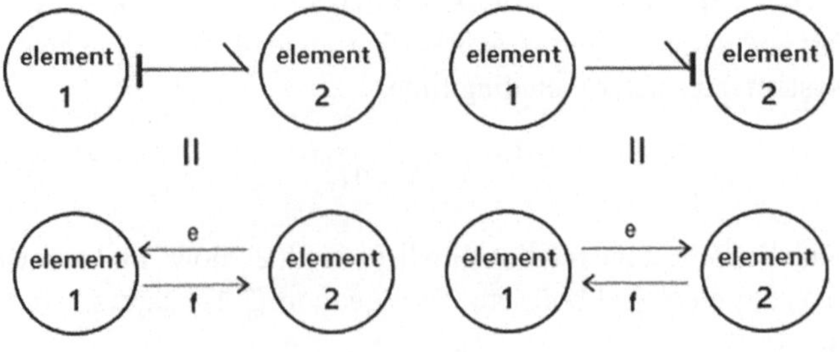

FIGURE 5.27 "Half-arrow" sign convention and meaning of the causal stroke.

Example 5.5

There is the damped mass–spring system for a mechanical system represented in the below figure. In mechanical diagrams, the port variables of the bond graph elements are the *force* on the element port and *velocity* of the element port. These two variables are associated with each other. The *power* being exchanged by a port with the rest of the system is the product of force and velocity. That is, $P = Fv$. The equations of a damper, spring, and mass are as follows (we use damping coefficient a, spring coefficient K_s, mass m, and applied force F_a):

$$F_d = \alpha v,\ F_s = K_s \int v dt = \frac{1}{C_s} \int v dt,\ F_m = m \frac{dv}{dt}$$

Because the loose ends of the example all have the same velocity, the common v is changed to a '1', a so-called 1-junction. This junction element also involves that the forces sum up to zero, considering the sign (related to the power direction). The force is depicted onto an *effort* and the velocity onto a *flow*.

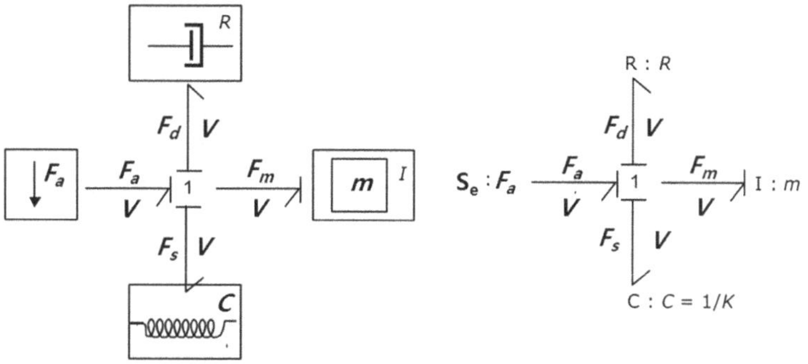

Example 5.6

To understand a transformer, a mass less ideal lever is considered. As seen in the figure below, the lever is rigid to establish a linear relationship between power variables at both its ends.

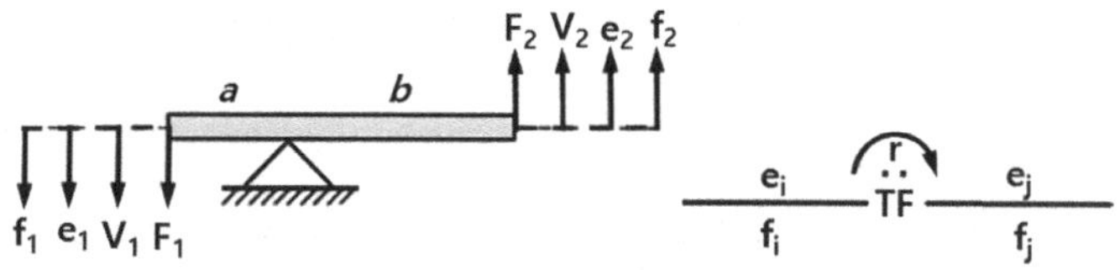

From the geometry, we have

$$V_2 = (b/a)V_1$$

The power transmission implies

$$F_2 = (a/b)F_1 \text{ so that } V_2F_2 = V_1F_1$$

In bond graphs, such a situation may be represented as TF (transformer). The 'r' above the transformer denotes the modulus of the transformer, which may be a constant ('b/a'):

$$f_j = rf_i \text{ and } e_j = \left(1/r\right)e_i$$

Thus, the following expression establishes the conservation of power:

$$e_j f_j = e_i f_i$$

5.5.3 Case Study: Failure Analysis and Redesign of a Helix Upper Dispenser

The icemaker in a domestic refrigerator comprises many mechanical parts. In the ice-making process, these components are subjected to a variety of mechanical loads. Ice making involves several mechanical processes: (1) the filtered water through a tap line is pumped to provide the tray; (2) the air chilled in the heat exchanger supplies the water tray; (3) ice cubes are harvested in the bucket until it is full; and (4) when the customer pushes the lever by force, cubed or crushed ice will be dispensed. In the United States, customers usually require an SBS refrigerator to harvest 10–200 cubes a day. Ice production might be affected by uncontrollable customer usage conditions such as ice consumption, water pressure, refrigerator notch settings, and the number of times the door is opened. When the refrigerator is plugged in, the cubed ice mode is chosen by itself without direct human control. A crusher shatters the cubed ice in the crushed mode. Usually, the mechanical load of the icemaker is small because it is operated without fused or webbed ice.

However, for Asian customers, fused or webbed ice will regularly crystallize in the tray because in cubed mode they dispense ice rarely. As ice is dispensed under these situations, a significant mechanical overload happens in the ice crusher. However, in Europe or the United States, the icemaker system operates continuously as it is repeatedly utilized in both crushed and cubed ice modes. This might produce a mechanical overload.

Figure 5.28 gives a general summary on the ice maker. Figures 5.29 and 5.30 show schematic diagrams of the mechanical load transfer in the ice bucket assembly and its bond graphs. An AC auger motor generates power. To obtain sufficient torque to crush

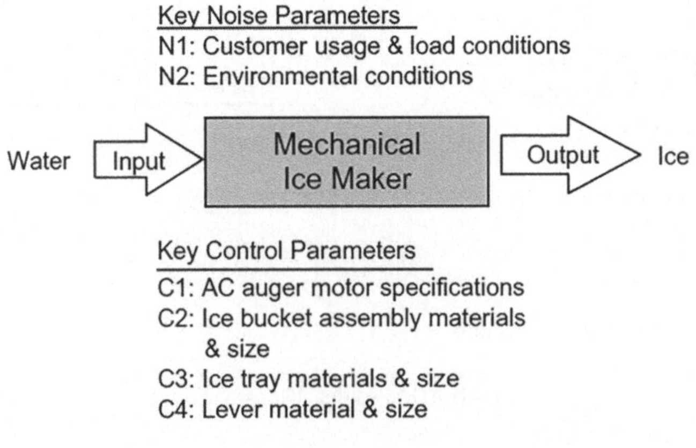

FIGURE 5.28 Robust design schematic of an ice maker.

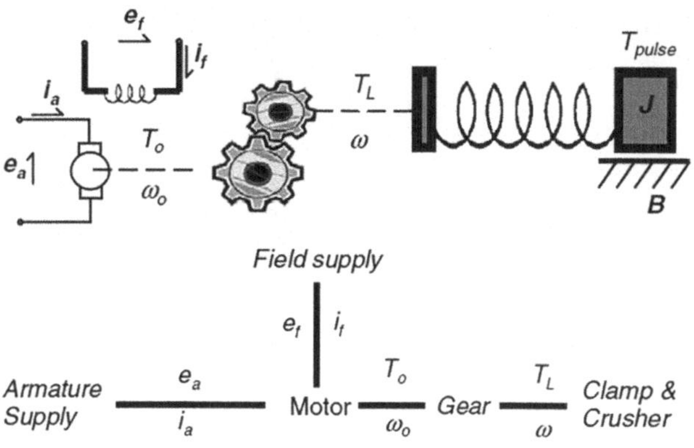

FIGURE 5.29 Schematic diagram for mechanical ice bucket assembly.

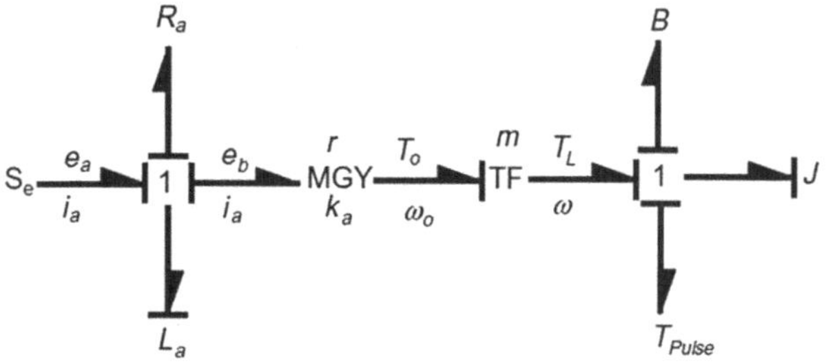

FIGURE 5.30 Bond graph of ice bucket assembly.

the ice, motor power is moved through the gear system to the ice bucket fabrication – in other words, to the helix upper dispenser, the blade dispenser, and the ice crusher.

The bond graph might be expressed as follows:

$$dfE_2/dt = 1/L_a \times eE_2 \tag{5.50}$$

$$dfM_2/dt = 1/J \times eM_2 \tag{5.51}$$

The junction from Equation (5.50)

$$eE_2 = e_a - eE_3 \tag{5.52a}$$

$$eE_3 = R_a \times fE_3 \tag{5.52b}$$

The junction from Equation (5.51)

$$eM_2 = eM_1 - eM_3 \tag{5.53a}$$

$$eM_1 = (K_a \times i) - T_{\text{Pulse}} \tag{5.53b}$$

$$eM_3 = B \times fM_3 \tag{5.53c}$$

Because $fM_1 = fM_2 = fM_3 = \omega$ and $i = fE_1 = fE_2 = fE_3 = i_a$.
From Equation (5.52a and b),

$$eE_2 = e_a - R_a \times fE_3 \tag{5.54}$$

$$fE_2 = fE_3 = i_a \tag{5.55}$$

If substituting Equations (5.54) and (5.55) into (5.50), then

$$di_a/dt = 1/L_a \times (e_a - R_a \times i_a) \tag{5.56}$$

And from Equation (5.53a–c), we can obtain

$$eM_2 = [(K_a \times i) - T_{\text{Pulse}}] - B \times fM_3 \tag{5.57a}$$

$$i = i_a \tag{5.57b}$$

$$fM_3 = fM_2 = \omega \tag{5.57c}$$

If substituting Equation (5.57a–c) into (5.51), then

$$d\omega/dt = 1/J \times [(K_a \times i) - T_{\text{Pulse}}] - B \times \omega \tag{5.58}$$

So, the governing equation (or state equation) might be obtained from Equations (5.56) and (5.58) as follows:

$$\begin{bmatrix} di_a/dt \\ d\omega/dt \end{bmatrix} = \begin{bmatrix} -R_a/L_a & 0 \\ mk_a & -B/J \end{bmatrix} \begin{bmatrix} i_a \\ \omega \end{bmatrix} + \begin{bmatrix} 1/L_a \\ 0 \end{bmatrix} e_a + \begin{bmatrix} 1 \\ -1/J \end{bmatrix} T_{Pulse} \tag{5.59}$$

When Equation (5.59) is integrated, the angular velocity of the ice bucket mechanical assembly is obtained as

$$y_p = \begin{bmatrix} 0 & 1 \end{bmatrix} \begin{bmatrix} i_a \\ \omega \end{bmatrix} \tag{5.60}$$

5.5.4 Case Study: Hydrostatic Transmission (HST) in Sea-Borne Winch

The winch structure is designed for throwing, towing, and holding the cable and array in ships. The operation conditions of a sea-borne winch can be varied such as operation states – sea state, ship speed, and towing cable length. Because its operation needs high torque, a sea-borne winch is often utilized by the hydrostatic transmission (HST). It consists of an electric motor, a pump, piping, a hydraulic motor, and loads. For modeling HST, tension and the response characteristics under the states of throwing, towing, and holding should be known before the design of HST. Tension data might be acquired from tension experiments. However, obtaining the exact time response characteristics from an experiment has many difficulties. And many previous design methods for HST involve extensive calculations because the energy type of HST alters from electrical and mechanical to hydraulic, and then mechanical system. Bond graph, therefore, can effortlessly model the HST system and obtain the dynamic response (Figure 5.31).

HST is often split into electric motor, hydraulic pump, piping system, safety switches, and hydraulic motor. A rotating electric motor works a hydraulic pump, which gives oil to pipe system. As cylinders in a hydraulic motor are filled with oil, it shall rotate the final load. Therefore, HST is a kind of closed-loop power transmission. The effort and flow in the rotating mechanical system are torque and angular velocity, respectively. If two elements are integrated, they became momentum and volume. No matter what systems in HST may be, we know that power does not change.

Bond graph of an electric motor and a hydraulic pump is displayed in Figure 5.32. Source flow SF_{11} designates an electric motor with a constant angular velocity. It is presumed that a 10% among total torque goes off by resistance element R_{12}.

(a)

(b)

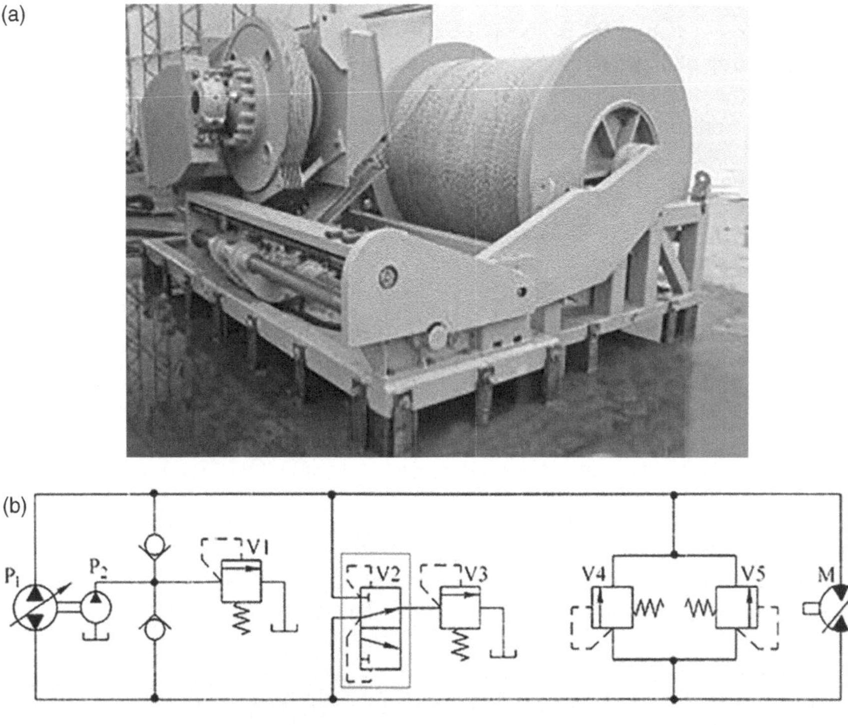

FIGURE 5.31 Ship-borne hydraulic-driven winch system: (a) winch system driven by hydraulic circuits and (b) hydrostatic transmission.

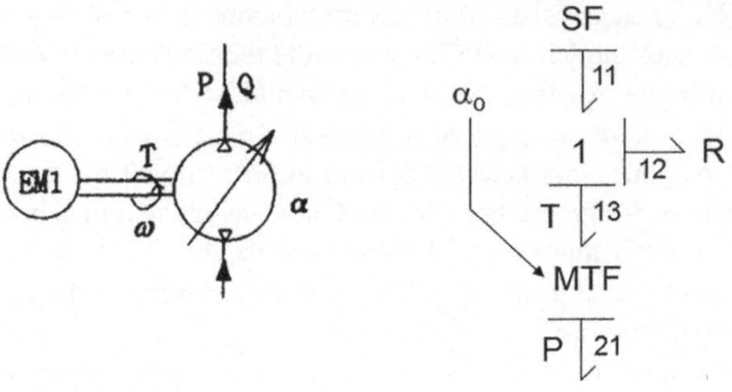

FIGURE 5.32 Electric motor and hydraulic pump modeling.

Transducer element MTF_{11} signifies the volume of a variable piston pump which can control volume with swivel angle α_o.

A bulk modulus B that implies the oil compressibility selects 10,000 bar among 6,000–12,000 bar. Fluid condensers $C_{23}=C_{21}$ are described as V/B. Fluid inertia I_{24} designates oil mass. Using the least square method, resistances R_{22} and R_{26} are calculated from the pump and motor leakage, respectively. Because pipe flow is laminar, fluid resistance R_{25} can be computed. Motor capacity TF_{3128} is determined from the number of filling cylinders. Moment of inertia of drum and flange I_{33} can be computed. It is presumed that torque loss of flange R_{32} is about 10%. When bond graph is drawn from top and bottom – starting with the electric motor and ending with the load, total bond graph and derivation of the state equations of a HST are represented in Figure 5.33.

To obtain non-dimensional state equations, non-dimensional variables are introduced as

$$\tilde{p}=\frac{P}{I\bar{Q}\, or\, \omega},\tilde{q}=\frac{Q}{C\bar{P}},\tilde{t}=\frac{t}{\omega_n^{-1}}=\frac{1}{\sqrt{IC}} \tag{5.61}$$

$$\frac{dQ}{dt}=\frac{d\left(C\bar{P}\tilde{q}\right)}{dt}=\frac{d\left(C\bar{P}\tilde{q}\right)}{d\tilde{t}}\frac{d\tilde{t}}{dt}=C\bar{P}\omega_n\dot{\tilde{q}} \tag{5.62}$$

$$\frac{dP}{dt}=\frac{d\left(I\bar{Q}\tilde{p}\right)}{dt}=\frac{d\left(I\bar{Q}\tilde{p}\right)}{d\tilde{t}}\frac{d\tilde{t}}{dt}=I\bar{Q}\omega_n\dot{\tilde{p}} \tag{5.63}$$

where P, Q, and t are dimensional integrals of pressure, volume, and time, and $\tilde{p}$, $\tilde{q}$, and $\tilde{t}$ are non-dimensional integrals of pressure, volume, and time, respectively. Therefore, non-dimensional state equations are derived as follows:

$$\frac{C_{23}\omega_n\bar{P}}{SF_{11}MTF_{2113}}\dot{\tilde{q}}_{23}+\frac{\bar{P}}{R_{23}SF_{11}MTF_{2113}}\tilde{q}_{23}+\frac{\bar{Q}}{SF_{11}MTF_{2113}}\tilde{p}_{24}=1 \tag{5.64}$$

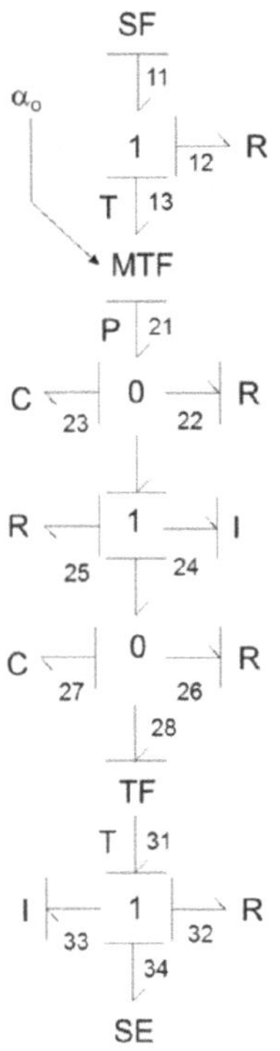

$F_{11} = SF_{11}$	$E_{34} = SE_{34}$
$E_{11} = E_{12} + E_{13}$	$F_{12} = F_{11}$
$F_{13} = F_{11}$	$E_{13} = E_{21} * MTF_{1321}$
$F_{21} = F_{13} * MTF_{1321}$	$F_{23} = F_{21} - F_{22} - F_{24}$
$E_{22} = E_{23}$	$E_{21} = E_{23}$
$E_{24} = E_{23} - E_{25} - E_{27}$	$F_{25} = F_{24}$
$E_{26} = E_{27}$	$F_{27} = F_{24} - F_{26} - F_{25}$
$E_{28} = E_{27}$	$E_{31} = E_{28} * TF_{3128}$
$F_{28} = F_{31} * TF_{3128}$	$F_{33} = F_{31} - F_{32} - F_{34}$
$F_{31} = F_{33}$	$F_{32} = F_{33}$
$F_{34} = F_{33}$	$E_{12} = F_{12} * R_{12}$
$\frac{dQ_{23}}{dt} = F_{23}$	$E_{23} = Q_{23}/C_{23}$
$F_{22} = E_{22}/R_{22}$	$E_{25} = F_{25} * R_{25}$
$\frac{dP_{24}}{dt} = E_{24}$	$F_{24} = P_{24}/I_{24}$
$\frac{dQ_{27}}{dt} = F_{27}$	$E_{27} = Q_{27}/C_{27}$
$\frac{dP_{33}}{dt} = E_{33}$	$F_{26} = E_{26}/R_{26}$
$E_{32} = F_{32} * R_{22}$	$F_{33} = P_{33}/I_{33}$

FIGURE 5.33 Bond graph and derivation of the state equations of the hydrostatic transmission in sea-borne winch.

$$\frac{I_{24}\omega_n}{R_{25}}\dot{\tilde{p}}_{24} - \frac{\bar{P}}{R_{25}\bar{Q}}\tilde{q}_{23} + \tilde{p}_{24} + \frac{\bar{P}}{R_{25}\bar{Q}}\tilde{q}_{27} = 0 \tag{5.65}$$

$$C_{27}\omega_n R_{26}\dot{\tilde{q}}_{27} - \frac{\bar{Q}R_{26}}{\bar{P}}\tilde{p}_{24} + \tilde{q}_{27} + \frac{\omega TF_{3128}R_{26}}{\bar{P}}\tilde{p}_{33} = 0 \tag{5.66}$$

$$\frac{I_{33}\omega\omega_n}{SE_{34}}\dot{\tilde{p}}_{33} - \frac{\bar{P}TF_{3128}}{SE_{34}}\tilde{q}_{27} + \frac{R_{32}\omega}{SE_{34}}\tilde{p}_{33} = 0 \tag{5.67}$$

To investigate the dynamic stability of the system, a simple asymptotic approach can be used and perturbations around stable points are expressed as follows:

$$\tilde{q}_{23} = \tilde{q}_{230} + \varepsilon^1 \tilde{q}_{231} + O(\varepsilon^2) \tag{5.68}$$

$$\tilde{p}_{24} = \tilde{p}_{240} + \varepsilon^1 \tilde{p}_{241} + O(\varepsilon^2) \tag{5.69}$$

$$\tilde{q}_{27} = \tilde{q}_{270} + \varepsilon^1 \tilde{q}_{271} + O(\varepsilon^2) \tag{5.70}$$

$$\tilde{p}_{33} = \tilde{p}_{330} + \varepsilon^1 \tilde{p}_{331} + O(\varepsilon^2) \tag{5.71}$$

where ε^1 is a very small value.

If we substitute Equations (5.64)–(5.67) into (5.68)–(5.71), then the terms of ε^0 are yielded:

$$\frac{\bar{P}}{R_{22} SF_{11} MTF_{2113}} (\tilde{q}_{230}) + \frac{\bar{Q}}{SF_{11} MTF_{2113}} \tilde{p}_{240} = 1 \tag{5.72}$$

$$-\frac{\bar{P}}{R_{25}\bar{Q}} \tilde{q}_{230} + \tilde{p}_{240} + \frac{\bar{P}}{R_{25}\bar{Q}} \tilde{q}_{270} = 0 \tag{5.73}$$

$$-\frac{\bar{Q}R_{26}}{\bar{P}} \tilde{p}_{240} + \tilde{q}_{270} + \frac{\omega TF_{3128} R_{26}}{\bar{P}} \tilde{p}_{330} = 0 \tag{5.74}$$

$$-\frac{\bar{P}TF_{3128}}{SE_{34}} \tilde{q}_{270} + \frac{R_{32}\omega}{SE_{34}} \tilde{p}_{330} = -1 \tag{5.75}$$

And then, the terms of ε^1 are yielded

$$\frac{C_{23}\omega_n \bar{P}}{SF_{11} MTF_{2113}} \dot{\tilde{q}}_{231} + \frac{\bar{P}}{R_{23} SF_{11} MTF_{2113}} \tilde{q}_{231} + \frac{\bar{Q}}{SF_{11} MTF_{2113}} \tilde{p}_{241} = 0 \tag{5.76}$$

$$\frac{I_{24}\omega_n}{R_{25}} \dot{\tilde{p}}_{241} - \frac{\bar{P}}{R_{25}\bar{Q}} \tilde{q}_{231} + \tilde{p}_{241} + \frac{\bar{P}}{R_{25}\bar{Q}} \tilde{q}_{271} = 0 \tag{5.77}$$

$$C_{27}\omega_n R_{26} \dot{\tilde{q}}_{271} - \frac{\bar{Q}R_{26}}{\bar{P}} \tilde{p}_{241} + \tilde{q}_{271} + \frac{\omega TF_{3128} R_{26}}{\bar{P}} \tilde{p}_{331} = 0 \tag{5.78}$$

$$\frac{I_{33}\omega\omega_n}{SE_{34}} \dot{\tilde{p}}_{331} - \frac{\bar{P}TF_{3128}}{SE_{34}} \tilde{q}_{271} + \frac{R_{32}\omega}{SE_{34}} \tilde{p}_{331} = 0 \tag{5.79}$$

If the perturbed Equations (5.76)–(5.79) are expressed as state space form $\left[dx/dt\right] = [A][X]$, then

$$\begin{bmatrix} \dot{\tilde{q}}_{231} \\ \dot{\tilde{p}}_{241} \\ \dot{\tilde{q}}_{271} \\ \dot{\tilde{p}}_{331} \end{bmatrix} = \begin{bmatrix} -\dfrac{1}{R_{22}C_{23}\omega_n} & -\dfrac{\bar{Q}}{C_{23}\omega_n\bar{P}} & 0 & 0 \\ \dfrac{-\bar{P}}{\bar{Q}I_{24}\omega_n} & -\dfrac{R_{25}}{I_{24}\omega_n} & -\dfrac{\bar{P}}{\bar{Q}\omega_n I_{24}} & 0 \\ 0 & \dfrac{1}{\bar{P}C_{27}\omega_n} & -\dfrac{1}{C_{27}\omega_n R_{26}} & -\dfrac{\omega TF_{3128}}{\bar{P}C_{27}\omega_n} \\ 0 & 0 & \dfrac{-\bar{P}TF_{3128}}{I_{33}\omega\omega_n} & -\dfrac{R_{32}}{I_{33}\omega_n} \end{bmatrix} \begin{bmatrix} \tilde{q}_{231} \\ \tilde{p}_{241} \\ \tilde{q}_{271} \\ \tilde{p}_{331} \end{bmatrix} \tag{5.80}$$

To investigate the dynamic stability of the non-dimensional state Equation (5.80), eigen-value of the bond graph can be represented as a state equation form $|A-\lambda I|[X]=0$. The system is unstable if the eigen-value is $\lambda>0$ and the system is stable $\lambda<0$. When the state equations are represented as state space form of

$$\begin{bmatrix} \dfrac{dQ_{23}}{dt} \\ \dfrac{dP_{24}}{dt} \\ \dfrac{dQ_{27}}{dt} \\ \dfrac{dP_{33}}{dt} \end{bmatrix} = \begin{bmatrix} -\dfrac{1}{C_{23}R_{22}} & -\dfrac{1}{I_{24}} & 0 & 0 \\ \dfrac{1}{C_{23}} & -\dfrac{R_{25}}{I_{24}} & -\dfrac{1}{C_{27}} & 0 \\ 0 & \dfrac{1}{I_{24}} & -\dfrac{1}{C_{27}R_{26}} & -\dfrac{TF_{3128}}{I_{33}} \\ 0 & 0 & \dfrac{TF_{3128}}{C_{27}} & -\dfrac{R_{32}}{I_{33}} \end{bmatrix} \begin{bmatrix} Q_{23} \\ P_{24} \\ Q_{27} \\ P_{33} \end{bmatrix} + \begin{bmatrix} MTF_{2113} \\ 0 \\ 0 \\ 0 \end{bmatrix} [SF_{11}] + \begin{bmatrix} 0 \\ 0 \\ 0 \\ -1 \end{bmatrix} [SE_{34}] \tag{5.81}$$

When Equation (5.81) is integrated, the pump pressure and motor pressure are obtained as

$$\begin{bmatrix} \bar{P}_{\text{pump}} \\ \bar{P}_{\text{motor}} \end{bmatrix} = \begin{bmatrix} 1/C_{23} & 0 \\ 0 & 1/C_{27} \end{bmatrix} \begin{bmatrix} Q_{23} \\ Q_{27} \end{bmatrix} \tag{5.82}$$

HST simulations are classified as models of low speed, high speed, and maximum tension. The tension values might be obtained by the drag force analysis of a cable. A steady solution of ε^0 equation and eigen-values from high speed, low speed, and maximum tension are calculated as stable. The values of (a) perturbed state Q_{23},

(b) perturbed state P_{24}, (c) perturbed state Q_{27}, and (d) perturbed state P_{33} from the high-speed mode are shown in Figure 5.24. State variables are converged after they perturbed around the steady-state value ε^0. It can figure out that simulation results with a big overshoot reach a stead-state value (Figures 5.34 and 5.35).

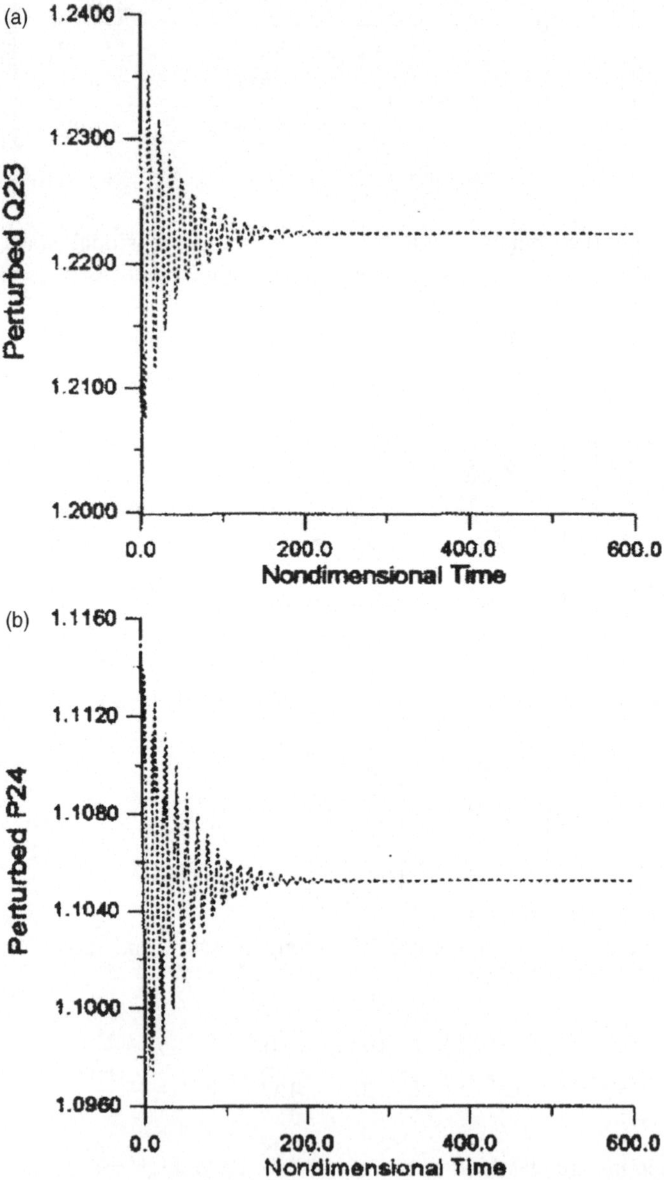

FIGURE 5.34 Perturbed state: (a) $\tilde{q}_{23}$, (b) $\tilde{p}_{24}$, (c) $\tilde{q}_{27}$, and (d) $\tilde{p}_{33}$.

(Continued)

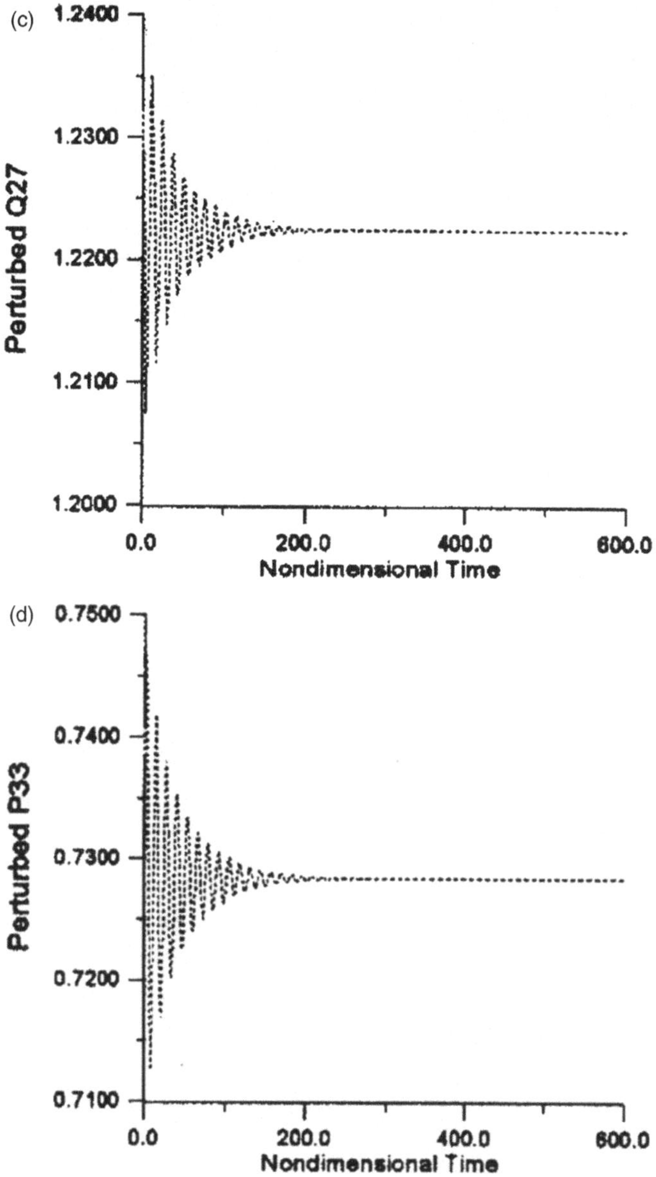

FIGURE 5.34 ***(CONTINUED)*** Perturbed state: (a) $\tilde{q}_{23}$, (b) $\tilde{p}_{24}$, (c) $\tilde{q}_{27}$, and (d) $\tilde{p}_{33}$.

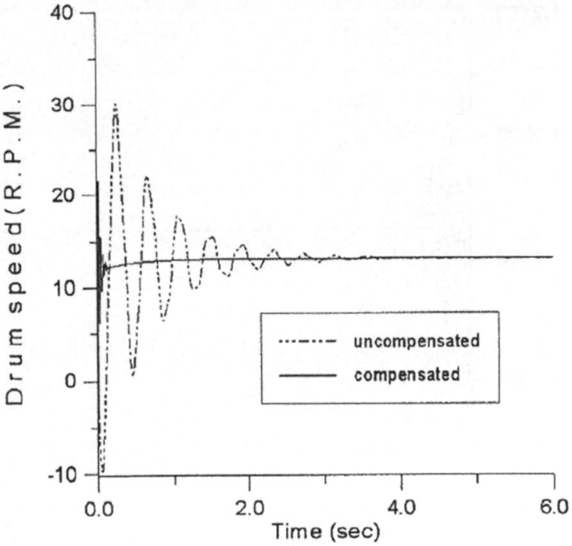

FIGURE 5.35 Simulation results for hydrostatic transmission

REFERENCES

1. Sonntag, R.E., Borgnakke, C., 2007, *Introduction to engineering thermodynamics*, York, PA: John Wiley & Sons, Inc.
2. Dijksterhuis, E.J. 1987, Archimedes (translated by C. Dikshoorn), Princeton, NJ: Princeton University Press.

6 Mechanical System Design (Strength and Stiffness)

6.1 INTRODUCTION

In major features of design, the mechanical system subjected to a variety of (random) loads is required to possess a fine quality such as sufficient stiffness or strength. Strength – the resistance against permanent deformation – is continually required to be high throughout the design of material or product shape because a permanent deformation of material due to loads may lead to crack, fracture, and eventually the loss of (intended) functions in a product. Requirements on stiffness –the resistance against reversible deformation – may change over a wide range based on the product's application.

If there is a design flaw in a structure, it can cause an inadequacy of strength (or stiffness), and when exposed to loads, the product structure can get permanent deformation, crack, or fracture at that location because it doesn't have enough strength to bear stress in the elasticity range. Hence, an engineer would prefer to make a better design of products that are subject to repeated loads. As a result, mechanical structure might be reshaped to have a better design that can robustly withstand repetitive loads in product lifetime.

The failure mechanics – especially, fatigue – of parts that might no longer be worked can be distinguished by two factors: (1) the loads (or stress) on the structure and (2) the pattern of product materials and their shape utilized in the structure including mechanism. To stop the mechanical failures such as fatigue or fracture, mechanical engineers reproduce them and modify product designs by selecting a proper material and product shape so that the products can have proper strength and stiffness. Consequently, the mechanical system can withstand the repetitive loads (or stress) in its lifetime (Figure 6.1).

The failure of a mechanical product is a structural problem that is produced when stress due to repetitive loads causes a permanent deformation, crack, or fracture. Failure mechanics try to understand the process of failure that is caused by system materials. The failure site in the problematic design of products' (or modules') structure might be noticeable when the failed products from the marketplace (or testing) are taken apart. To recognize these failure phenomena, engineers therefore should comprehend the design concepts of mechanical system mentioned in Chapter 5 and elasticity or solid mechanics that will be discussed in this chapter.

Almost all mechanical systems are composed of multi-module structures with mechanical or electrical components. If the product structure is preferably designed

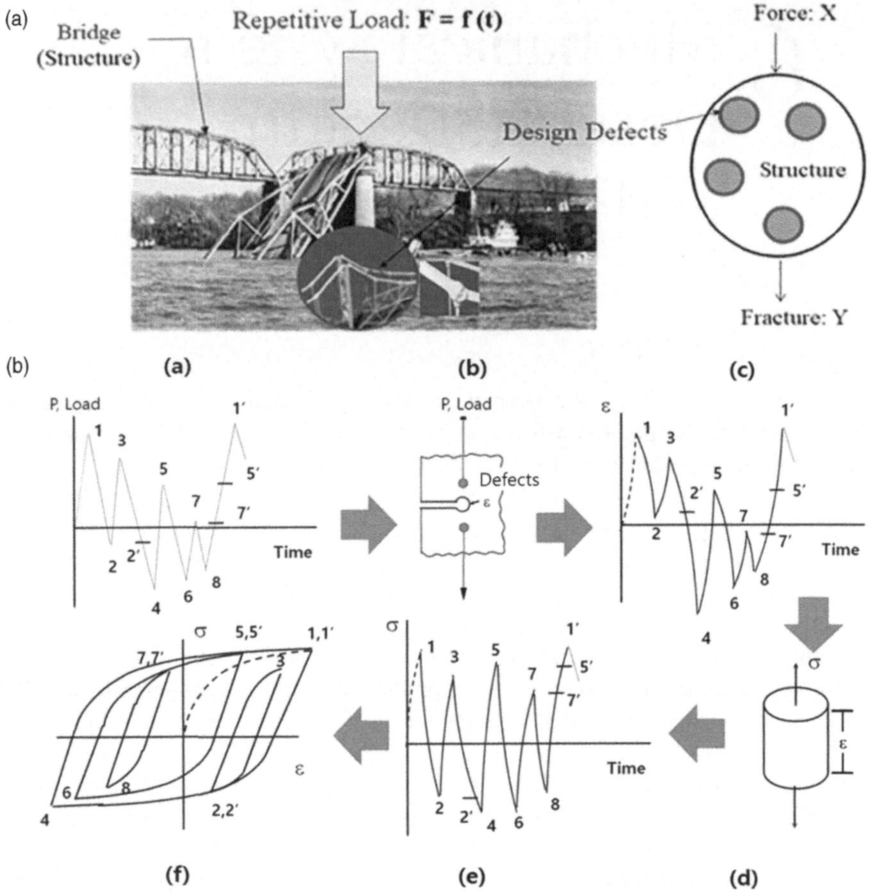

FIGURE 6.1 Failure mechanics created by a load on a component made from a specific material. (a) Failure of the structure caused by repetitive loads and its design defects. (b) A tensile and compression load-time history repeatedly applied to a problematic component and the stress–strain response: (a) load-time history, (b) faulty (or notched) component, (c) its strain history, (d) strain history applied to a smooth specimen, (e) its stress response obtained from a smooth specimen, and (f) its stress–strain hysteresis loops [1].

to possess sufficient strength and stiffness against loads, there should be no troubles in product lifetime. But though the mechanical design is developed with an optimal design using conventional design technics such as finite element analysis (FEA), it may have design flaws that will suddenly happen in the field. Especially, fatigue failure often arises not due to impractical stresses in an ideal component, but rather due to the existence of a tiny crack or pre-existing fault on the exterior of a part, which is nearly unnoticeable and plainly impractical to model using conventional finite element techniques. A product structure including a mechanism completed through the design process, should be experimentally verified if it possess a fine quality – sufficient strength and stiffness.

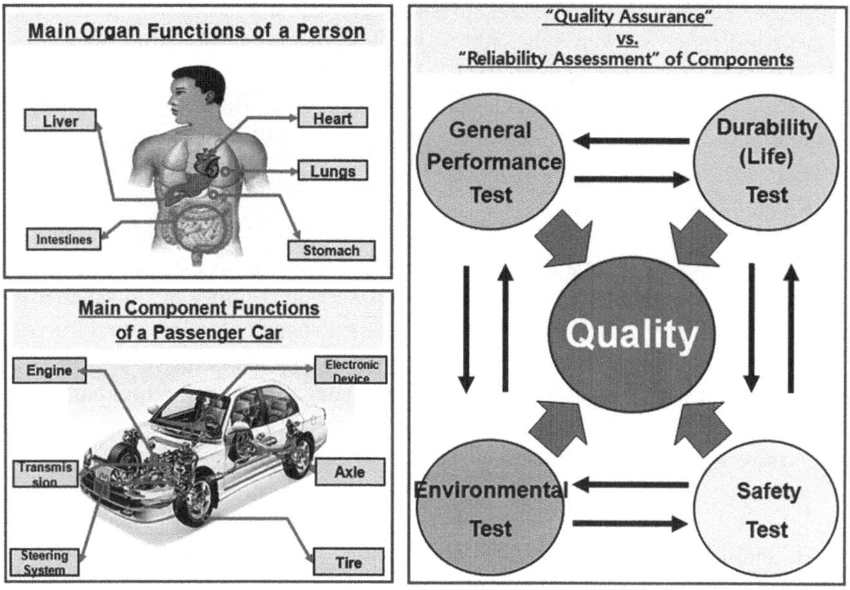

FIGURE 6.2 Assessment concepts of product quality.

To do this, the product design might be successfully connected with test engineering and design engineering in the design process. Based on the customer requirements (or specifications), design engineer tries to implement the intended functions in the structure. On the other hand, test engineer tries to affirm whether last designs of product achieve the lifetime target through a variety of testing – performance, reliability, safety, etc. Engineer therefore might develop parametric ALT as reliability testing in the design process, which might confirm if product have a good quality – strength and stiffness (Figure 6.2).

6.2 STRENGTH OF MECHANICAL PRODUCT – ELASTICITY

6.2.1 Introduction

Mechanics of materials (or strength of mechanics) discuss the strength and physical effectiveness of mechanical/civil structures subjected to various patterns of (random) loading. A complete understanding of a mechanical behavior such as displacement/deformation, strain, and stress from a standpoint of elasticity is crucial for a sound design of all patterns of structures, regardless of airplanes, automobiles, refrigerators, construction machines such as excavator, bridges and buildings, or ships and spacecraft. These ideas are situated under the design and analysis of an enormous variety of structural and mechanical systems for strength. Engineers should recognize these fundamental design concepts of structural material as well as numerous physical laws – especially, mechanical loads.

We will firstly discuss how to obtain the displacements/deformation, stresses, and strains in structures due to the loads in the range of elasticity. We can explain the fatigue mechanism of mechanical parts that accounts for about 80%–95% of all structural failures when subjected to variable, fluctuating loading (Chapter 7). Fatigue failure arises not due to impractical stresses in an ideal component, but rather due to the existence of a tiny crack or pre-existing fault on the exterior of a part that the stress and strain can become plastic when subjected to repetitive stress. Product engineers therefore determine whether a mechanical system endures the applied stresses in the design process.

There are a variety of solid bodies in the mechanical system, which comprise bars with axial loads, beams and plate in bending, shafts in torsion, and columns in compression. For example, let's imagine a beam, seen in Figure 6.3a, on which lots of cars are continuously moving on the bridge. As time changes, loads exerted in system will continuously change. If load is simply supposed to be subjected to the end of the beam, the beam will bend downward. While the materials in the bottom parts of the beam shrink, those in the top part expand. Those that are compressed are subject to compressive stresses, while those that are lengthened have tensile stresses acting on the materials.

As the deformation of the beam fibers consequently sets up strains and stresses, we can find the loop of the displacements, strains, and stresses in structures due to the repetitive loads from the standpoint of elasticity – plasticity because the existence

(a)

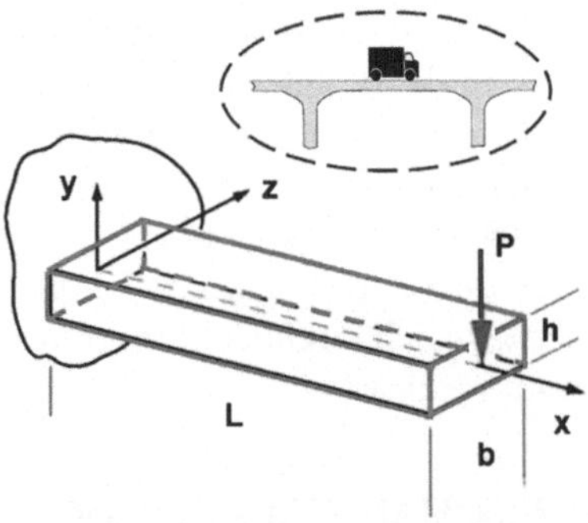

(b)

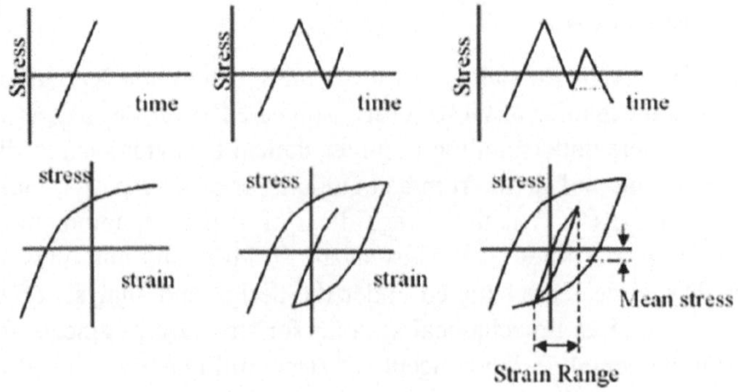

FIGURE 6.3 Cantilever beam subjected to random loads in transit: (a) tip-loaded cantilever beam and (b) loop of elastic-plastic stress and strain due to loads.

of a tiny crack or pre-existing flaw on the surface of a part, not due to ideal stresses in a perfect component (Figure 6.3b).

6.2.1.1 A Brief History

Galileo Galilei (1564–1642) and Leonardo da Vinci (1452–1519) made the earliest endeavors at finding a beam theory. Although Leonardo da Vinci made the major findings, to finish his theory by the help of calculus in mathematics and Hooke's law, he still needed to be developed and compensated. Galileo was also held back by an insufficient presumption he proposed. They carried out experiments to decide the strength of wires, beams, and bars, though they did not suggest proper theories to describe their experimental outputs.

The well-known mathematician Leonhard Euler (1707–1783) suggested the mathematical concept of columns and computed the major load of a column in 1744. With inappropriate tests to support his theories, the Euler–Bernoulli beam theory endured unemployed for over a hundred years. Buildings and bridges were continually designed by exemplars till the latter of part of the nineteenth century, while the Ferris wheel and Eiffel Tower explained the effectiveness of the theory on big scales. Now, they are the foundation for the analysis and design of most columns such as tip-loaded cantilever beam.

6.2.2 Elasticity

Elasticity is a property of (solid) materials to become distorted when exposed to an outer force and to recover their initial shape after the force is eliminated. To understand the design of the mechanical product, we will discuss the elasticity. After starting one-dimensional elasticity, we can expand three-dimensional one.

We can define a simple body of volume, V, and surface, S. The action of deforming at local point $x = [x, y, z]^T$ is given by the three components of its displacement $u = [u, v, w]^T$, where $u = u(x, y, z)$, i.e., each displacement component is a function of position. In this case, there are two basic patterns of outer forces that are acted on a body: (1) body force (force per unit volume) e.g., weight, and (2) surface traction (force per unit surface area), e.g., friction (Figure 6.4).

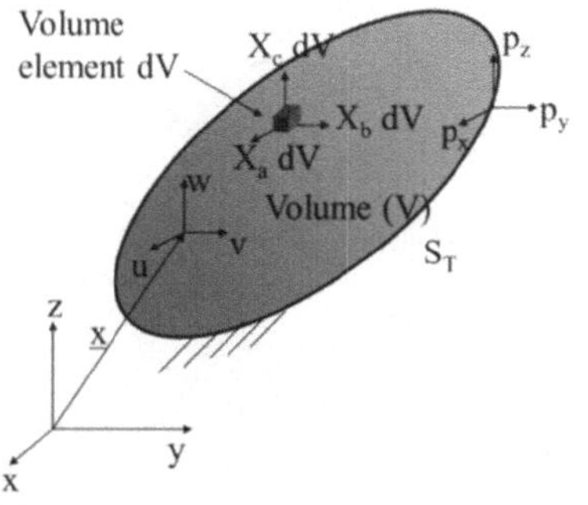

FIGURE 6.4 A simple body subjected to body force and surface traction.

6.2.2.1 Body Force

For volume element dV, body force can be defined as $\underline{X} = [X_a, X_b, X_c]^T$. If the body is accelerating, the inertia force might be contemplated as part of $\underline{X}$. That is,

$$\underline{X} = \underline{\tilde{X}} - \rho\underline{\ddot{u}} \tag{6.1}$$

where $\rho\underline{\ddot{u}} = \begin{Bmatrix} \rho\ddot{u} \\ \rho\ddot{v} \\ \rho\ddot{w} \end{Bmatrix}$.

6.2.2.2 Surface Traction

As distributed force per unit surface area, surface traction can be defined as follows:

$$\underline{T}_S = \begin{Bmatrix} p_x \\ p_y \\ p_z \end{Bmatrix} \tag{6.2}$$

And there is an internal force that is reacted in a body. If it is expressed as the internal reaction forces per unit area, we can say it stress.

6.2.2.3 Internal Forces

If a thick, solid piece of material from the body is taken out, there are reaction forces due to the outer forces applied to it. For the hexahedron, on each surface, the internal reaction force per unit area (black arrows) is known as stress and may be decomposed into three orthogonal components (Figure 6.5).

Assume a body aligned in the Cartesian coordinate system with a number of forces applying on it, such that the vector total of all the forces is zero. Take a piece orthogonal to the x direction and explain a small area on this slice as ΔA_x. Let the total force applying on this small area be

$$\underline{\Delta F} = \Delta F_x \cdot \hat{i} + \Delta F_y \cdot \hat{j} + \Delta F_z \cdot \hat{k} \tag{6.3}$$

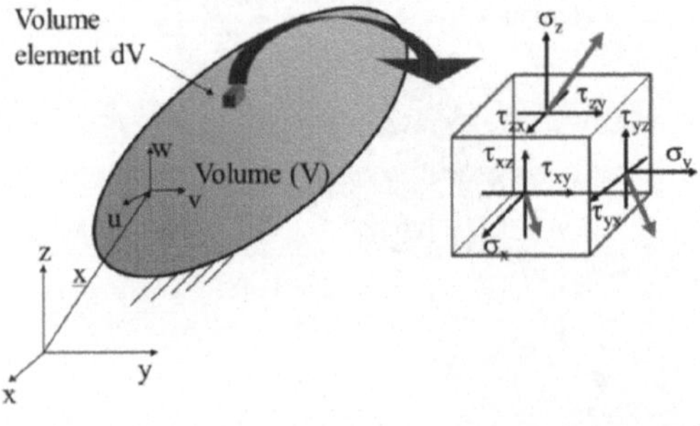

FIGURE 6.5 A stress notation on an infinite body when subjected to forces.

We can explain the following scalar amounts:

$$\sigma_{xx} = \lim_{\Delta A_x \to 0} \frac{\Delta F_x}{\Delta A_x}, \tau_{xy} = \lim_{\Delta A_x \to 0} \frac{\Delta F_y}{\Delta A_x}, \tau_{xz} = \lim_{\Delta A_x \to 0} \frac{\Delta F_z}{\Delta A_x} \tag{6.4}$$

Equivalently, considering slices orthogonal to the y and z directions, we find

$$\tau_{yx} = \lim_{\Delta A_y \to 0} \frac{\Delta F_x}{\Delta A_y}, \sigma_{yy} = \lim_{\Delta A_y \to 0} \frac{\Delta F_y}{\Delta A_y}, \tau_{yz} = \lim_{\Delta A_y \to 0} \frac{\Delta F_z}{\Delta A_y} \tag{6.5}$$

and

$$\tau_{zx} = \lim_{\Delta A_z \to 0} \frac{\Delta F_x}{\Delta A_z}, \tau_{zy} = \lim_{\Delta A_z \to 0} \frac{\Delta F_y}{\Delta A_z}, \sigma_{zz} = \lim_{\Delta A_z \to 0} \frac{\Delta F_z}{\Delta A_z} \tag{6.6}$$

We can say that σ_x, σ_y, and σ_z are normal stresses. On the other hand, the rest six are the shear stresses. Because $\tau_{xy} = \tau_{yx}$, $\tau_{yz} = \tau_{zy}$, and $\tau_{zx} = \tau_{xz}$, we can say that only six stress components are independent.

The stress vector is described as follows:

$$\underline{\sigma} = \left[\sigma_x \, \sigma_y \, \sigma_z \, \tau_{xy} \, \tau_{yz} \, \tau_{zx}\right]^T \tag{6.7}$$

On the other hand, the quantity of deformation is defined as the *strain* and may be decomposed into three orthogonal components. Strains have six independent strain components. That is,

$$\underline{\varepsilon} = \left[\varepsilon_x \, \varepsilon_y \, \varepsilon_z \, \gamma_{xy} \, \gamma_{yz} \, \gamma_{zx}\right]^T \tag{6.8}$$

Assume the equilibrium of a differential volume element to get the three equilibrium equations of elasticity:

$$\begin{aligned}
&\frac{\partial \sigma_x}{\partial x} + \frac{\partial \tau_{xy}}{\partial y} + \frac{\partial \tau_{xz}}{\partial z} + X_a = 0 \\
&\frac{\partial \tau_{xy}}{\partial x} + \frac{\partial \sigma_y}{\partial y} + \frac{\partial \tau_{yz}}{\partial z} + X_b = 0 \\
&\frac{\partial \tau_{xz}}{\partial x} + \frac{\partial \tau_{yz}}{\partial y} + \frac{\partial \sigma_z}{\partial z} + X_c = 0
\end{aligned} \tag{6.9}$$

Concisely, we can describe equilibrium equations as follows:

$$\underline{\partial}^T \underline{\sigma} + \underline{X} = \underline{0} \tag{6.10}$$

$$\text{where } \underline{\partial} = \begin{bmatrix} \frac{\partial}{\partial x} & 0 & 0 \\ 0 & \frac{\partial}{\partial y} & 0 \\ 0 & 0 & \frac{\partial}{\partial z} \\ \frac{\partial}{\partial y} & \frac{\partial}{\partial x} & 0 \\ 0 & \frac{\partial}{\partial z} & \frac{\partial}{\partial y} \\ \frac{\partial}{\partial z} & 0 & \frac{\partial}{\partial x} \end{bmatrix}.$$

6.2.2.4 Strong Formulation for 3D Elasticity Problem

If the external applied loads (on S_T and in V) and the stated displacements (on S_u) are given, we can resolve for the resultant displacements, strains, and stresses required to keep equilibrium of the body.

Equilibrium equations in a certain volume are described as follows:

$$\underline{\partial}^T \underline{\sigma} + \underline{X} = \underline{0} \quad \text{in } V \tag{6.11}$$

As displacement boundary conditions, displacements are stated on the boundary of S_u:

$$\underline{u} = \underline{u}^{\text{specified}} \quad \text{on } S_u \tag{6.12}$$

If we set $\sin\theta = \frac{dx}{ds} = n_y$ and $\cos\theta = \frac{dy}{ds} = n_x$ in Figure 6.6, elasticity traction in 2D can be derived as follows:

$$p_x ds = \sigma_x dy + \tau_{xy} dx \tag{6.12a}$$

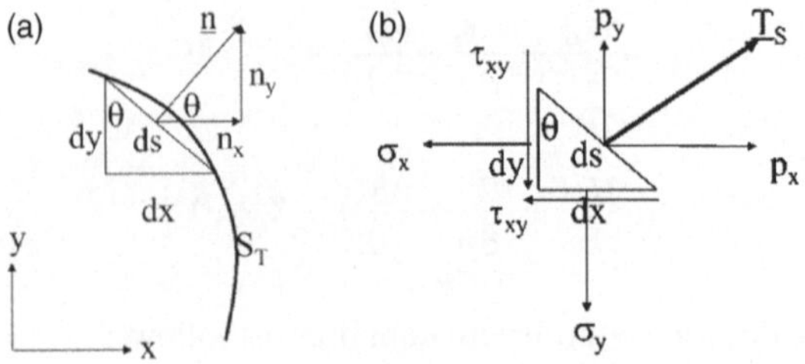

FIGURE 6.6 A simple wedge body subjected to 2D surface traction (a) and (b).

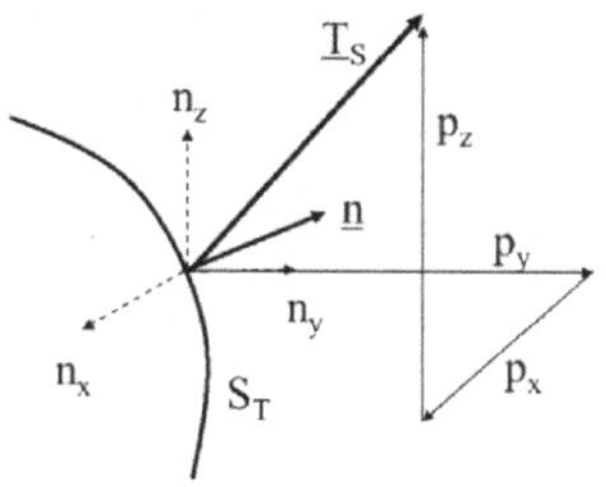

FIGURE 6.7 A simple wedge body subjected to 3D surface traction.

$$p_x = \sigma_x \frac{dy}{ds} + \tau_{xy} \frac{dx}{ds} \tag{6.12b}$$

$$p_x = \sigma_x n_x + \tau_{xy} n_y \tag{6.12c}$$

As traction (force) boundary conditions, tractions are stated on the boundary of S_T. Traction can be described as a distributed force per unit area. That is,

$$\underline{T}_S = \begin{Bmatrix} p_x \\ p_y \\ p_z \end{Bmatrix} \tag{6.13}$$

If $\underline{n}$ is the unit external normal to S_T in Figure 6.7, traction can be defined as follows:

$$\begin{aligned} p_x &= \sigma_x n_x + \tau_{xy} n_y + \tau_{xz} n_z \\ p_y &= \tau_{xy} n_x + \sigma_y n_y + \tau_{yz} n_z \\ p_z &= \tau_{xz} n_x + \tau_{zy} n_y + \sigma_z n_z \end{aligned} \tag{6.14}$$

6.2.2.5 Strain–Displacement Relationships

As seen in Figure 6.8, in 2D elasticity, the strain–displacement relationships from small fragment can be obtained as follows:

$$\varepsilon_x = \frac{A'B' - AB}{AB} = \frac{\left(dx + \left(u + \frac{\partial u}{\partial x} dx\right) - u\right) - dx}{dx} = \frac{\partial u}{\partial x} \tag{6.15a}$$

$$\varepsilon_y = \frac{A'C' - AC}{AC} = \frac{\left(dy + \left(v + \frac{\partial v}{\partial y} dy\right) - v\right) - dy}{dy} = \frac{\partial v}{\partial y} \tag{6.15b}$$

$$\begin{aligned} \gamma_{xy} &= \frac{\pi}{2} - \text{angle}\,(C'A'B') = \beta_1 + \beta_2 \approx \tan\beta_1 + \tan\beta_2 \\ &\approx \frac{\partial v}{\partial x} + \frac{\partial u}{\partial x} \end{aligned} \tag{6.15c}$$

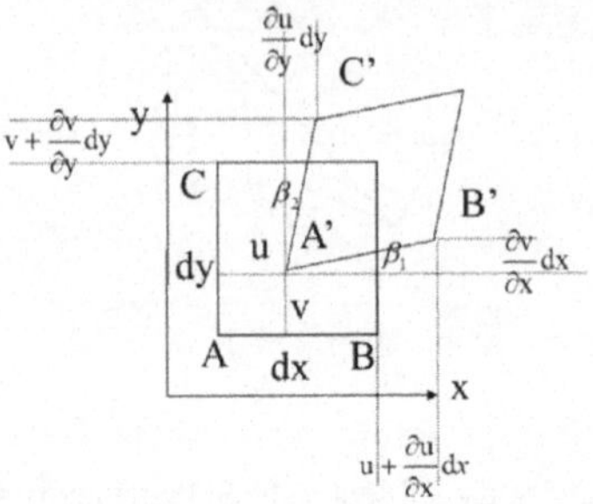

FIGURE 6.8 Strain–displacement relationships of a small fragment in 2D elasticity.

For 3D elasticity, strain–displacement relationships can be expanded as follows:

$$\begin{aligned}
\varepsilon_x &= \frac{\partial u}{\partial x} \\
\varepsilon_y &= \frac{\partial v}{\partial y} \\
\varepsilon_z &= \frac{\partial w}{\partial z} \\
\gamma_{xy} &= \frac{\partial u}{\partial y} + \frac{\partial v}{\partial x} \\
\gamma_{yz} &= \frac{\partial v}{\partial z} + \frac{\partial w}{\partial y} \\
\gamma_{zx} &= \frac{\partial u}{\partial z} + \frac{\partial w}{\partial x}
\end{aligned} \tag{6.16}$$

If compactly defined, it will be described as follows:

$$\varepsilon = \partial u \tag{6.17}$$

where $\underline{\varepsilon} = \begin{Bmatrix} \varepsilon_x \\ \varepsilon_y \\ \varepsilon_z \\ \gamma_{xy} \\ \gamma_{yz} \\ \gamma_{zx} \end{Bmatrix}$ $\underline{\partial} = \begin{bmatrix} \frac{\partial}{\partial x} & 0 & 0 \\ 0 & \frac{\partial}{\partial y} & 0 \\ 0 & 0 & \frac{\partial}{\partial z} \\ \frac{\partial}{\partial y} & \frac{\partial}{\partial x} & 0 \\ 0 & \frac{\partial}{\partial z} & \frac{\partial}{\partial y} \\ \frac{\partial}{\partial z} & 0 & \frac{\partial}{\partial x} \end{bmatrix}$ $\underline{u} = \begin{Bmatrix} u \\ v \\ w \end{Bmatrix}$.

6.2.2.6 Stress–Strain relationship

If general Hooke's law is applied, linear elastic material can be defined as follows:

$$\underline{\sigma} = \underline{D}\,\underline{\varepsilon} \tag{6.18}$$

where

$$\underline{D} = \frac{E}{(1+\nu)(1-2\nu)} \begin{bmatrix} 1-\nu & \nu & \nu & 0 & 0 & 0 \\ \nu & 1-\nu & \nu & 0 & 0 & 0 \\ \nu & \nu & 1-\nu & 0 & 0 & 0 \\ 0 & 0 & 0 & \frac{1-2\nu}{2} & 0 & 0 \\ 0 & 0 & 0 & 0 & \frac{1-2\nu}{2} & 0 \\ 0 & 0 & 0 & 0 & 0 & \frac{1-2\nu}{2} \end{bmatrix}$$

6.2.2.7 Principle of Minimum Potential Energy

For a linear elastic frame subjected to surface tractions $\underline{T}_S = \left[p_x, p_y, p_z\right]^T$ and body forces $\underline{X} = \left[X_a, X_b, X_c\right]^T$, leading to displacements $u = \left[u, v, w\right]^T$, strains $\underline{e}$, and stresses $\underline{s}$, the possible energy P is expressed as the strain energy minus the potential energy of the loads requiring $\underline{X}$ and $\underline{T}_S$:

$$\Pi = U - W \tag{6.19}$$

where $U = \frac{1}{2}\int_V \underline{\sigma}^T \underline{\varepsilon}\, dV$ and $W = \int_V \underline{u}^T \underline{X}\, dV + \int_{S_T} \underline{u}^T \underline{T}_S\, dS$.

The one that fulfils the equilibrium equations amidst all allowable displacement fields also gives the potential energy P to a minimum. The 'allowable displacement field' has the following properties:

1. The first derivative of the displacement components prevails.
2. It fulfils the boundary conditions on S_u.

6.2.3 Beam

A beam in mechanical/civil engineering is a structural constituent piece whose length is longer compared to its cross-sectional area. In other words, it is a bar-like structural member whose main function is to bear transverse loading and convey it to the supports. The Euler–Bernoulli beam theory plays a critical role in structural analysis because it helps mechanical engineers to compute the load-carrying and deflection characteristics of beams in a straightforward way.

When a beam is bent downward, the materials in the bottom part of the beam shrink, but those in the top part expand. These changes in the lengths of the materials produce stresses in the materials. Those that are enlarged possess tensile stresses applying on the materials in the orientation of the longitudinal axis of the beam, while those that are compressed are subjected to compressive stresses. There always exists the neutral surface of the beam containing materials that do not experience any compression or extension, and thus are not subjected to any compressive or tensile stress. It, therefore, can be deduced that from the neutral axis, the value of the bending stress will change linearly with distance. Now we start the analysis of beam from the standpoint of elasticity.

6.2.3.1 Euler–Bernoulli Beam Theory

As shown in Figure 6.9, before bending:

$$AB = CD = EF = \Delta x \tag{6.20}$$

As bending line segment *AB* compressed, line segment *CD* elongated and line segment *EF* does not vary. Hence, line segment *EF* is referred to as the neutral surface. The junction of the neutral surface with the longitudinal plane of symmetry is defined as the neutral axis.

From calculus and analytic geometry in Figure 6.6a, we can find the curvature

$$\kappa = \frac{1}{\rho} = \frac{\Delta\theta}{\Delta x} \tag{6.21}$$

where κ is the curvature and ρ is the radius of the curvature.

6.2.3.2 Strain–Displacement Relationships

Because $\rho = \Delta x/\Delta\theta$ and $\Delta x = \rho\Delta\theta$, along line segment *AB*, the normal stain in the longitudinal direction can be expressed as follows:

$$\varepsilon_{AB} = \frac{\delta}{L} = \frac{L_f - L_i}{L_i} = \frac{(\rho - y)\Delta\theta - \rho\Delta\theta}{\rho\Delta\theta} = \frac{-y}{\rho} \tag{6.22}$$

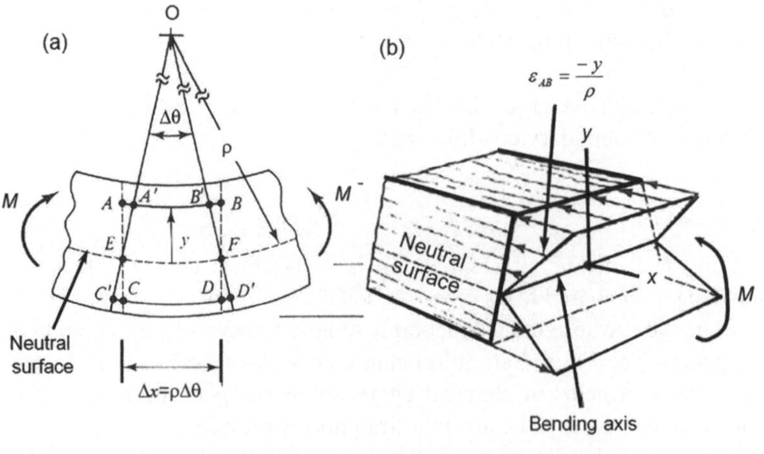

FIGURE 6.9 A deformed section in beam after bending: (a) and (b).

6.2.3.3 Stress–Strain Relationship

Since Hooke's law holds, the stresses are proportionate to the interval y from the neutral axis:

$$\sigma = E\varepsilon = \frac{-Ey}{\rho} \tag{6.23}$$

6.2.3.4 Equilibrium Equations

From Figure 6.10, we can find that internal stresses are statically identical to the external forces and moment. That is,

$$\sum F_{x(\mathrm{I})} = \sum F_{x(\mathrm{II})} \Rightarrow \int \sigma dA = 0 \tag{6.24}$$

$$\sum M_{z(\mathrm{I})} = \sum M_{z(\mathrm{II})} \Rightarrow -\int y\sigma dA = M \tag{6.25}$$

If Equation (6.23) is substituted, we can yield:

$$-\int y\frac{-Ey}{\rho}dA = M \tag{6.26}$$

$$M = \frac{E}{\rho}\int y^2 dA = \frac{EI}{\rho} \tag{6.27}$$

where moment of inertia $I = \int y^2 dA$.

Because curvature, κ, equals $1/\rho$, the bending momentum can be redefined as follows:

$$M = \frac{EI}{\rho} = EI\kappa = EIy'' \tag{6.28}$$

where $y = f(x)$, y' = slope, and y'' = curvature.

Finally, if combining Equations (6.23) and (6.27), we can find the bending equation:

$$\sigma = \frac{My}{I} \tag{6.29}$$

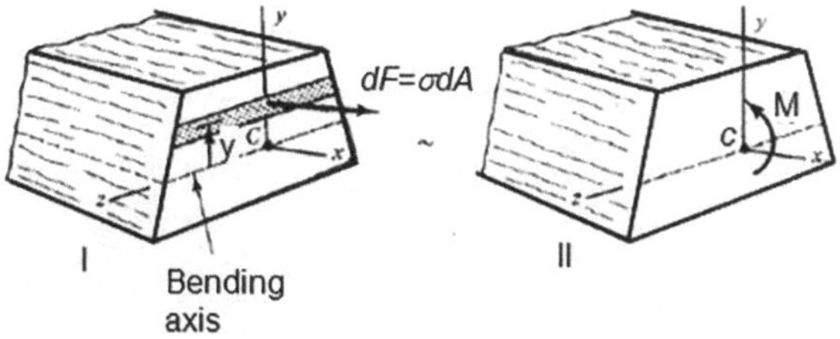

FIGURE 6.10 Bending stresses on a beam cross section.

6.2.4 Flat Plate

At first, plates are flat structural members possessing thicknesses much smaller than the other sizes. To ascertain the distribution of displacement and stress for a plate subjected to a given set of forces, the basic assumptions such as the Kirchhoff hypothesis are required as follows:

1. The deflecting action of the mid-surface is tiny compared with the thickness of the plate.
2. The mid-plane lasts unstrained following to bending.
3. At first plane segments normal to the mid-surface and normal to that surface after the bending.
4. The stress component normal to the mid-plane is tiny compared with the other stress components.

Based on these assumptions, a three-dimensional plate problem can be reduced to a two-dimensional one. Consider a plate with no load, in which the *xy* plane coincides with the mid-plane (Figure 6.11). The components of displacement at a point are denoted by *u*, *v*, and *w*.

According to the preceding presumptions, the displacement field could be expressed as follows:

$$u(x,y,z) = -z\frac{\partial w}{\partial x} \tag{6.30a}$$

$$v(x,y,z) = -z\frac{\partial w}{\partial y} \tag{6.30b}$$

$$w(x,y,z) = w(x,y) \tag{6.30c}$$

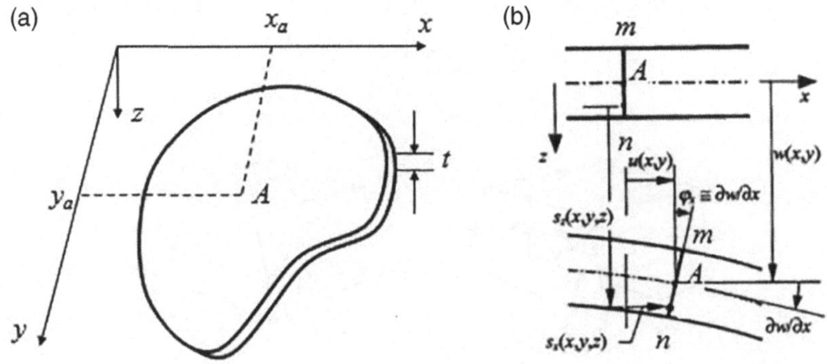

FIGURE 6.11 A plate of constant thickness: (a) plate and (b) plate part before and after deflection.

6.2.4.1 Strain–Displacement Relationships

The non-zero linear strains connected with the displacement field are defined as follows:

$$\varepsilon_x = \frac{\partial u}{\partial x} = -z\frac{\partial^2 w}{\partial x^2} \tag{6.31a}$$

$$\varepsilon_y = \frac{\partial v}{\partial y} = -z\frac{\partial^2 w}{\partial y^2} \tag{6.31b}$$

$$\gamma_{xy} = \frac{\partial u}{\partial y} + \frac{\partial v}{\partial x} = -2z\frac{\partial^2 w}{\partial x \partial y} \tag{6.31c}$$

where γ_{xy} is the shear strain and $\left(\varepsilon_x, \varepsilon_y\right)$ are the normal strains.

6.2.4.2 Stress–Strain Relationship

From Hooke's law, we can obtain stress because $\varepsilon_z = \gamma_{yz} = \gamma_{xz} = 0$. That is,

$$\sigma_x = \frac{E}{1-\nu^2}\left[\varepsilon_x + \nu\varepsilon_y\right] \tag{6.32a}$$

$$\sigma_y = \frac{E}{1-\nu^2}\left[\varepsilon_y + \nu\varepsilon_x\right] \tag{6.32b}$$

$$\tau_{xy} = G\gamma_{xy} \tag{6.32c}$$

From Equation (6.31), we can simplify as follows:

$$\sigma_x = -\frac{E}{1-\nu^2}z\left[\frac{\partial^2 w}{\partial x^2} + \nu\frac{\partial^2 w}{\partial y^2}\right] \tag{6.33a}$$

$$\sigma_x = -\frac{E}{1-\nu^2}z\left[\frac{\partial^2 w}{\partial y^2} + \nu\frac{\partial^2 w}{\partial x^2}\right] \tag{6.33b}$$

$$\tau_{xy} = -\frac{E}{1+\nu}z\frac{\partial^2 w}{\partial x \partial y} \tag{6.33c}$$

6.2.4.3 Equilibrium Equations

The stresses dispersed over the thickness of the plate yield bending moments, twisting moments, and vertical shear forces. These moments and forces per unit length are also defined as stress resultants and described as follows:

$$\left\{\begin{matrix} M_x \\ M_y \\ M_{xy} \end{matrix}\right\} = \int_{-t/2}^{t/2}\left\{\begin{matrix} \sigma_x \\ \sigma_y \\ \tau_{xy} \end{matrix}\right\} z\,dz \tag{6.34}$$

Substituting Equation (6.33) into Equation (6.34), we can derive the following formulas for the bending and twisting moments:

$$M_x = -D\left(\frac{\partial^2 w}{\partial x^2} + v\frac{\partial^2 w}{\partial y^2}\right)$$

$$M_y = -D\left(\frac{\partial^2 w}{\partial y^2} + v\frac{\partial^2 w}{\partial x^2}\right) \tag{6.35}$$

$$M_{xy} = -D(1-v)\frac{\partial^2 w}{\partial x \partial y}$$

where $D = \dfrac{Et^3}{12\left(1-v^2\right)}$.

So, for bending of a thin plate, the differential equation of equilibrium can be described as follows:

$$\frac{\partial^2 M_x}{\partial x^2} + 2\frac{\partial^2 M_{xy}}{\partial x \partial y} + \frac{\partial^2 M_y}{\partial y^2} = -p \tag{6.36}$$

Substituting Equation (6.35) into Equation (6.36), the governing equation for plate can be derived as follows:

$$\frac{\partial^4 w}{\partial x^4} + 2\frac{\partial^4 w}{\partial x^2 \partial y^2} + \frac{\partial^4 w}{\partial y^4} = \frac{p}{D} \tag{6.37}$$

6.2.4.4 Boundary Conditions

The boundary conditions that put in the edge $x = a$ of the rectangular plate with edges parallel to the x and y axes (Figure 6.12) are as follows:

1. Clamped, fixed, or built-in edge: $w = 0, \dfrac{\partial w}{\partial x} = 0 \ (x = a)$
2. Simply supported edge: $w = 0, M_x = -D\left(\dfrac{\partial^2 w}{\partial x^2} + v\dfrac{\partial^2 w}{\partial y^2}\right) = 0 \ (x = a)$
3. Free edge: $\dfrac{\partial^2 w}{\partial x^2} + v\dfrac{\partial^2 w}{\partial y^2} = 0, \dfrac{\partial^3 w}{\partial x^3} + (2-v)\dfrac{\partial^3 w}{\partial x \partial y^2} = 0 \ (x = a)$

6.2.5 Torsion Member

Torsion is a form of distorted shape of an object due to an applied torque. Because the shear stresses for non-circular sections are no longer denoting to the circumference of a curved geometry, the torsional theory of circular sections cannot be applied to the torsion of noncircular sections. Moreover, plane cross-sections do not possess plane and not distort on the implementation of torque, and in fact, warping of the cross section happens.

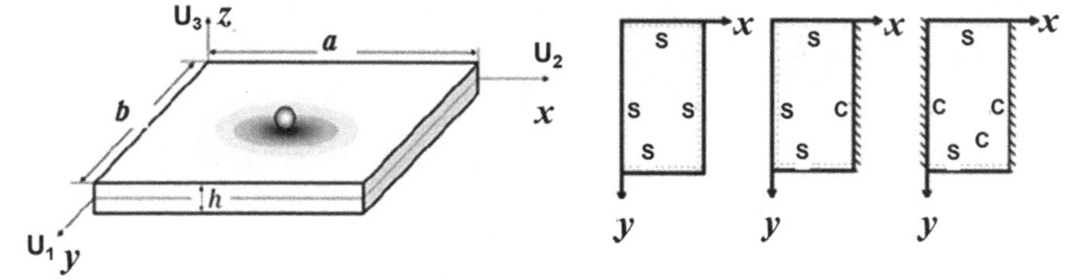

FIGURE 6.12 Various boundary conditions of a plate (S – simply supported edge, F – free edge, C – clamped or built-in edge).

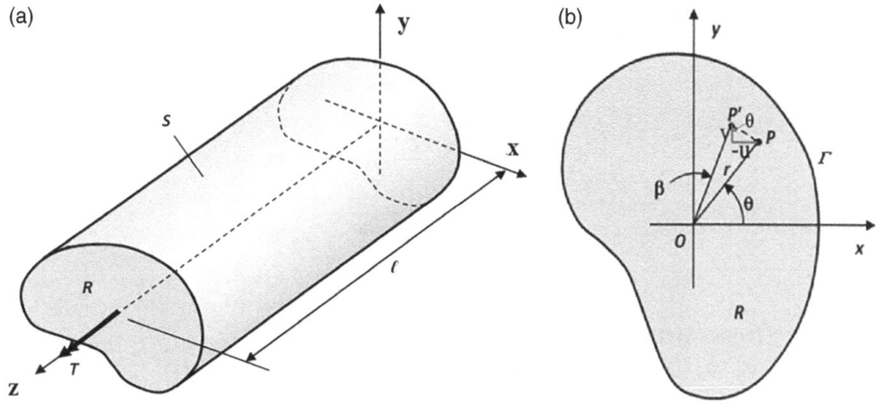

FIGURE 6.13 Non-circular section under twist: (a) torsion member under twist and (b) cross section.

As shown in Figure 6.13a, assume a prismatic bar of isotropic homogeneous non-circular segment subjected to twisting action. Based on the mechanics convention, that is, the axis of the bar corresponds with z-axis, the cross section is a set in the x, y plane. The bar is attached at $z = 0$ base, and the opposite base $z = l$ is twisted by angle $l\,\alpha$. We can acquire the following assumptions:

- The cross sections in the x, y plane revolve like a rigid body. In the case of a non-circular shape, the cross section is not planar, but it is diverted in the z direction.
- The deflection and the twist rate α are constant along the whole length of the bar. Therefore, the problem is lessened to a two-dimensional one.

Presume the following geometrical behavior: based on the Saint-Venant hypothesis, consider any point P in the section, which will rotate and warp owing to the application of T, as shown in Figure 6.13b. The displacements u, v, and w in directions x, y, and z under these presumptions can be expressed as follows:

$$u = -r\beta \sin\theta = -\beta y = -\alpha yz$$

$$v = -r\beta\cos\theta = \beta x = \alpha xz \tag{6.38}$$

$$w = \alpha\psi(x,y)$$

where $\beta = \alpha z$, and $\psi(x,y)$ is an unknown function defining the deflection.

6.2.5.1 Strain–Displacement Relationships

It is supposed to be differentiable. A simple computation produces the equivalent strain (small deformation). Thus, we can assume the following:

$$\varepsilon_x = \varepsilon_y = \varepsilon_z = \gamma_{xy} = 0 \tag{6.39}$$

and, therefore, the only shearing strains that exist are γ_{xz} and γ_{yz}, which are defined as follows:

$$\gamma_{xz} = \frac{\partial w}{\partial x} + \frac{\partial u}{\partial z} = \alpha\left(\frac{\partial\psi}{\partial x} - y\right) \tag{6.40}$$

$$\gamma_{yz} = \frac{\partial w}{\partial y} + \frac{\partial v}{\partial z} = \alpha\left(\frac{\partial\psi}{\partial y} + x\right) \tag{6.41}$$

6.2.5.2 Stress–Strain Relationship

From Hooke's law, we can obtain stress from Equations (6.39)–(6.41). That is,

$$\sigma_x = \sigma_y = \sigma_z = \tau_{xy} = 0 \tag{6.42}$$

$$\tau_{xz} = G\alpha\left(\frac{\partial\psi}{\partial x} - y\right) \tag{6.43}$$

$$\tau_{yz} = G\alpha\left(\frac{\partial\psi}{\partial y} + x\right) \tag{6.44}$$

6.2.5.3 Equilibrium Equations

$$\frac{\partial\tau_{xz}}{\partial x} + \frac{\partial\tau_{yz}}{\partial y} = 0 \tag{6.45}$$

Let introduce the Prandt stress function $\phi = \phi(x,y)$. We can define shear stress as $\tau_{xz} = \frac{\partial\phi}{\partial y}$, $\tau_{yz} = -\frac{\partial\phi}{\partial x}$. By differentiating Equations (6.43) and (6.44) with respect to y and x, respectively, we can get the following equation:

$$\frac{\partial\tau_{xz}}{\partial y} - \frac{\partial\tau_{yz}}{\partial x} = \frac{\partial^2\varphi}{\partial y^2} - \left(-\frac{\partial^2\varphi}{\partial x^2}\right) = G\alpha\left(\frac{\partial^2\psi}{\partial x\partial y} - 1\right) - G\alpha\left(\frac{\partial^2\psi}{\partial x\partial y} + 1\right) = -2G\alpha \tag{6.46}$$

The previous equality can be restated into an inhomogeneous second-order partial differential equation, which is called as *Poisson equation*:

$$\nabla^2\phi = \frac{\partial^2\phi}{\partial y^2} + \frac{\partial^2\phi}{\partial x^2} = -2G\alpha \tag{6.47}$$

6.2.5.4 Boundary Conditions

Since zero surface forces are examined, the traction vector $T = \left(T_x, T_y, T_z\right)$ on the boundary has zero components. Inserting τ_{xy} and τ_{xz} from Equations (6.43) and (6.44) to equality $T_z = 0$, we obtain

$$T_z = \tau_{xz}n_x + \tau_{yz}n_y = \frac{\partial\phi}{\partial y}n_x - \frac{\partial\phi}{\partial x}n_y = \frac{\partial\phi}{\partial y}\frac{dy}{ds} + \frac{\partial\phi}{\partial x}\frac{dx}{ds} = \frac{\partial\phi}{\partial s} = 0 \tag{6.48}$$

Thus, Equation (6.48) involves the tangent derivative of ϕ which is equal to zero, and therefore, ϕ is constant along each component of the boundary.

6.2.5.5 Torque, Section Moment, and Shear Stresses

The torsional moment or torque, M, is calculated as the double integral over the cross section:

$$M = \iint_{R^2} \left(-\tau_{xz}y + \tau_{yz}x\right)dxdy \tag{6.49}$$

where R^2 is the cross section of the bar.

$$M = -\iint_{R^2} \left(\frac{\partial\phi}{\partial y}y + \frac{\partial\phi}{\partial x}x\right)dxdy \tag{6.50}$$

Using the method of integration by parts, we can get the following results:

$$\iint_{R^2} \frac{\partial\phi}{\partial y} y\,dx\,dy = \int_{\Gamma} \phi y n_y\,ds - \iint_{R^2} \phi(x,y)\,dx\,dy = -\iint_{R^2} \phi(x,y)\,dx\,dy \tag{6.51}$$

$$\iint_{R^2} \frac{\partial\phi}{\partial x} x\,dx\,dy = -\iint_{R^2} \phi(x,y)\,dx\,dy \tag{6.52}$$

So, we can get the torsional moment. That is,

$$M = 2\iint_{R^2} \phi(x,y)\,dxdy = J \tag{6.53}$$

where $J = 2\iint_{R^2} \phi(x,y)\,dxdy$.

6.2.6 Fluid Mechanics

Fluids on unlimited control volumes are subject to two definite types of forces: a surface force and a body (volume) force. While increasing the distance between the interacting elements, volume force like gravity decreases at a slow speed and is superbly piercing into the inside of a fluid. One result of the comparatively slow variation of volume forces with position is that they move equally on fluid in an enough small volume. The net force acting on the element becomes straight corresponding to its control volume.

Assume that there is a fluid possessed in a small volume element *dV*, and the whole volume force applying at time *t* on it is expressed as *F*(*r*, *t*) *dV*. Surface forces are modeled as momentum transport in the fluid. Such transport comes from a combination of reciprocal forces applied by adjacent molecules, and momentum fluxes are raised by a comparative molecular motion (Figure 6.14).

The net surface force applied by the fluid on the planar surface element, dS_j, can be stated as follows:

$$\sigma_{ij} dS_j \tag{6.54}$$

where σ_{ij} is the local stress tensor and dS_j is the components of an random first-order tensor as surface elements.

The *i*-component of the whole force applying on a minute surface element is a combination of body force and surface force. It is expressed as follows:

$$f_i = f_V + f_S = \iiint_V F_i\, dV + \oint\!\!\oint_S \sigma_{ij}\, dS_j \tag{6.55}$$

where f_i is the force per volume, isolated into those forces appearing from volume force (f_V, long range) and from surface force (f_S, short range) with a minute contribution *dF*.

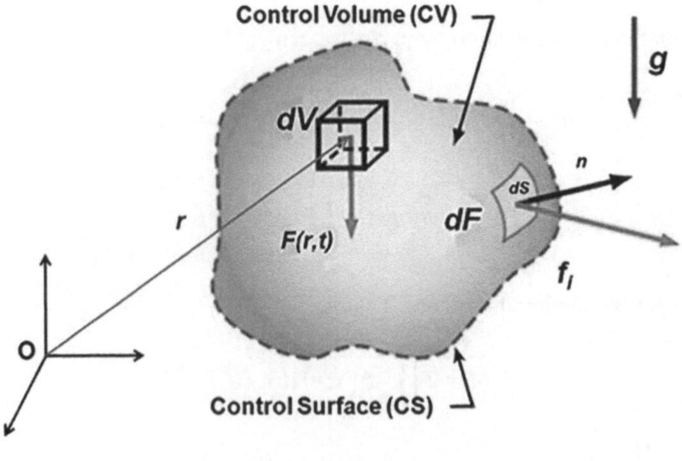

FIGURE 6.14 Fluid forces of a surface element on infinite control volume (or control surface).

If the tensor divergence theorem is used, Equation (6.55) becomes

$$f_i = \iiint_V F_i \, dV + \iiint_V \frac{\partial \sigma_{ij}}{\partial x_j} dV \tag{6.56}$$

As we can take the limit where $V \to 0$, F_i and $\frac{\partial \sigma_{ij}}{\partial x_j}$ converge to steady across the element. Both things on the right-hand side in Equation (6.56) are $O(V)$. According to Newtonian dynamics, the i-component of the net force applying on the volume element is identical to the i-component of the rate of change of its linear momentum. However, if volume converges to extremely smaller, the mass density and linear acceleration of the fluid are both about constant across the element. Thus, the rate of change of the element's linear momentum also is $O(V)$. We know that the linear equation of motion on a tiny fluid element places no particular reductions on the stress tensor.

Assuming that there is a fluid element surrounded by a surface S in a fixed volume V, the i-component of the whole torque in the origin O of the coordinate system can be stated as follows:

$$\tau_i = \iiint_V \xi_{ijk} x_j F_k \, dV + \oiint_S \xi_{ijk} x_j \sigma_{kl} \, dS_l \tag{6.57}$$

where the first and second terms on the right-hand side are due to volume and surface forces.

If the tensor divergence theorem is utilized, Equation (6.57) becomes

$$\tau_i = \iiint_V \xi_{ijk} x_j F_k \, dV + \iiint_V \xi_{ijk} \frac{\partial (x_j \sigma_{kj})}{\partial x_j} dV \tag{6.58}$$

It might be restated as

$$\tau_i = \iiint_V \xi_{ijk} x_j F_k \, dV + \iiint_V \xi_{ijk} \sigma_{kj} \, dV + \iiint_V \xi_{ijk} x_j \frac{\partial \sigma_{kj}}{\partial x_j} dV \tag{6.59}$$

As the fluid particle is lessened to a point where $V \to 0$, we can presume that F_i, σ_{ij} and $\frac{\partial \sigma_{ij}}{\partial x_j}$ are all roughly constant across the element. Because $x \approx V^{1/3}$, the first, second, and third terms on the right-hand side of Equation (6.59) are $O(V^{1/3})$, $O(V)$, and $O(V^{1/3})$, respectively.

Based on Newtonian mechanics, the i-component of the whole torque applying on the fluid element is identical to the i-component of the rate of change of its net angular momentum about O. Presuming that the linear acceleration and density of the fluid are roughly constant across the element, we conclude that the rate of change of its angular momentum is $O(V^{4/3})$.

We recognize that the rotational equation of motion of a fluid element relies on the second term on the right-hand side of Equation (6.59). Otherwise, the second term is zero. So, we can be relevant if

$$\xi_{ijk}\sigma_{kj} = 0 \tag{6.60}$$

From Equation (6.60), we recognize that the stress tensor should be exactly similar:

$$\sigma_{ji} = \sigma_{ij} \tag{6.61}$$

These six stress tensor (i.e., σ_{11}, σ_{22}, σ_{33}, σ_{12}, σ_{13}, and σ_{23}) are independent components of an amount. According to the principal axes, the diagonal components of the stress tensor σ_{ij} are the principal stresses – σ'_{11}, σ'_{22}, and σ'_{33}.

Generally, the positioning of the principal axes changes with position. The normal stress σ'_{11} applying across a surface element vertical to the first principal axis is identical to a tension in the way of the axis. It is identical for σ'_{22} and σ'_{11}. Consequently, the common state of the fluid can be considered as a superposition of compressions or tensions at a particular point in space. Regardless of the positioning of the principle axes, because the trace of the stress tensor, $\sigma_{ii} = \sigma_{11} + \sigma_{22} + \sigma_{33}$, is having only magnitude, the trace of the stress tensor at a stated point is identical to the summation of the principal stresses:

$$\sigma_{ii} = \sigma'_{11} + \sigma'_{22} + \sigma'_{33} \tag{6.62}$$

Assume that there is the surface forces applied on the tiny cubic volume element of a static fluid. The components of the stress tensor are roughly constant across the element. The sides of the cube are placed parallel to the principal axes of the local stress tensor. This tensor, which has zero non-diagonal components, is the summation of two tensors:

$$\begin{pmatrix} \frac{1}{3}\sigma_{ii} & 0 & 0 \\ 0 & \frac{1}{3}\sigma_{ii} & 0 \\ 0 & 0 & \frac{1}{3}\sigma_{ii} \end{pmatrix} \tag{6.63}$$

and

$$\begin{pmatrix} \sigma'_{11} - \frac{1}{3}\sigma_{ii} & 0 & 0 \\ 0 & \sigma'_{22} - \frac{1}{3}\sigma_{ii} & 0 \\ 0 & 0 & \sigma'_{33} - \frac{1}{3}\sigma_{ii} \end{pmatrix} \tag{6.64}$$

Because the tensor (6.63) is the same value when measured in different directions, it is identical to the normal force per unit area bound inward on each face of the volume element. This constant compression changes the fluid element's volume without reforming its shape, which can be endured by the fluid element. The tensor (6.64) signifies the departure of the stress tensor from an isotropic form. The diagonal components of this tensor have zero sum in view of Equation (6.62). It represents equal and opposite forces per unit area, applying on opposing faces of the volume element. The forces on at least one pair of opposing faces form a tension, and the forces on at least one pair form a compression. Such forces react to alter the shape of the volume element: (1) a pure translation, (2) a pure strain along the main axes, and (3) a rotation (connected with the vorticity).

Furthermore, this likelihood cannot be nullified by any volume force applying on the element, because such forces become randomly small compared to surface forces ($V \rightarrow 0$; in other words, in the limit that the element's volume converges to zero).

A fluid might be called a material that is continuously reshaped by applied forces. It follows that if the diagonal components of the tensor (6.64) are non-zero in any place in the fluid, it is going at that point. Thus, we deduce that the principal stresses, σ'_{11}, σ'_{22}, and σ'_{33}, should be equal to one another at all points in a static fluid. It signifies that the stress tensor has the isotropic form (6.63) in a motionless fluid. Regardless of the positioning of the coordinate axes, the components of an isotropic tensor are not rotationally changing.

As resting fluids are in a condition of compression with no shear forces, the stress tensor should have only diagonal terms. The stress tensor of fluid can be expressed as follows:

$$\sigma_x = \sigma_y = \sigma_z = -p \tag{6.65}$$

Based on Pascal's law, in a motionless fluid, the force per unit at a given point is the same in all directions. The pressure is normal to the surface on which it acts. Its magnitude is unconstrained of the surface positioning. So, this gives rise to the comparatively straightforward form of the equation of motion for an inviscid flow. The stress tensor is expressed as follows:

$$\sigma_{ij} = -p\delta_{ij} \tag{6.66}$$

where $p = -\sigma_{ij}/3$ is the static pressure: that is, minus the normal stress acting in any direction.

As the fluid is subjected to the pressure, expressed as the mean normal force on a fluid element, the mean normal stress is expressed as $(\sigma_{11} + \sigma_{22} + \sigma_{33}) = \sigma_{ij}/3$. This is taken to be $-p$ in some otherwise fine texts, but it is strictly applicable only for mere mono atomic gases. Generally, there is an unpredictability between the pressure and the mean normal stress. It is easy to explain pressure in a moving fluid as minus the mean normal stress:

$$p = -\frac{1}{3}\sigma_{ii} \tag{6.67}$$

We can break the stress tensor in a moving fluid into two parts: an isotropic part, $p\delta_{ij}$, which tends to change the volume of the body in a static fluid, and a present non-isotropic part, τ_{ij}, which comprises any shear stresses. Thus, the deviator tensor can be defined as follows:

$$\sigma_{ij} = -p\delta_{ij} + \tau_{ij} \tag{6.68}$$

Since δ_{ij} and σ_{ij} are symmetric tensors, we recognizes that τ_{ij} is symmetric:

$$\tau_{ji} = \tau_{ij} \tag{6.69}$$

The deviatoric stress tensor τ_{ij} is a result of fluid motion. It can be explained as the difference between the whole stress tensor and the pressure. If a static fluid is at rest, the fluid arises stationary and the deviatoric stress tensor will be zero. If fluid is stably moving, it has a spatially uniform velocity field and the deviatoric stress tensor is also zero. Consequently, we are aware that the deviatoric stress tensor is operated by velocity gradients in the fluid.

For Newtonian fluids, the relationship between rate of strain $\partial v_i/\partial x_j$ and stress τ_{ij} is linear. In other words,

$$\tau_{ij} = C_{ijkl}\frac{\partial v_k}{\partial x_l} \tag{6.70}$$

where C_{ijkl} is a coefficient tensor of rank 4.

In fact, there are $3^4 = 81$ coefficients. As independent of positing in space, the fourth-order tensor C_{ijkl} is isotropic, i.e., it has the same property in all rotated coordinate frames. The problem is how to get its fourth-order equivalent. For an isotropic fourth-order tensor, the most common expression is

$$C_{ijkl} = \alpha\delta_{ij}\delta_{kl} + \beta\delta_{ik}\delta_{jl} + \gamma\delta_{il}\delta_{jk} \tag{6.71}$$

where α, β, and γ are arbitrary scalars.

From Equations (6.70) and (6.71), we are aware that it follows

$$\tau_{ij} = \alpha\frac{\partial v_k}{\partial x_k}\delta_{ij} + \beta\frac{\partial v_i}{\partial x_j} + \gamma\frac{\partial v_j}{\partial x_i} \tag{6.72}$$

From Equation (6.69), τ_{ij} is a symmetric tensor that $\beta = \gamma$,

$$\tau_{ij} = \alpha e_{kk}\delta_{ij} + 2\beta e_{ij} \tag{6.73}$$

where $e_{ij} = \frac{1}{2}\left(\frac{\partial v_i}{\partial x_j} + \frac{\partial v_j}{\partial x_i}\right)$ is defined as the *rate of strain tensor.*

As τ_{ij} is a traceless tensor, we get $3\alpha = -2\beta$.

$$\tau_{ij} = 2\mu\left(e_{ij} - \frac{1}{3}e_{kk}\delta_{ij}\right) \tag{6.74}$$

where $\mu = \beta$.

From Equation (6.68), the stress tensor in an isotropic Newtonian fluid is expressed as follows:

$$\sigma_{ij} = -p\delta_{ij} + 2\mu\left(e_{ij} - \frac{1}{3}e_{kk}\delta_{ij}\right) \tag{6.75}$$

where p and μ are arbitrary scalars.

For a simple shearing motion, the viscosity coefficient μ can be obtained from Equation (6.74). With $\frac{\partial v_1}{\partial x_2}$ as the only non-zero velocity derivative, all of the components of τ_{ij} are zero apart from the shear stresses:

$$\tau_{12} = \tau_{21} = \mu\frac{\partial v_1}{\partial x_2} \tag{6.76}$$

As aligned plane layers of fluid slide over one another, the coefficient μ is the constant of proportionality between the tangential force per unit area and the rate of shear. The force applying between layers of fluid experiencing slide motion tends to be against the motion, which suggests that $\mu > 0$.

Think about a property γ (e.g. density, velocity component, temperature) of the fluid element in space. Generally, at that time, this will rely on the time, t, and on the position (x, y, z) of the fluid element. In a tiny time δt, assume that the element goes from (x, y, z) to ($x+\delta x$, $y+\delta y$, $z+\delta z$). It has precisely the same material at the two times. But there will be a corresponding small change in γ, signified by $\delta\gamma$. For instance, the rate of change of the density $\rho = \rho(x, t)$ of a particle instantly at x is

$$\frac{D\rho}{Dt} = \left(\frac{\partial\rho}{\partial t}\right) + \left(\frac{\partial\rho}{\partial x}\right)\left(\frac{\partial x}{\partial t}\right) + \left(\frac{\partial\rho}{\partial y}\right)\left(\frac{\partial y}{\partial t}\right) + \left(\frac{\partial\rho}{\partial z}\right)\left(\frac{\partial z}{\partial t}\right) = \frac{\partial\rho}{\partial t} + v\cdot\nabla\rho \tag{6.77}$$

If the continuity equation is united, it can be restated as follows:

$$\frac{1}{\rho}\frac{D\rho}{Dt} = \frac{D\ln\rho}{Dt} = -\nabla\cdot v \tag{6.78}$$

Assume that there is a volume element V that is moving with the fluid. As the fluid element in volume is convected, we can use the divergence theorem. That is,

$$\frac{DV}{Dt} = \oint\!\!\oint_S v\cdot dS = \oint\!\!\oint_S v_i\, dS_i = \iiint_V \frac{\partial v_i}{\partial x_i}dV = \iiint_V \nabla\cdot v\, dV \tag{6.79}$$

where S is the bounding surface of the fluid element.

As we take the limit where $V \to 0$, $\nabla\cdot v$ will be roughly constant across the element. Then, we obtain

$$\frac{1}{V}\frac{DV}{Dt} = \frac{D\ln V}{Dt} = \nabla\cdot v \tag{6.80}$$

Thus, we deduce that the divergence of the fluid velocity at a designated point in space states the fractional rate of increase in the volume of a minute co-moving fluid element at that point. If a surface S encloses a fixed volume V, the i-component of whole linear momentum included in V is

$$P_i = \iiint_V \rho v_i dV \tag{6.81}$$

The flux of i-momentum across S is

$$\Phi_i = \oiint_S \rho v_i v_j \, dS_j \tag{6.82}$$

Momentum conservation supports that the increasing rate of the net i-momentum of the fluid included in V, joining the flux of i-momentum out of V, is equal to the rate of i-momentum generation in V. From Newton's second law of motion, the latter quantity is identical to the i-component of the net force acting on the fluid bounded in V. Consequently, we get

$$\frac{dP_i}{dt} + \Phi_i = f_i \tag{6.83}$$

Lastly, the i-component of the net force applying on the fluid in V is restated as follows:

$$\iiint_V \frac{\partial(\rho v_i)}{\partial t} dV + \oiint_S \rho v_i v_j \, dS_j = \iiint_V F_i \, dV + \oiint_S \sigma_{ij} \, dS_j \tag{6.84}$$

Because the volume V is non-time-varying and the divergent theorem is utilized, Equation (6.84) becomes

$$\iiint_V \left[\frac{\partial(\rho v_i)}{\partial t} + \frac{\partial(\rho v_i v_j)}{\partial x_j} \right] dV = \iiint_V \left[F_i + \frac{\partial \sigma_{ij}}{\partial x_j} \right] dV \tag{6.85}$$

Reorganizing Equation (6.85), we get

$$\left(\frac{\partial \rho}{\partial t} + v_j \frac{\partial \rho}{\partial x_j} + \rho \frac{\partial v_j}{\partial x_j} \right) v_i + \rho \left(\frac{\partial v_i}{\partial t} + v_j \frac{\partial v_i}{\partial x_j} \right) = F_i + \frac{\partial \sigma_{ij}}{\partial x_j} \tag{6.86}$$

From the continuity equation, the first term in tensor notation is

$$\frac{\partial \rho}{\partial t} + v_j \frac{\partial \rho}{\partial x_j} + \rho \frac{\partial v_j}{\partial x_j} = 0 \tag{6.87}$$

So, we get the following fluid equation of motion:

$$\rho \left(\frac{\partial v_i}{\partial t} + v_j \frac{\partial v_i}{\partial x_j} \right) = F_i + \frac{\partial \sigma_{ij}}{\partial x_j} \tag{6.88}$$

If Equations (6.75) and (6.88) are incorporated, the equation of fluid motion for an isotropic and Newtonian fluid can be stated as follows:

$$\rho\frac{Dv_i}{Dt} = F_i - \frac{\partial p}{\partial x_i} + \frac{\partial}{\partial x_j}\left[\mu\left(\frac{\partial v_i}{\partial x_j} + \frac{\partial v_j}{\partial x_i}\right)\right] - \frac{\partial}{\partial x_i}\left(\frac{2}{3}\mu\frac{\partial v_j}{x_j}\right) \tag{6.89}$$

This equation is normally familiar as the Navier-Stokes equation. If there are no strong temperature gradients in the fluid, the Navier-Stokes equation can be restated as follows:

$$\rho\frac{Dv_i}{Dt} = F_i - \frac{\partial p}{\partial x_i} + \mu\left(\frac{\partial^2 v_i}{\partial x_j \partial x_j} + \frac{1}{3}\frac{\partial^2 v_j}{\partial x_i \partial x_j}\right) \tag{6.90}$$

And the vector form for Equation (6.90) becomes

$$\rho\frac{Dv}{Dt} \equiv \rho\left[\frac{\partial v}{\partial t} + (v\cdot\nabla)v\right] = F - \nabla p + \mu\left[\nabla^2 v + \frac{1}{3}\nabla(\nabla\cdot v)\right] \tag{6.91}$$

6.3 STIFFNESS OF MECHANICAL PRODUCT – VIBRATION

6.3.1 Introduction

This section we will discuss vibrations which are the fluctuations of a mechanical system. We know that all human beings involve vibrations in one form or the other. For instance, we hear through eardrum vibration, see by light wave vibration, breath by lung vibration, and speak through tongues vibration. Vibration can be defined as a time-dependent displacement of a system of particles or a particle with respect to an equilibrium location.

Vibration in a mechanical system is a restoring force that continually pulls the system toward its equilibrium position by elastic or springing property of the system. That is, when the spring is stretched, energy is stored as a potential energy and a spring force is developed. When the block released, the potential energy will be converted to a kinetic energy so that the block will be pulled toward equilibrium position, and it continues oscillations about equilibrium position (Figure 6.15).

What else is essentially required in the vibration components of a mechanical system? We can suggest an inertia or mass that is associated with the kinetic energy from an energy perspective. And there is deformation such as compliance or stiffness that is associated with strain or potential energy. The other vibration component also involves dissipation or damping as the resistance supplied by a body to the motion of a vibratory system is damping. It just controls the amplitude of vibration so that the mechanical failure happening because of resonance may be evaded.

Uncontrolled vibrations in many common engineering systems can lead to catastrophic failure. For example, vibrations due to the rotating unbalance in helicopter blades can lead to loss of the pilot's control and crash of the helicopter. Structural vibrations during earthquake lead to large displacements/stresses and might result in

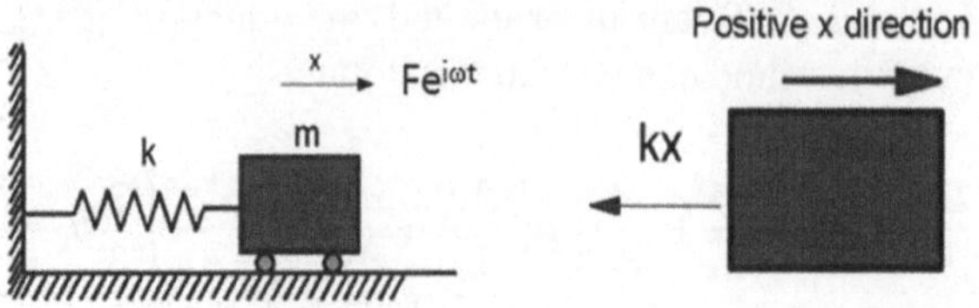

FIGURE 6.15 Vibration of a mechanical system.

structural failure. Vibrations of machine tools lead to improper machining of a part and noise.

From the definition of vibration what causes it? The answer is simple. That is, it comes from dynamic loads. For internal load, we can give an example as unbalance forces in engines/turbines. On the other hand, for external load, we can give an instance as wind loads/earthquake excitations, and road bumps.

There are different techniques that can reduce or eliminate the undesirable vibrations. Some of the techniques are as follows: (1) if possible, eliminating external excitation, (2) using shock absorbers, (3) dynamic observers, and (4) resting the system on right vibration isolator. Vibration theory is developed by applying basic laws of nature and appropriate constitutive equations to dynamic systems. The study of vibrations requires a synthesis of basic engineering sciences and mathematics, dynamics, strength of materials, and fluid mechanics.

To estimate vibrations, there are two methods: (1) measurements and (2) mathematical modeling. The measurements can be classified as (1) experimental modal analysis – natural frequencies and damping and (2) running vibration measurements – actual levels. On the other hand, mathematical modeling can be classified as (1) lumped parameter models – single and multi-degree-of-freedom and (2) distributed parameter models – infinite degrees-of-freedom (simple structures).

A vibratory system is a dynamic system for which the variables such as the input excitations and response outputs are time dependent. It depends on the external excitation as well as the initial conditions. Because most practical vibrating systems are very complicated, it is not possible to examine all the details for a mathematical analysis. Only the most major features are examined in the analysis to estimate the system behavior under specified input conditions. The analysis of a vibrating system includes (1) mathematical modeling, (2) derivation of the governing equation, (3) solution of the equation, and (4) interpretation of the result.

6.3.1.1 Terminologies

There are basic terminologies to understand the vibration as follows: (1) *periodic motion* is a motion that repeats itself after equal intervals of time, (2) *time period* is a time that is required to finish one cycle, (3) *frequency* is the number of cycles per unit time, (4) *amplitude* is the maximum displacement of a vibrating body from its equilibrium position, (5) *natural frequency* is the frequency of free vibration without external excitation, and (6) *degree of freedom* is the least number of independent coordinates needed to state the motion of a system at any instant.

6.3.2 Mathematical Modeling of Mechanical Systems

As the first step of vibration study, a mechanical system is modeled as single-degree-of-freedom systems or the others. Regardless of lumped systems or distributed systems, a product can be typically modeled as mass in the form of the body, damper in the form of shock absorbers, and spring in the form of suspension when the external forces like the skin drag of airplane wing are applicable.

For instance, suppose that there is an airplane wing that can be modeled as the distributed mass of the wing. Because of a variety of loads in field, the wing distorts the shape so that it can be modeled as a spring. The deflection of the wing arises damping because of relative motion between components such as connections, joints, and support. And to assess the ride quality of a product mounted on a truck, the most effective mathematical model of a vehicle suspension system is a quarter car model [2]. The presumed model of the vehicle consists of the un-sprung mass and the sprung mass, separately. The un-sprung mass, m_{us}, represents one wheel of the vehicle, and the sprung mass, m_s, represents one fourth of the body of the vehicle. The principal suspension is modeled as a spring k_s and a damper c_s in parallel, which connects the un-sprung mass to the sprung mass. The tire (or rail) is modeled as a spring k_{us} and represents the transfer of the road force to the un-sprung mass (Figure 6.16).

Step 1: Mathematical Modeling for Deriving Governing Equations

To derive the governing equations, the common approaches used are D'Alembert's principle and Newton's second law mentioned in Chapter 5. After mathematical

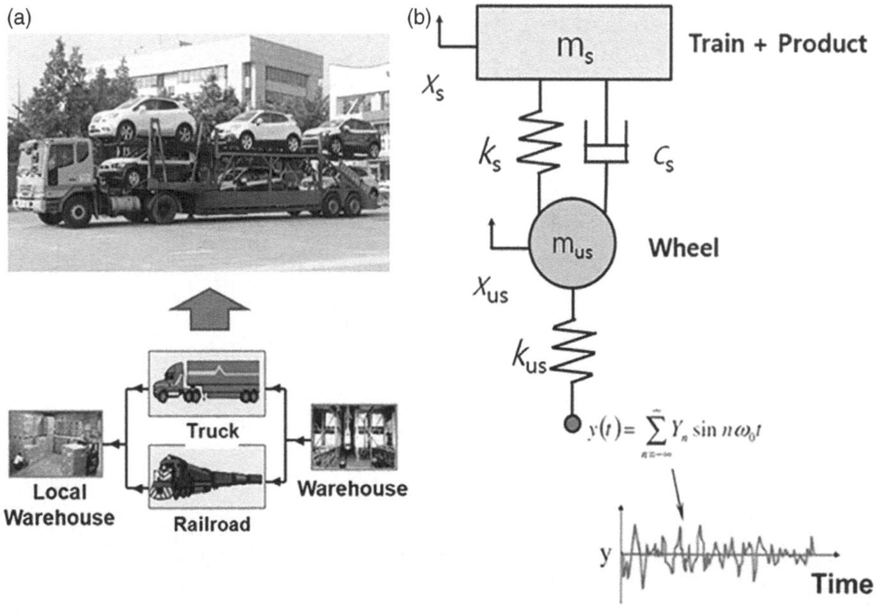

FIGURE 6.16 Automobile subjected to random vibrations in transportation: (a) automobile in transportation and (b) its modeling.

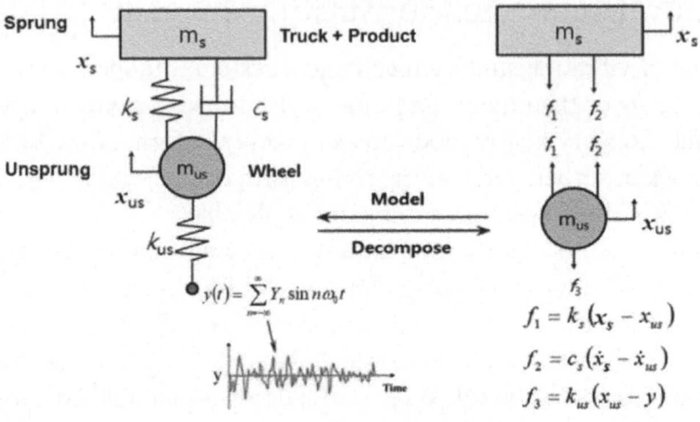

FIGURE 6.17 A quarter car model subjected to random loads from base (or road).

modeling of a vibrating system, we can find the governing equations, results of the equations, and interpretation of the results. The mathematical modeling is to represent all the major features – mass, damping, and spring – of the mechanical system. To explain the procedure of refinement utilized in the mathematical modeling, for example, we consider the automobile subjected to (random) vibrations in transportation (Figure 6.17).

The governing differential equations of motion for the quarter car model might be represented as follows:

$$m_s\ddot{x}_s + c_s(\dot{x}_s - \dot{x}_{us}) + k_s(x_s - x_{us}) = 0 \tag{6.92}$$

$$m_{us}\ddot{x}_{us} - c_s(\dot{x}_s - \dot{x}_{us}) - k_s(x_s - x_{us}) = -k_{us}(x_{us} - y) \tag{6.93}$$

So, we can concisely represent the above equations of motion as follows:

$$\left[\begin{bmatrix} m_s & 0 \\ 0 & m_{us} \end{bmatrix}\begin{bmatrix} \ddot{x}_s \\ \ddot{x}_{us} \end{bmatrix} + \begin{bmatrix} c_s & -c_s \\ -c_s & c_s \end{bmatrix}\begin{bmatrix} \dot{x}_s \\ \dot{x}_{us} \end{bmatrix}\right]$$

$$+\begin{bmatrix} k_s & -k_s \\ -k_s & k_{us}+k_s \end{bmatrix}\begin{bmatrix} x_s \\ x_{us} \end{bmatrix} = \begin{bmatrix} 0 \\ k_{us}y \end{bmatrix} \tag{6.94}$$

As a result, Equation (6.32) can be expressed in a matrix form:

$$[M]\ddot{X} + [C]\dot{X} + [K]X = F \tag{6.95}$$

Step 2: Solution of the Governing Equations and Its Interpretation

By analysis of mechanical system, the governing Equation (6.95) can be solved by the following methods: (1) standard methods for solving differential equations, (2) Laplace transform methods, (3) matrix methods, and (4) numerical methods. Lastly, when Equation (6.95) is numerically integrated, we can find the time response of

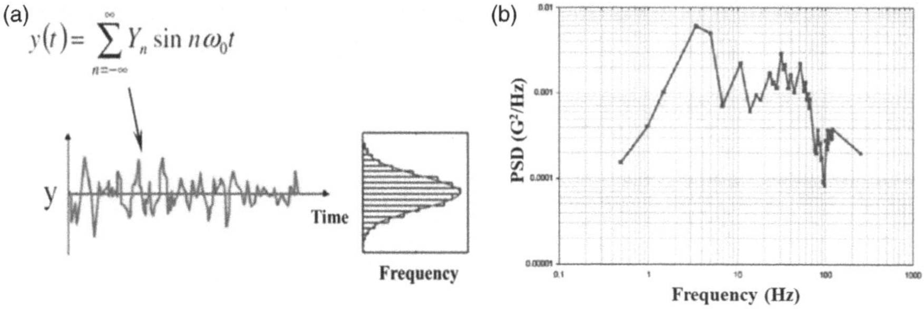

FIGURE 6.18 Intermodal random vibration for base random vibrations: (a) base random vibrations and (b) typical intermodal random vibration.

the state variables due to random vibration. By the Laplace transformation, we can obtain the power spectral density (PSD) in the frequency domain (Figure 6.18).

After engineers understand the results, they determine whether the action plans such as shock absorbers or vibration isolation are needed. If required, engineers will redesign the system. Generally, whenever the frequency of the external excitation like wind matches with the natural frequency of a product, a resonance will happen, which leads to excessive deflections and lastly failure.

6.3.3 Characteristics of the Vibratory Motion Due to the (Internal/External) Force

We will deal with the response of mechanical systems subjected to excitations due to (internal/external) force. After we model mechanical system as a single-degree-of-freedom system subjected to excitation, we can present the equation of motion and its solution. For example, consider a reciprocating compressor as shown in Figure 6.19.

Supposing that a compressor runs at a constant speed, the motion equation of this system subjected to a harmonic force $F(t) = F_0 \cos \omega t$ can be expressed as follows:

$$m\ddot{x} + c\dot{x} + kx = F_0 \cos \omega t \tag{6.96}$$

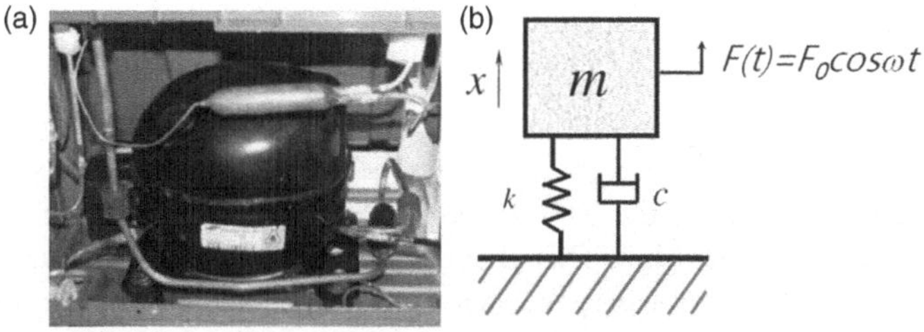

FIGURE 6.19 A reciprocating compressor model: (a) reciprocating compressor and (b) its dynamic model.

The particular solution of Equation (6.96) is also expected to be harmonic; we assume it in the following form:

$$x_p(t) = X\cos(\omega t - \phi) \tag{6.97}$$

where X and ϕ denote the amplitude and phase angle of the response, respectively, which are constants to be determined.

By substituting Equation (6.97) into Equation (6.96), we arrive at

$$X\left[\left(k - m\omega^2\right)\cos(\omega t - \phi) - c\omega\sin(\omega t - \phi)\right] = F_0\cos\omega t \tag{6.98}$$

Utilizing the trigonometric relations in Equation (6.98) and equating the coefficients of $A = \cos\omega t$ and $\sin\omega t$ on both sides of the resulting equation, we obtain

$$X\left[\left(k - m\omega^2\right)\cos\phi + c\omega\sin\phi\right] = F_0 \tag{6.99a}$$

$$X\left[\left(k - m\omega^2\right)\sin\phi - c\omega\cos\phi\right] = 0 \tag{6.99b}$$

The solution of Equation (6.99a and b) gives

$$X = \frac{F_0}{\left[\left(k - m\omega^2\right)^2 + \omega^2 c^2\right]^{1/2}} \tag{6.100}$$

$$\phi = \tan^{-1}\left(\frac{c\omega}{k - m\omega^2}\right) \tag{6.101}$$

By substituting the equations of X and ϕ from Equations (6.100) and (6.101) into Equation (6.99a and b), we get the particular solution of Equation (6.96). Dividing both the numerator and denominator of Equation (6.100) by spring k and making the following substitutions, we get the following equation and Figure 6.20 that describes the magnification factor in accordance with damping ζ and frequency ratio r:

$$M = \frac{X}{\delta_{st}} = \frac{1}{\left[\left\{\left[1 - \left(\frac{\omega}{\omega_n}\right)^2\right]^2 + \left[2\zeta\frac{\omega}{\omega_n}\right]^2\right\}\right]^{1/2}} = \frac{1}{\sqrt{\left(1 - r^2\right)^2 + \left(2\zeta r\right)^2}} \tag{6.102}$$

The quantity M is called the 'magnification factor'. In a lightly damped system when the forcing frequency nears the natural frequency ($r \approx 1 r \approx 1$), the amplitude of the vibration can be extremely high. This phenomenon is defined as resonance. In rotor-bearing systems, any rotational speed that excites a resonant frequency is referred to as a critical speed. If resonance happens in a mechanical system, it can be very dangerous – leading to ultimate failure of the system. Thus, one of the crucial reasons for vibration analysis is to forecast when this type of resonance may happen and

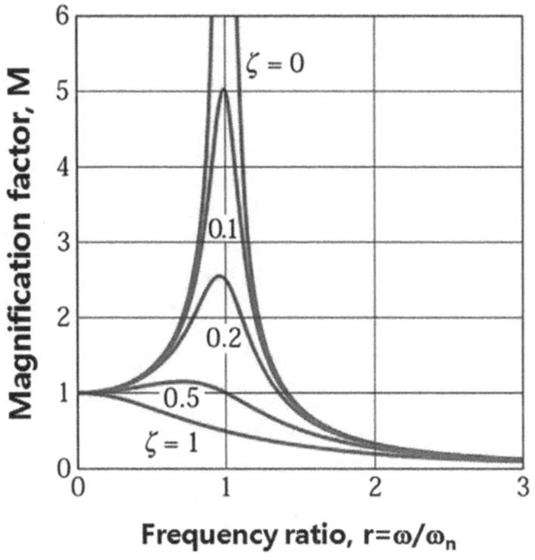

FIGURE 6.20 Magnification factor in accordance with frequency ratio r and damping factor ζ.

then to decide what steps to take to prevent it from occurring. Adding on damping can notably lessen the magnitude of the vibration. Also, the magnitude can be lessened if the natural frequency might be shifted away from the forcing frequency by modifying the mass or stiffness of the system. The magnification factor relies on the frequency ratio and the damping factor as follows.

In many applications, the isolation is required to lessen the motion of the mass (machine) under the applied force. The displacement amplitude of the mass m due to the force $F(t)$ can also be expressed as follows:

$$T_d = \frac{X}{\delta_{st}} = \frac{1}{\sqrt{\left(1-r^2\right)^2 + \left(2\zeta r\right)^2}} \tag{6.103}$$

It is the amplitude ratio (or displacement transmissibility) and designates the ratio of the amplitude of the mass, X, to the static deflection under the constant force δ_{st}. The form of the displacement transmissibility with the frequency ratio r for some values of the damping ratio is displayed in Figure 6.18. The displacement transmissibility becomes greater to a maximum value at

$$r = \sqrt{1-2\zeta^2} \tag{6.104}$$

Equation (6.103) manifests that the displacement transmissibility (or the amplitude of the mass) for small values of damping ratio will possess maximum at $r \approx 1$. Thus, the value of $r \approx 1$ is to be avoided in application. In most instances, the excitation frequency is set, and as a consequence, we can prevent from happening by changing the value of the natural frequency which can be achieved by altering the value of either or both of m and k.

The amplitude of the mass, X, comes near to zero as r makes greater to a big value. The reason is that at large values of r, the applied force $F(t)$ changes very quickly and the inertia of the mass prevents it from following the fluctuating force.

6.3.4 Vibration Isolation of a Mechanical Product

Vibration design of a mechanical product is causally connected to the vibration isolation which makes less the unpleasant results of vibration. So, action plans for a mechanical design require the introduction of a mounting rubber between the source of vibration and the mechanical product so that a lessening in the dynamic response of the system is attained under the stated conditions of vibration excitation.

Vibration isolation can be utilized when (1) the base of vibrating machine is secured against its big unstable forces and (2) the product is secured from environmental conditions like transportation. The first pattern of isolation is utilized when a mechanical system is subjected to an excitation force. On the other hand, the second type of isolation can be found as the following instance: when a domestic refrigerator is transported, it will be subjected to random vibration from the refrigerator basement. Vibration should be isolated as refrigerator uses the mounting rubber. Vibration (or noise) can also cause uneasiness to customers.

In these approaches, vibration isolation is to make less the vibratory motion of the system under the applied force. Thus, transmissibility becomes crucial for this kind of isolators. When a product is mounted in a train, it will be subjected to a random vibration in transportation from the base. Hence, a mounting rubber between the base and the product is set to reduce the transmitted force. The mounted rubber is presumed to have both damping and elasticity, which is modeled as a spring k and a dashpot c. Refrigerator compressor can then be modeled as a single-degree-of-freedom system (Figure 6.21).

To find the transmitted force from excitation, first of all, it is presumed that the operation of the product gives rise to a harmonically varying force $F = F_0 \cos \omega t$. The equation of motion of the product is given by

$$m\ddot{x} + c\dot{x} + kx = F_0 \cos \omega t \tag{6.105}$$

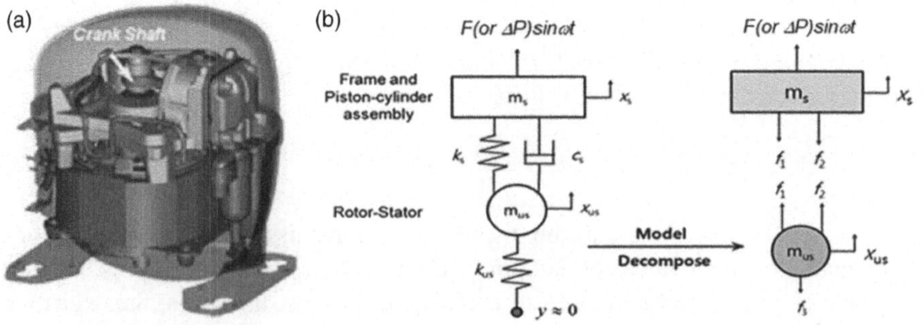

FIGURE 6.21 A simplified compressor model subjected to excitation: (a) refrigerator compressor and (b) single-degree-of-freedom model.

Since the transient solution dies out after some time, only the steady-state solution will be left. The steady-state solution of Equation (6.105) is given by

$$x(t) = X\cos(\omega t - \phi) \tag{6.106}$$

where

$$X = \frac{F_0}{\left[\left(k - m\omega^2\right)^2 + \omega^2 c^2\right]^{1/2}} \tag{6.107}$$

and

$$\phi = \tan^{-1}\left(\frac{\omega c}{k - m\omega^2}\right) \tag{6.108}$$

The force transmitted due to piston movement through the dashpot and the spring, F_T is given by

$$F_T(t) = kx(t) + c\dot{x}(t) = kX\cos(\omega t - \varphi) - c\omega X\sin(\omega t - \phi) \tag{6.109}$$

The magnitude of the whole transmitted force (F_T) is given by

$$\begin{aligned} F_t(t) &= \left[(kx)^2 + (c\dot{x})^2\right]^{1/2} = X\sqrt{k^2 + \omega^2 c^2} \\ &= \frac{F_0\left(k^2 + \omega^2 c^2\right)^{1/2}}{\left[\left(k - m\omega^2\right)^2 + \omega^2 c^2\right]^{1/2}} \end{aligned} \tag{6.110}$$

The transmissibility, the ratio of the magnitude of the force transmitted to that of the exciting force, is defined as the following equation and Figure 6.22 that can be described as damping ζ and frequency ratio r:

$$\begin{aligned} T_f = \frac{F_T}{F_0} &= \left\{\frac{k^2 + \omega^2 c^2}{\left(k - m\omega^2\right)^2 + \omega^2 c^2}\right\}^{1/2} \\ &= \left\{\frac{1 + (2\zeta r)^2}{\left(1 - r^2\right)^2 + (2\zeta r)^2}\right\}^{1/2} \end{aligned} \tag{6.111}$$

where $r = \omega/\omega_n$ is the frequency ratio.

The change of T_f with the frequency ratio $r = \omega/\omega_n$ is displayed in Figure 6.20. In order to attain isolation, the force transmitted to the base is required to be less than the excitation force. It can be seen that the forcing frequency has to be greater than the natural frequency of the system in order to attain isolation of vibration.

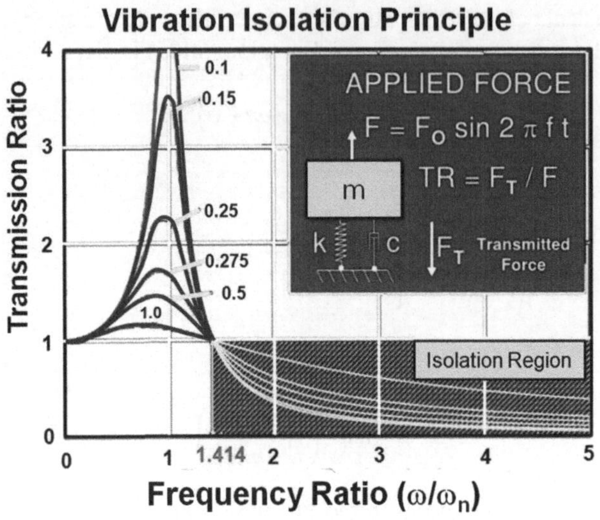

FIGURE 6.22 A simplified refrigerator model subjected to random vibrations.

For small values of damping ratio and for frequency ratio, the force transmissibility, given by Equation (6.55), can be approximated as

$$T_f = \frac{F_T}{F_0} \approx \frac{1}{r^2 - 1} \tag{6.112}$$

The following observations can be made from Figure 6.20:

1. The magnitude of the force transmitted to the foundation can be reduced by decreasing the natural frequency of the system (ω_n).
2. The force transmitted to the foundation can also be reduced by decreasing the damping ratio. However, since vibration isolation requires $r > \sqrt{2}$, the machine should pass through resonance during stopping and start-up. Hence, some damping is crucial to avoid infinitely big amplitudes at resonance.
3. Although damping reduces the amplitude of the mass (X) for all frequencies, it reduces the maximum force transmitted to the foundation only if $r < \sqrt{2}$. Above that value, the addition of damping increases the force transmitted.
4. If the speed of the machine (forcing frequency) changes, we must compromise in selecting the amount of damping to reduce the force transmitted. The quantity of damping should be adequate to limit the amplitude X and the force transmitted while passing through the resonance, but not so much to increase unnecessarily the force transmitted at the operating speed.

Example 6.1

An exhaust fan, rotating at 1,000 RPM, is to be borne by four springs, each having a stiffness of *K*. If only 10% of the unbalanced force of the fan is to be transmitted to the base, what should be the value of *K*? Assume the mass of the exhaust fan to be 40 kg.

Since the transmissibility has to be 0.1, the forcing frequency is given by $\omega = (1{,}000 \times 2\pi)/60 = 104.72$ rad/s and the natural frequency of the system by

$$\omega_n = \left(\frac{k}{m}\right)^{1/2} = \left(\frac{4K}{40}\right)^{1/2} = \frac{\sqrt{K}}{3.1623}.$$

By presuming the damping ratio to be $\zeta = 0$, we get from Equation (6.112)

$$T_f = 0.1 = \frac{1}{r^2 - 1} = \frac{1}{\left(\frac{104.72 \times 3.1623}{\sqrt{K}}\right)^2 - 1}$$

This leads to

$$\frac{331.1561}{\sqrt{K}} = 3.3166 \quad K = 9.97 \text{ N/m}$$

REFERENCES

1. Sonsino, C.M., 2007, Fatigue testing under variable amplitude loading, *International Journal of Fatigue*, 29 (6), 1080–1089.
2. Jazar, R., 2008, *Vehicle dynamics: theory and application*, New York: Springer, 931–975.

7 Mechanical System Failure

7.1 INTRODUCTION

A typical failure mechanism of a mechanical system due to repetitive or varying loads is fatigue and fracture. As coined by the French engineer Jean-Victor Poncelet in the midst of the nineteenth century, the term 'fatigue' means that the material becomes weak, which eventually leads to disintegration. It evolves in an imperfect part of a product as a consequence of cyclic plastic deformation in a limited area on the outside of the part where there is a tiny crack or a pre-existing defect. Based on individualistic studies conducted by Battelle in 1982 [1], approximately 80%–90% of all structural failures happen through a fatigue mechanism. The value for failures due to fracture is also approximated to be $1.5 billion per year. The required costs are crucial, and also injury and hazards to human life due to many failures are infinitely more so. Hence, Battelle suggested that these can be lessened by 29% by implementation of modern fatigue technology.

7.1.1 (Static) Loads

Applied static forces include tension, compression, bending, torsion, and shear (Figure 7.1). Tensile signifies that the material is under the state of being stretched tight. The forces applied on it are pulling the material. In the case of compression, the forces applied on an object are squashing it.

- Axial loading (compression/tension) – The acted forces are lying in the same straight line with the longitudinal axis of the part. The forces give rise to the member to either shorten or stretch.
- Transverse loading (shear) – Forces applied upright to the longitudinal axis of a member. Transverse loading gives rise to the member to bend and divert from its first position, with internal tensile and compressive strains going along with the change in curvature of the part. Transverse loading also causes shear forces that induce shear deformation of the material and increase the transverse turning of the part.
- Torsional loading – Twisting action induced by a pair of outer acted identical and oppositely directed force couples acting on parallel planes or by a single outer force acted to a member that has one end attached against rotation.

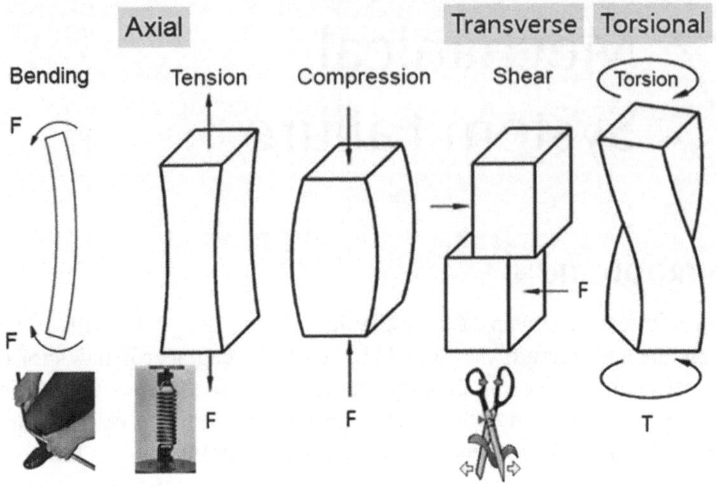

FIGURE 7.1 A variety of stresses due to four kinds of static loads.

7.1.2 Stress

Stress is a bodily quantity that indicates the response of the matter on the unit area (A) applied in the outer (or inner) forces (F):

$$\sigma = \frac{F}{A} \tag{7.1}$$

Deformation includes any changes in the shape of an object because of an acted force. Strain is a physical deformation response of a material to stress. A result of stresses in the vertical axis is the corresponding strains along the horizontal axis. When an external load is applied, it will cause deforms on the body of product. Once the load is eliminated, the body will regain its original shape without permanent deformation in the member. We call this 'elastic deformation'. This pattern of deformation includes stretching of the atomic bonds.

Hooke's law governs linear elastic deformation. That is, it states

$$\sigma = E\varepsilon \tag{7.2}$$

where σ is the applied stress, E is Young's modulus, and ε is the strain.

By Hook's law, the stresses are directly proportional to the strains within the elastic limit. The linear part of the stress–strain curve is the elastic area, and the gradient is Young's modulus (or stiffness). However, if the acted stress is beyond the elastic limit, the body will fail with a permanent deformation (or failure). Mild steel goes proportional limit, yield point, ultimate stress point into fracture. We know that a failure happens when the material holds out its yield strength – the limit of the elastic range (Figure 7.2).

After the yield point, the curve usually decreases somewhat, but the deformation continues. Strain hardening and plastic deformation continue until it reaches the

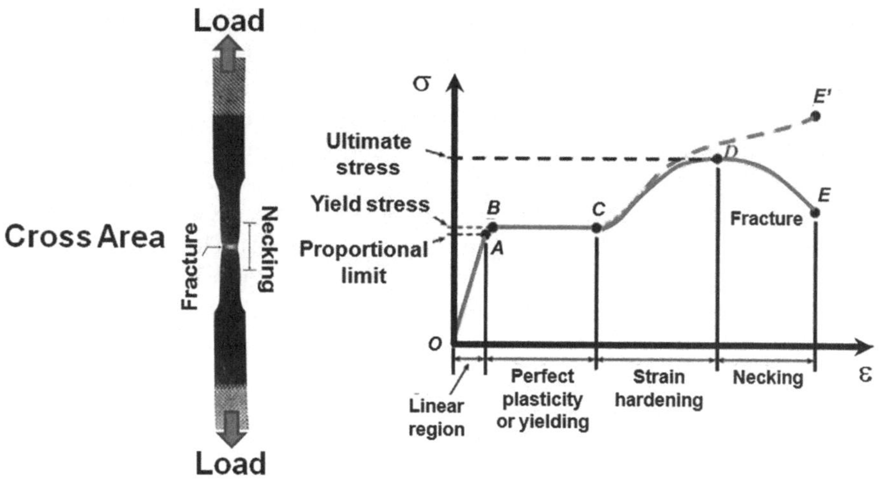

FIGURE 7.2 Definition of stress and stress–strain curve (e.g. mild steel).

ultimate tensile stress. During strain hardening, the material becomes stronger due to the occurrence of atomic dislocations. The necking phase is specified by a reduction in the cross-sectional area of the specimen. After the ultimate strength is reached, necking starts. During necking, the material can no longer endure the maximum stress and the strain in the specimen quickly increases. Plastic deformation ends with the fracture of the material.

Many materials are composed of lots of grains, each of which has grain boundaries and second-phase particles. Thus, it is easier to study plastic deformation in a sole crystal to eliminate the effects of grain boundaries and second-phase particles. If a single crystal in a metal is stressed in tension beyond its elastic limit, it elongates somewhat, which is defined as plastic deformation (Figure 7.3).

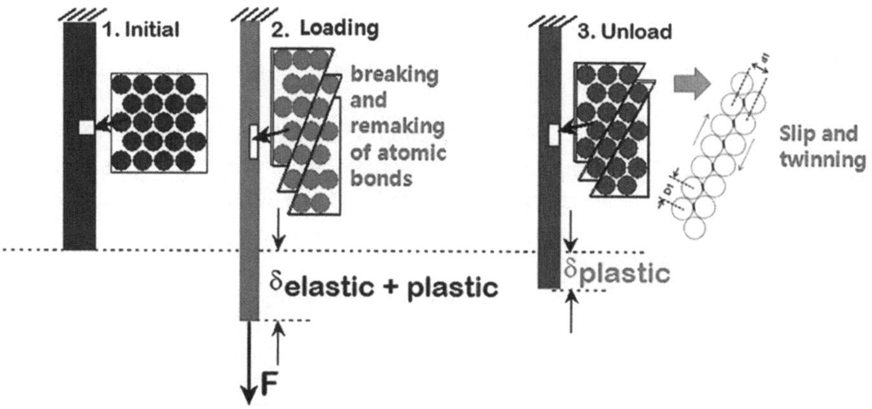

FIGURE 7.3 Plastic deformation (metals) for axial force.

Plastic deformation requires breaking and remaking of atomic bonds. Plastic deformation may occur by a slip, twinning, or a combination of both. Plastic deformation cannot be restored to its initial state by changes, i.e. irreversible process. Under a tensile stress, plastic deformation is distinguished by a strain-hardening region, necking region, and lastly fracture (also called 'rupture').

Fracture is the breakdown of a single body into pieces by an acted stress. Studies in the field of fracture mechanics cover the response in materials for (dynamic) loading and, in results, occurrence and propagation of its cracks. The process of crack formation in a material includes the following fracture mechanisms: (1) fatigue, (2) brittle (cleavage) fracture, (3) ductile (shear) fracture, (4) de-adhesion, and (5) crazing. The branch of fracture mechanics obviously depends on the material's behavior. Brittle fracture is the dominant failure mechanism of low-strength materials. At intermediate strength levels, there is a transition between brittle ductile overload and fracture under linear elastic conditions. Fracture failure at very high strength values is governed by the flow properties of the material.

The earliest fracture study focused only on linear elastic materials under quasi-static states. Integrating other types of material behavior, elastic-plastic fracture mechanics considers plastic deformation under quasi-static states. Especially, dynamic, viscoelastic, and viscoplastic fracture mechanics include time as a variable. Fracture mechanics also utilizes the methods of analytical solid mechanics to compute the dominant force on a crack and the methods of experimental solid mechanics to distinguish the material's resistance to fracture. When subjected to a variety of loadings, fractures happen due to inappropriate designs. An efficient design of products to avoid all fractures due to fatigue is still a challenge and an ultimate goal in the modern research.

7.2 BUCKINGHAM PI THEOREM

Dimensional analysis is a crucial tool for evolving mathematical models of physical phenomena. Assume that q_1, q_2, q_3, ..., q_n be n dimensional variables that are physically pertinent in a demonstrated problem and that are correlated by an unspecified dimensionally homogeneous set of equations, these might be defined via a functional relation of the form

$$F(q_1,q_2,\ldots,q_n)=0 \text{ or equivalently } q_1=f(q_2,\ldots,q_n) \tag{7.3}$$

If k is the number of crucial dimensions required to define the n variables, there will be k main variables and the present variables might be defined as $(n-k)$ dimensionless and independent quantities or 'Pi groups', $\Pi_1,\Pi_2,\ldots,\Pi_{n-k}$. The functional relation might thus be reduced to the much more concise form:

$$\Phi(\Pi_1,\Pi_2,\ldots,\Pi_n)=0 \text{ or equivalently } \Pi_1=\Phi(\Pi_2,\ldots,\Pi_{n-k})=0 \tag{7.4}$$

Example 7.1

Consider a simple edge-cracked plate subjected to a tensile stress σ_∞ (see Figure 7.4). Assume that this is a two-dimensional problem that the thickness, compared to

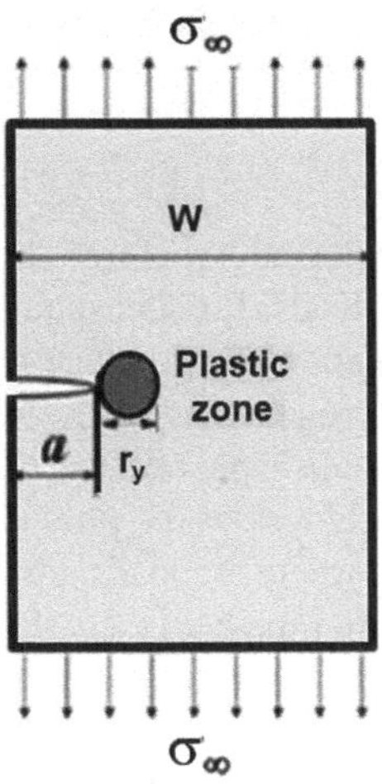

FIGURE 7.4 Edge-cracked plates with a plastic zone.

wide and height in plates, can be negligible. We would like to know how one of the stress components σ_{ij} changes with position. A generalized functional relation can be stated as

$$f\left(\sigma_{ij},\sigma_{\infty},E,v,\sigma_{kl},\varepsilon_{kl},a,r,\theta\right)=0 \tag{7.5}$$

where ν is Poisson's ratio, σ_{kl} are other stress components, ε_{kl} are all nonzero components of the strain tensor.

For infinite plate, we can eliminate σ_{kl} and ε_{kl} as per Hooke's law. Let σ_{∞} and a be the main quantities. Calling on the Buckingham Pi Theorem gives

$$\frac{\sigma_{ij}}{\sigma_{\infty}}=f_1\left(\frac{E}{\sigma_{\infty}},\frac{r}{a},v,\theta\right) \tag{7.6}$$

When the plate width is finite, the above equation is redefined as

$$\frac{\sigma_{ij}}{\sigma_{\infty}}=f_2\left(\frac{E}{\sigma_{\infty}},\frac{r}{a},\frac{W}{a},v,\theta\right) \tag{7.7}$$

When a plastic zone sets up ahead of the crack tip and the material does not strain harder, the yield strength can express the stress field

$$\frac{\sigma_{ij}}{\sigma_{\infty}}=f_3\left(\frac{E}{\sigma_{\infty}},\frac{\sigma_{ys}}{\sigma},\frac{r}{a},\frac{W}{a},\frac{r_y}{a},v,\theta\right) \tag{7.8}$$

We know that two functions, f_1 and f_2, are identical to linear elastic fracture mechanics (LEFM). On the other hand, f_3 is the elastic–plastic relationship. So, we know that LEFM is valid only when $r_y \ll a$ and $\sigma_{\infty} \ll \sigma_{ys}$.

7.3 FAILURE MECHANICS AND DESIGN FOR MECHANICAL PRODUCTS

7.3.1 Introduction

Critical design features of mechanical structures should have enough stiffness and strength to endure a variety of loads. Product requirements on stiffness, the resistance against reversible deformation, may change over a wide span. Strength, the defense against irreversible deformation, is always required to be high through the design of product such as material/shape, because this permanent deformation may cause cracks, fractures, and lastly loss of the product's intended functions. If there is a design fault in the structure where the loads are applied, the structure will fracture at that location in product lifetime because it doesn't have enough stiffness and strength to bear stress. Engineers would want to make products with better designs without faults to withstand when subjected to repeated loads. Thus, the mechanical system might be redesigned to be a good design in product lifetime.

The failure mechanics of mechanical parts can be distinguished by two factors: (1) the stress (or loads) on the structure including mechanism and (2) the pattern of materials and shape used in the structure. To prevent the failure in field, mechanical engineers modify product designs by choosing a proper product material and shape before the products are released. Thus, the mechanical system of the products can withstand the loads (or stress) in its lifetime with enough strength and stiffness (Figure 7.5).

Product failure – fatigue – in a mechanical system is a physical problem created when stress due to loads causes a fracture. Failure mechanics pursues to understand the process how product materials subjected to stress cause the failure. That is, if there are design defects in a product that can cause inadequacy of strength or stiffness, then the product will collapse in its lifetime when subjected to stress. Hence, the failure site in the product (or module) structure must be identified when the failed (or reproduced) products are taken apart in the field or from the results of reliability testing such as parametric accelerated life testing (ALT).

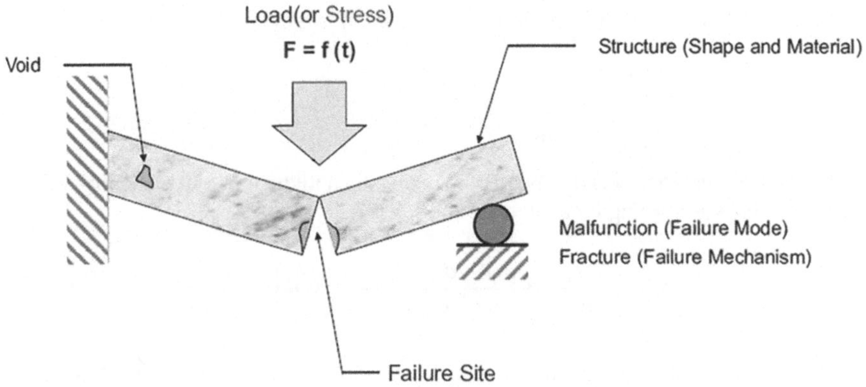

FIGURE 7.5 Failure mechanics produced by a load on a part made from a specific material.

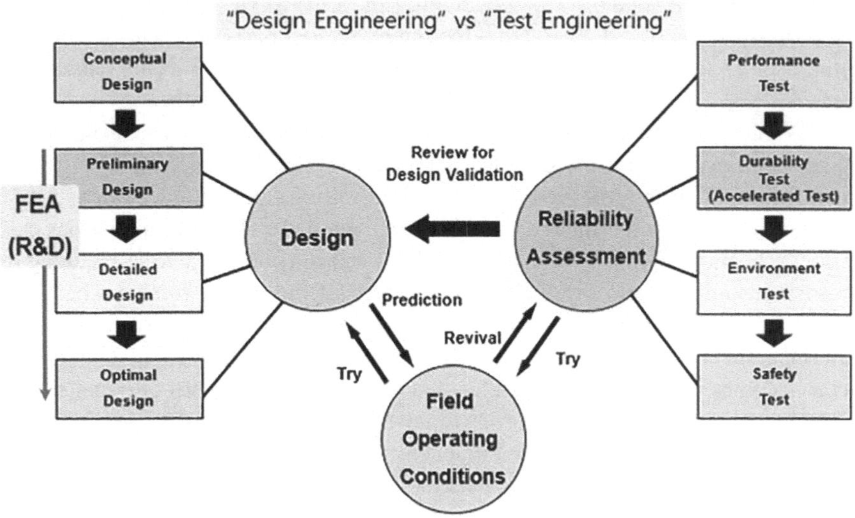

FIGURE 7.6 Optimal designs completed together by test engineering and design engineering.

Most mechanical systems are composed of multi-module structures. If the product structure, as a design result, is ideally designed in the developing process, it can robustly endure its own loading. That is, there would be no problems or failures in the products' lifetime. Although the product is optimally designed using a design tool such as finite element analysis (FEA), it might have design flaws that will unexpectedly show up in the field. Especially, fatigue failure arises due to the existence of a tiny crack or pre-existing defect on the exterior of a part when subjected to repeated stresses and decide the lifetime of the product because it is practically unnoticeable using the conventional finite element techniques. To prevent this problem, the product structure should be experimentally confirmed if it has a good quality – sufficient stiffness and strength. The design problems, therefore, should be revealed and corrected by a reliability testing such as parametric ALTs.

As seen in Figure 7.6, to complete this design mission, the products should successfully undergo test engineering and design engineering in the design process. Design engineers implement the structure of the products to perform intended functions in accordance with the customer requirements. On the other hand, test engineers confirm whether the products' final design can achieve the reliability target.

7.3.2 Failure Mechanics – Fatigue

The designs of a mechanical system involve components subjected to cyclic or fluctuating loads. Such loadings cause fluctuating or cyclic stresses which usually result in fatigue failure if improperly designed. Approximately 90% of all structural failures happen through a fatigue mechanism. The idea of fatigue that describes the failure of an (imperfect) structural system having a tiny crack or pre-existing defect when subjected to repeated loadings was started from the mid-nineteenth century in the railroad industry. At that time, fatigue failures of railway axles had become an extensive problem. This was the first time that many related parts had been subjected to millions of cycles at stress levels well below the monotonically varying tensile yield stress.

A. Wöhler initially started the modern study of fatigue. He, a German engineer on the Lower Silesia-Brandenberg Railroad, worked for the railroad system in the mid-nineteenth century and was the head manager of rolling stock. Wöhler worried about the causes of fracture in railcar axles after lengthened use. A railcar axle is basically a circular beam with four-point bending, which yields a compressive stress along the top surface and a tensile stress along the bottom. If the axle is rotated a half turn, the bottom becomes the top and vice versa, so the stresses on a specific region of the material at the surface change repetitively from compression to tension. Although the metal becomes weak, fatigue was specified to explain this pattern of damage. This is just now known as completely reversed fatigue loading. Some of Wöhler's data are for Krupp axle steel and are plotted in terms of nominal stress (S) vs. the number of cycles to failure (N), which is known as the S–N diagram. Each curve on such diagram is still referred to as a Wöhler line (Figure 7.7).

Since 1830, it has been acknowledged that metal under a fluctuating or repeated load will fail at a stress level lower than required to lead failure under an individual implementation of the same load. Figure 7.8 represents a bar-shaped part subjected to a constant sinusoidal changing force. After a period of time, a crack might start to form on the perimeter of the hole. This crack then grows throughout the part until the whole section is unable to endure the applied stresses, and finally the part fails.

The physical growth of a crack can be usually split into two different phases. These are the crack beginning phase (Phase I) and the crack growth phase (Phase II). Fatigue cracks start by the release of shear strain energy. The above diagram represents how a tiny crack or pre-existing defect starts and the shear stresses result in internal plastic deformation along slip planes. As the sinusoidal loading is cycled, the slip planes pass back and forth, resulting in small extrusions and intrusions on

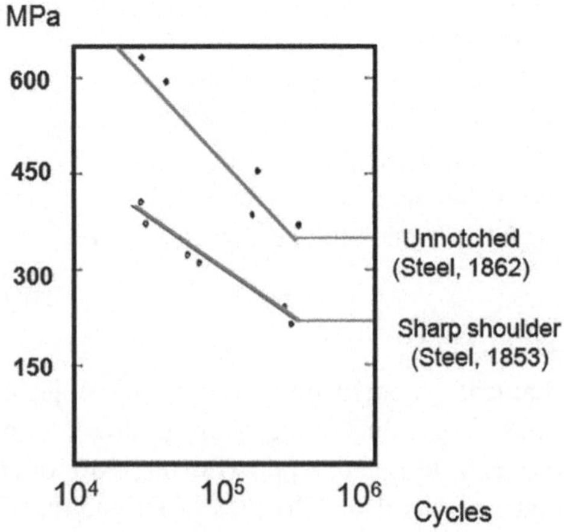

FIGURE 7.7 Some of Wöhler's data for railcar axles steel on the S–N diagram [2].

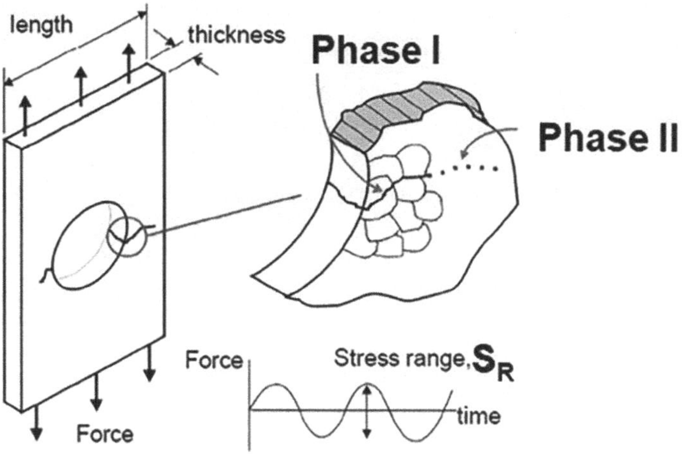

FIGURE 7.8 Bar-shaped part subjected to a constant sinusoidal changing force.

the crystal surface. These surface disturbances are about 1–10 μm in height and add up to embryonic cracks.

Phase I growth pursuits the orientation of the maximum shear plane, or 45° to the direction of loading. A crack starts in this way and extends to the grain boundary. The mechanism at this spot is slowly moved to the adjoining grain. Once the crack has grown through about three grains, it is seen to change its orientation of propagation. The physical mechanism for fatigue moves to Phase II. The crack is enough larger to generate a geometrical stress concentration. A tensile plastic zone is created at the crack tip as shown in the following Figure 7.13. After this stage, the crack moves perpendicular to the direction of the acted load.

7.3.3 Classification of Failures

The meaning of product failure is clear when there is a whole loss of its (intended) functions that can be differentially discerned by specifications or from the viewpoints of the customers. If a part breaks while using a product unintentionally, it may fail through failure mechanism. However, if only a partial loss of (intended) function happens, it will be complex to explain the product failure. In such instances, defining the failure may be incorrect as only an intermittent loss is observed or performance can be regained over time gradually. Although the activity is completed successfully, a person may still feel unsatisfactory if the underlying process is recognized to be below expected specification.

For example, as seen in Figure 7.9, automobile failures can be categorized as follows:

A Class indicates that the failure will damage the body of a passenger or will lead to the loss of the car's control, failure of brake equipment, and fire risk. Instances are too many to be supposed: (1) differential gear fixation, (2) accident due to loss of acceleration control, (3) air exposure of pump, (4) damaged flywheel, (5) overheat

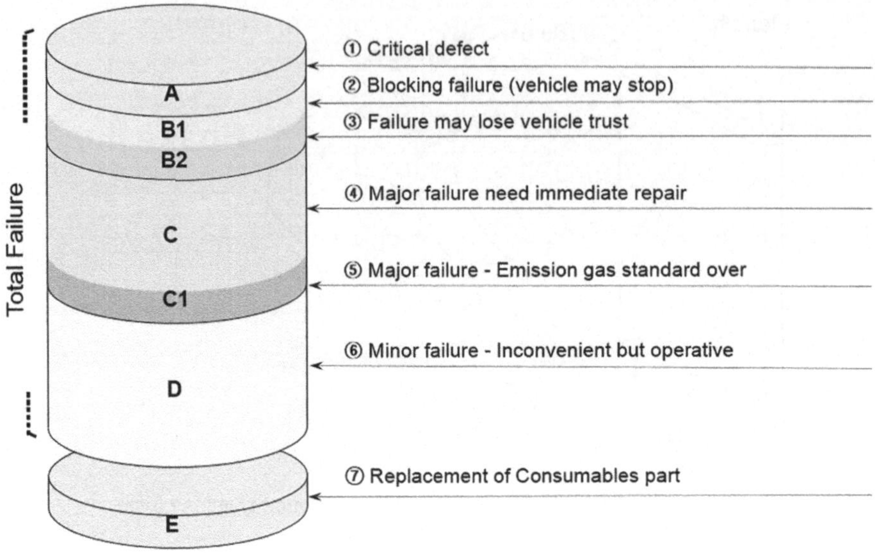

FIGURE 7.9 Definition for failure class (e.g. automobile).

or disconnection of cable, (6) malfunctioned clutch, and (7) malfunctioned injection pump.

B1 Class indicates that the failure will cease the car. Examples are too many to be supposed: (1) engine stop or no starting from injection, pump, computing engine, common rail, engine control, engine fixation, ignition coil, car starting, distribution chain, (2) transmission stop, and (3) stop of gear box, no opposite operation,

B2 Class indicates that the failure may stop the car. There are lots of instances: (1) overheat of engine, (2) abnormal noise of engine or gear box, and (3) vibration of engine or gearbox.

C Class indicates that the car can be operated, but it needs excessive cost to recover it. Instances also are too many to be supposed: (1) it makes the car not working, (2) it influences hearing, visual, and smell, (3) crucial motor surges and power loss, (4) unusual noise, oil spill, cooling water spill, smell, and excessive oil leakage, (5) abnormal smell, and (6) clutch malfunction, not working for gear transmission. C1 Class is also unpredictable to the quality of discharge gas. An instance is not to satisfy the standard of emission gas.

D Class indicates that there is no effect on the car's usage but there is a minor operational failure. Instances are too many to be supposed: (1) affect hearing and visual, (2) driving the car is difficult, (3) slow acceleration, (4) excessive fuel consumption, (5) engine is not starting, (6) idle speed is instable, (7) vibration, and (8) noise (vibration noise, discharge noise, cooling pump noise, cracking noise, noise in gear transmission, starting noise, erosion of engine part).

E Class is related to wearable parts such as spark plug, filter, and timing belt which need to be replaced occasionally. Therefrom, failures can be expressed as A, B, C, and D classes.

7.4 MECHANICAL FAILURE

7.4.1 Theories of Failure

A failure happens when a material begins to display inelastic behavior (or permanent deformation). Ductile and brittle materials rely on loading. Ductile materials that display yielding have plastic deformation before failure. Brittle materials – doesn't display yielding – are susceptible to a sudden failure. The failure pattern of the material might be obtained from uniaxial tensile test, in which the stress–strain characteristics of the material are shown (Figure 7.10).

7.4.1.1 Maximum Principal Stress Theory

Generally, the maximum principal stress theory is applicable to brittle materials. A product failure happens when the maximum principal stress is equal to tensile yield stress (σ_Y). The yield function can be described as follows:

$$f = \max\left(|\sigma_1|,|\sigma_2|,|\sigma_3|\right) - Y \tag{7.9}$$

If $f < 0$, there is no yielding. When $\sigma_1 = \sigma_Y$, $\sigma_2 = 0$, $\sigma_3 = 0$, and $f = 0$, we can start to yield the product.

7.4.1.2 The Maximum Shear Stress Theory

Generally, the maximum shear stress theory is applicable to ductile materials. Here, a failure happens when the maximum shear stress from a mix of principal stresses equals or surpasses the value obtained for the shear stress at yielding in the uniaxial tensile test.

At yielding, the principal stresses in a uniaxial test are $\sigma_1 = S_y$; $\sigma_2 = 0$, and $\sigma_3 = 0$. Therefore, the shear strength at yielding is $\tau_{sy} = [\sigma_1 - (\sigma_2 \text{ or } \sigma_3 = 0)]/2 = S_y/2$. If the maximum shear stress to the failure criterion is compared, it can be defined as follows:

$$\tau_{max} \le \tau_{sy} \tag{7.10}$$

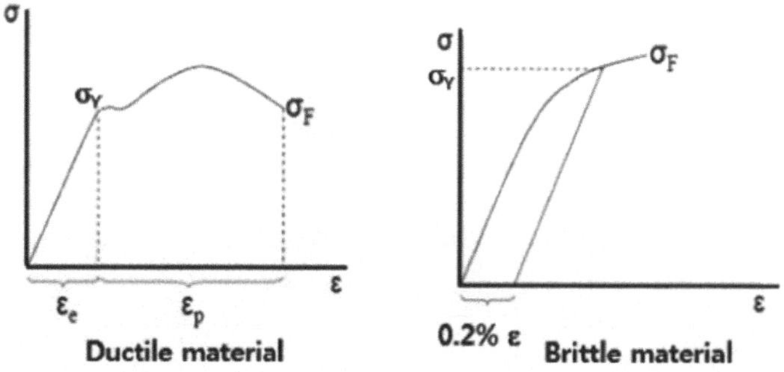

FIGURE 7.10 Typical stress–strain curves for (a) ductile and (b) brittle materials.

7.4.1.3 Maximum Principal Strain

A failure happens at a point in a body when the maximum strain at that point surpasses the value of the maximum strain in a uniaxial test of the material at yield point. The yield function can be defined as follows:

$$f = \max_{i \neq j \neq k} \left| \sigma_i - U\sigma_j - U\sigma_k \right| - \sigma_Y = \sigma_e - \sigma_Y \quad i,j,k = 1,2,3 \tag{7.11}$$

We know that yielding happens when $f = 0$. Hence, the material is safe when $f < 0$.

7.4.1.4 Strain Energy Theory

"Failure at any point in a body subjected to a state of stress starts only when the energy density taken up at that point is equal to the energy density taken up by the material when subjected to elastic limit in a uniaxial stress state."

Strain energy density from Hooke's law ($\sigma = E\varepsilon$) can be defined as follows:

$$U = \int \sigma_{ij}\, d\varepsilon_{ij} \Rightarrow U = \int_0^{\varepsilon_y} \sigma\, d\varepsilon = \frac{1}{2}\sigma_Y \varepsilon_Y = \frac{1}{2}\frac{\sigma_Y^2}{E} \tag{7.12}$$

When the body is subjected to external loads, principal stresses will be generated. Strain energy connected with principal stresses can be defined as follows:

$$U = \frac{1}{2}(\sigma_1\varepsilon_1 + \sigma_2\varepsilon_2 + \sigma_3\varepsilon_3) = \frac{1}{2E}\left(\sigma_1^2 + \sigma_2^2 + \sigma_3^2 - 2v(\sigma_1\sigma_2 + \sigma_3\sigma_1 + \sigma_2\sigma_3)\right) \tag{7.13}$$

So, the yield function is

$$f = \sigma_1^2 + \sigma_2^2 + \sigma_3^2 - 2v(\sigma_1\sigma_2 + \sigma_3\sigma_1 + \sigma_2\sigma_3) - \sigma_Y^2 = \sigma_e^2 - \sigma_Y^2 \tag{7.14}$$

We know that yielding happens when $f = 0$. Hence, the material is safe when $f < 0$.

7.4.1.5 Maximum Shear Stress Theory (Tresca)

"Yielding starts when the maximum shear stress at a point equals the maximum shear stress at yield in a uniaxial tension."

$$\tau_{\max} = K_T = \left| \frac{\sigma_1 - \sigma_2}{2} \right| \Rightarrow \tau_{\max} = \frac{\sigma_Y}{2} = K_T \tag{7.15}$$

If maximum shear stress $\tau_{\max} < \sigma_Y/2$, no failure occurs.

So, the yield function is defined as follows:

$$f = \max.\left\{ \left| \frac{\sigma_1 - \sigma_2}{2} \right|, \left| \frac{\sigma_1 - \sigma_2}{2} \right|, \left| \frac{\sigma_1 - \sigma_2}{2} \right| \right\} - K_T\left(= \frac{\sigma_Y}{2} \right) \tag{7.16}$$

If $f < 0$, no yielding.

If $f = 0$, we know that yielding will start.

So, we can summarize all theories of failure as shown in Table 7.1.

TABLE 7.1
Relationship between Uniaxial and Pure Shear in Failure Theories

	Loading		
Failure Theory	**Uniaxial**	**Pure Shear**	**Relationship**
Maximum principal stress	$\sigma_{max} = \sigma_{YP}$	$\sigma_{max} = \tau_{YP}$	$\tau_{YP} = \sigma_{YP}$
Maximum shear stress	$\tau_{max} = \sigma_{YP}/2$	$\tau_{max} = \tau_{YP}$	$\tau_{YP} = 0.5\ \sigma_{YP}$
Maximum principal strain	$\varepsilon_{max} = \sigma_{YP}/E$	$\varepsilon_{max} = 5\tau_{YP}/4E$	$\tau_{YP} = 0.8\sigma_{YP}$
Maximum octahedral shear stress (Von Mises)	$\tau_{oct} = 0.471\ \sigma_{YP}$	$\tau_{oct} = 0.816\ \tau_{YP}$	$\tau_{YP} = 0.577\ \sigma_{YP}$

7.4.2 Mechanism of Slip

If a mechanical product (or a part) is subjected to a repeated stress (or loads), permanent deformation (or failure) happens at early stage. In material science, a dislocation is a linear crystallographic defect that possesses a sudden change in the positioning of atoms. The movement of dislocations (or slip) allows atoms to slide over each other at low stress levels. Slip happens on planes that have the highest planer density of atoms and in the direction with the highest linear density of atoms. In other words, slip happens in the directions in which the atoms are packed with little space since this requires the lowest quantity of energy. Therefore, they can slip past each other with force. The slip flow relies upon the repetitive structure of the crystal, which allows the atoms to shear away from their original neighbors. It, therefore, passes along the face and joins up with the atom of new crystals.

A slip happens as a result of a straightforward shearing stress. Resolution of axial tensile load F gives two loads. One F_s is shear load along the slip plane, and the other F_N is a normal tensile load perpendicular to the plane. By analysis and experiment, maximum shear stress happens at 45°. Figure 7.11 represents the packing of atoms on a slip plane. We know that there are three ways in which the atoms are close-packed, and these would be the uncomplicated slip directions.

Parts of the crystal on either side of a particular slip plane move in opposing directions and come to rest with the atoms in almost equilibrium locations, so that there is very little change in the lattice positioning. Thus, the external shape of the crystal is changed without destroying it. Simply, a slip can be described in a face-centered cubic (FCC) lattice. The (111) plane is the slip plane possessing the maximum number of atoms (densest plane). It bisects the (001) plane in the line AC, (110) direction possessing the maximum number of atoms on it. A slip is seen as a movement along the (111) planes in the close-packed (110) direction (Figure 7.12a).

From the schematic diagram of slip in a FCC crystal, one may presume that the atoms slip one after another without interruption, beginning at one place or at a few places in the slip plane, and then moving outward over the rest of the plane. For example, if one tries to slide the whole rug as one piece, the resistance is too much. What one can do is to make a fold in the rug and then slide the entire rug a little at a time by pushing the fold along. A similar analogy to the fold in the rug is the movement of an earthworm. It moves in a direction by advancing a part of its body at a time.

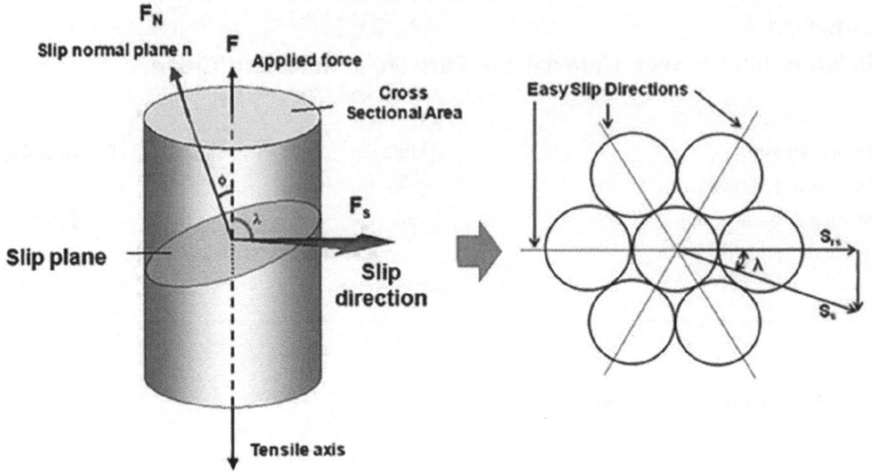

FIGURE 7.11 Components of force on a slip plane.

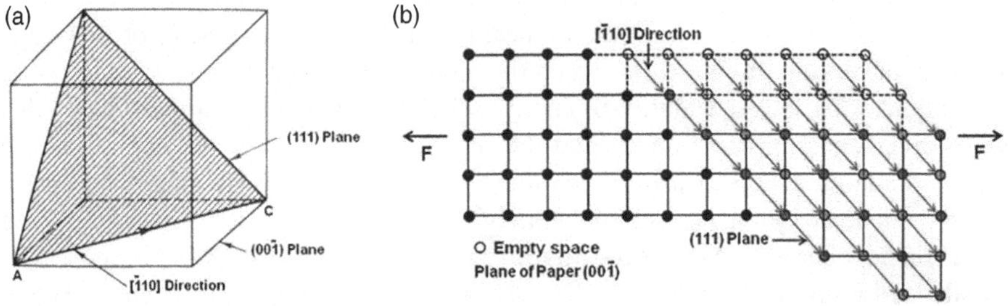

FIGURE 7.12 Schematic diagram of slip plane in FCC lattice: (a) slip plane in FCC lattice and (b) its schematic diagram.

By implementation of the shear force, first an additional plane of atoms (called a dislocation) is created above the slip plane. Thus, the bond between atoms breaks, which leads to the production of a new bond between atoms and a dislocation. On continued application of force, this dislocation proceeds by breaking old bonds and making new bonds. In the next move, as the bond between atoms is broken and a new bond is made between atoms repeatedly, resulting in a dislocation. As a result, this dislocation moves across the slip plane and leaves a step when it shows up at the surface of the crystal. Each time the dislocation moves across the slip plane, the crystal moves one atom spacing (Figure 7.12b).

7.4.3 Stress Concentration at Crack Tip

Although most mechanical products are designed such that the nominal stress remains elastic ($S_n < \sigma_{ys}$), stress concentrations in a mechanical system often give rise to plastic strains to occur in the neighborhood of design faults or stress raisers such as holes, grooves, notches, and fillets where the stress is increased.

The fracture strength of a material is connected with the cohesive forces between atoms. One can approximate that the theoretical cohesive strength of a material should be one-tenth of the elastic modulus (E). However, the exploratory fracture strength for a brittle material is usually $E/100$–$E/10{,}000$ below this conceptual value. This much lower fracture strength comes from the stress concentration due to the existence of microscopic defects or cracks found on the exterior of the material. Stress profile along the x-axis is concentrated on an elliptically shaped inner-crack (Figure 7.13).

Stress has a maximum value at the crack tip, and it decreases to the nominal applied stress with increasing distance away from the crack. Defects such stress raisers or stress concentrators have the ability to enlarge the stress at a designated point. The magnitude of amplification relies on crack orientation and geometry.

Inglis's solution (1913) not only used elliptical coordinates to solve the elliptical hole problem, but it also utilized complex numbers [3]. He derived the results that the confined stresses around a corner or hole in a stressed plate could be many times bigger than the mean acted stress. The maximum stress at the ending of the ellipse is related to its size and shape by

$$\sigma_{\max} = \sigma_{\infty}\left[1 + 2\frac{a}{b}\right] \tag{7.17}$$

It is obvious that Inglis's elliptical result lessens to the known $\sigma_{\max} = 3\sigma_{\infty}$ for the particular case of a hole when $a = b$. That is, a familiar result might be seen in most strength-of-materials books. On the other hand, the max stress is forecasted to approach to infinity as the ellipse makes flat to generate a crack ($b \rightarrow 0$).

The radius of curvature, ρ, at the tip of an ellipse is related to its length and width by

$$\rho = \frac{b^2}{a} \tag{7.18}$$

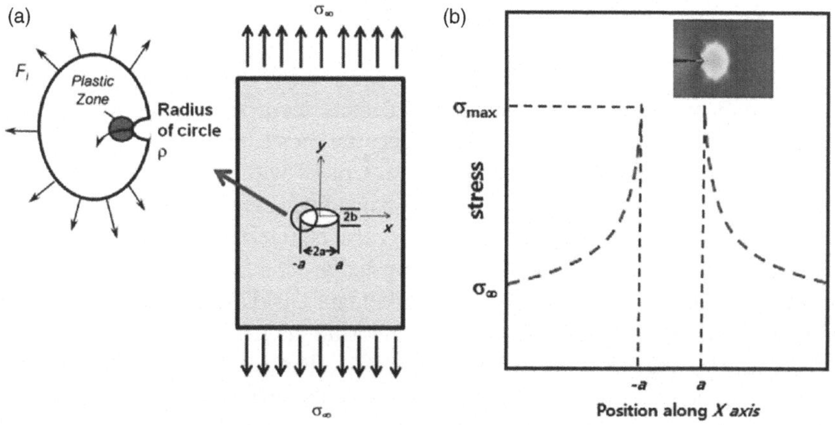

FIGURE 7.13 Stress concentration at crack tip positions: (a) geometry of internal cracks and (b) schematic stress profile along x-axis at crack tip.

As solving this for b and substituting into the a/b ratio in Equation (7.17), it can be estimated by Equation (7.19):

$$\sigma_{\max} = \sigma_{\infty}\left[1 + 2\sqrt{\frac{a}{\rho}}\right] \tag{7.19}$$

where ρ is the radius of curvature, σ_{∞} is the applied stress, $\sigma_{\max}$ is the stress at crack tip, and a is the half-length of internal crack or the full length for a surface flaw.

If the crack is similar to an elliptical hole through plate and is oriented perpendicular to applied stress, the maximum stress $\sigma_{\max}$ happens at a crack tip. The quantity of the theoretical acted tensile stress is σ_{∞}; the radius of the curvature of the crack tip is ρ; and a represents the length of a surface crack, or half-length of an internal crack. For a comparatively lengthy micro-crack, the factor $(a/\rho)^{1/2}$ may be very big. So, Equation (7.19) can be restated as follows:

$$\sigma_{\max} \cong 2\sigma_{\infty}\left(\frac{a}{\rho}\right)^{1/2} \tag{7.20}$$

The ratio between the maximum stress and the nominal acted tensile stress is defined as the stress concentration factor K_t. The stress concentration factor is a straightforward measure of the degree to which an external stress is expanded at the tip of a tiny crack and defined as follows:

$$K_t = \frac{\sigma_{\max}}{\sigma_o} \approx 2\left(\frac{a}{\rho}\right)^{1/2} \tag{7.21}$$

Because an external stress is expanded at the tip of a crack, Equation (7.21) can be restated as follows:

$$\sigma_{\max} = 2\sigma_{\infty}\left(\frac{a}{\rho}\right)^{1/2} = K_t\sigma_{\infty} \tag{7.22}$$

Stress amplification not only happens at tiny defects or cracks on a microscopic range of material but can also occur in stress concentration such as sharp corners, fillets, holes, and notches on the macroscopic level. Cracks with sharp tips spread easier than cracks having blunt tips. Because of an expanding acted stress, stress concentration may happen at microscopic flaws, sharp corners, internal discontinuities (voids/inclusions), notches, and scratches that are usually defined as stress raisers. Stress raisers are usually more devastating in brittle materials. Ductile materials have the ability to plastically deform in the area neighboring the stress raisers, which in turn evenly disperses the stress load around the defect. The maximum stress concentration factor results in a value less than that found for the theoretical value. Since brittle materials cannot plastically deform, the stress raisers will produce the conceptual stress concentration circumstances.

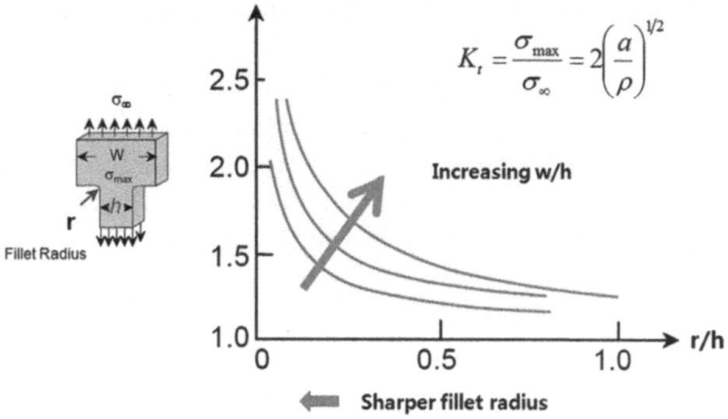

FIGURE 7.14 Stress concentration at sharp corners in accordance with fillet radius [4].

The quantity of this amplification relies on micro-crack positioning, dimensions, and geometry. For instance, stress concentration at sharp-edged corners relies on fillet radius (Figure 7.14).

7.4.4 Fracture Toughness and Crack Propagation

Cracks with sharp-edged tips grow and propagate easier than cracks possessing blunt-edged tips. In ductile materials, plastic deformation at a crack tip 'blunt' evolves to the crack. Elastic strain energy is stored in a material as it is elastically deformed. This energy is released when the crack grows. And formation of new surfaces needs energy.

As a crack has grown into a solid to a depth a, an area of material adjoining the free surfaces is unloaded and its strain energy is released. A straightforward way of seeing this energy release is to consider two triangular regions near the crack flanks, of width a and height πa, as being totally unloaded, while the remaining material continues to feel the full stress σ. The whole strain energy U released is then the strain energy per unit volume in both triangular regions:

$$U^* = -\frac{\sigma^2}{2E} \cdot \pi a^2 \tag{7.23}$$

At this time, the area normal to the x-y plane is taken to be unity, so U is the strain energy released per unit thickness of specimen. This strain energy is released by crack growth. But in shaping the crack, bonds must be destroyed, and the necessary bond energy is in effect absorbed by the material. The surface energy S related with a crack of length a (and unit depth) is

$$S = 2\gamma a \tag{7.24}$$

where γ is the surface energy and the factor 2 is needed since two free surfaces have been formed.

The whole energy related to the crack is then the sum of the (positive) energy absorbed to produce the new surfaces, plus the (negative) strain energy released by allowing the regions near the crack flanks to be unloaded.

As the crack propagates, the strain energy depend on the surface energy. Beyond a critical crack length a_c, the system can become its low energy by letting the crack grow still longer. Up to the point where $a = a_c$, the crack will grow only if the stress increases. Beyond that point, crack growth is unforced and catastrophic (Figure 7.15).

The quantity of the crucial crack length can be obtained by setting the derivative of the whole energy $S + U^*$ to zero:

$$\frac{\partial\left(S + U^*\right)}{\partial a} = 2\gamma - \frac{\sigma_f^2}{E} \cdot \pi a = 0 \tag{7.25}$$

Since fast fracture is approaching when this condition is satisfied, we can solve Equation (7.25). The required crucial stress for crack propagation is expressed as follows:

$$\sigma_c = \left(\frac{2E\gamma_s}{\pi a}\right)^{1/2} \tag{7.26}$$

where γ_s is the specific surface energy.

When the tensile stress at the tip of crack surpasses the crucial stress value, the crack grows and results in fracture. Most metals and polymers have plastic deformation. For ductile materials, specific surface energy γ_s should be substituted into $\gamma_s + \gamma_p$, where γ_p is the plastic deformation energy. So, Equation (7.26) can be expressed as follows:

$$\sigma_c = \left(\frac{2E\left(\gamma_s + \gamma_p\right)}{\pi a}\right)^{1/2} \tag{7.27}$$

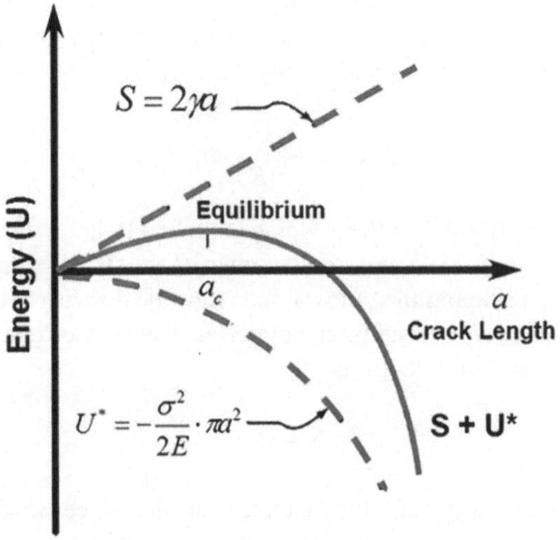

FIGURE 7.15 Total energy associated with the crack.

For highly ductile materials, $\gamma_p \gg \gamma_s$ is valid. So, Equation (7.27) can be restated as follows:

$$\sigma_c = \left(\frac{2E\gamma_p}{\pi a}\right)^{1/2} \tag{7.28}$$

Most brittle materials possess a population of tiny defects that are in a variety of sizes. When the magnitude of the tensile stress at the tip of crack surpasses the crucial stress value, the crack grows and results in fracture. Only very small and almost defect-free metallic and ceramic materials have been produced with facture strength that approximates their conceptual values.

Example 7.2

There is a long plate of glass subjected to a tensile stress of 30 MPa. If the modulus of elasticity and specific surface energy for this glass are 70 GPa and 0.4 J/m², respectively, find out the crucial length of a surface flaw that can have no fracture.

From Equation (7.18), $E = 70$ GPa, $\gamma_s = 0.4$ J/m², and $\sigma = 40$ MPa. So, the critical length can be obtained as

$$a_c = \left(\frac{2E\gamma_s}{\pi\sigma^2}\right) = \left(\frac{2 \cdot 70\text{ GPa} \cdot 0.4\text{ J/m}^2}{\pi \cdot (30\text{ MPa})^2}\right) = 2.0 \times 10^{-6}\text{ m}$$

Because of the catastrophic nature of fracture, much attempt has been expended to figure out fracture mechanics. From a collaboration of basic and empirical causes, brittle fracture will happen when the fracture toughness (K_c) of a material is surpassed. That is, fracture toughness K_c is a material's resistance to fracture when a crack exists. It, therefore, denotes the amount of stress required to grow a flaw. It can be expressed as follows:

$$K_c = Y(a/w)\sigma_c\sqrt{\pi a} \tag{7.29}$$

Here, $Y(a/w)$ is a geometrical factor that relies on the crack dimensions, where a is the crack length, w is the sample thickness, σ_c is the crucial stress for the crack to grow, and a is again the length of a surface crack of half the length of an interior crack.

If $a \to 0$ or $w \to \infty$, then $Y \to 1$. As the sample thickness increases, the fracture toughness decreases until the plane strain region is obtained. Fracture toughness depends on temperature, strain rate, and microstructure. Its magnitude diminishes with increasing strain rate and decreasing temperature. If yield strength is made better by alloying and strain hardening, fracture toughness will increase with reduction in grain size.

7.4.5 Crack Growth Rates

The metal fatigue starts at a surface (or an internal) flaw by concentrated stresses, and progression at the start of shear flow along slip planes. Slip can occur (111) plane in a FCC lattice because the atoms are most compactly packed. Over a number of (random) loading cycles in field, this slip produces intrusions and extrusions that start to resemble a crack. A true crack running inward from an intrusion region may propagate initially along one of the first slip planes, but in the end turns to move transversely to the principal normal stress.

After repeated loadings, the slip bands can grow into tiny shear-driven micro-cracks. These Stage I cracks can be explained as a back and forth slip on a series of adjacent crystallographic planes to produce a band. In this slip bands the pores nucleation and coalescence happen. The process eventually leads to micro-cracks formation. Often, extrusion and intrusions may also appear which, being a much localized discontinuity, results in a much faster micro-crack formation.

Micro-cracks join to form a macro-crack in Stage II of fatigue. Now, the crack is already long enough to escape shearing stress control and be driven by normal stress which produces a continuous growth, cycle by cycle, on a plane that is no longer crystallographic, but rather normal to external loads. Ahead of this macro-crack, two plastic lobes are generated by stress concentration. The cracks grow perpendicular to the dominant stress and increases dramatically by plastic stresses at the crack tip as seen in Figure 7.16.

It is crucial that engineers be able to predict the rate of crack growth during load cycling in the engineering structures, so that the problematic parts can be modified before the crack reaches a critical length. A great deal of experimental verification assists the view that the crack growth rate can be corrected with the cycle variation in the stress intensity factor [5]:

$$\frac{da}{dN} = A\Delta K^m \tag{7.30}$$

where da/dN is the fatigue crack rate per cycle, $\Delta K = K_{min} - K_{max}$ is the stress intensity factor range during the cycle, and A and m are parameters that depend on the material, environment, frequency, temperature, and stress ratio.

The rate of fatigue crack propagation during Stage II depends on stress level, crack size, and materials. This is sometimes known as the 'Paris law', which leads to plots similar to that shown in Figure 7.17.

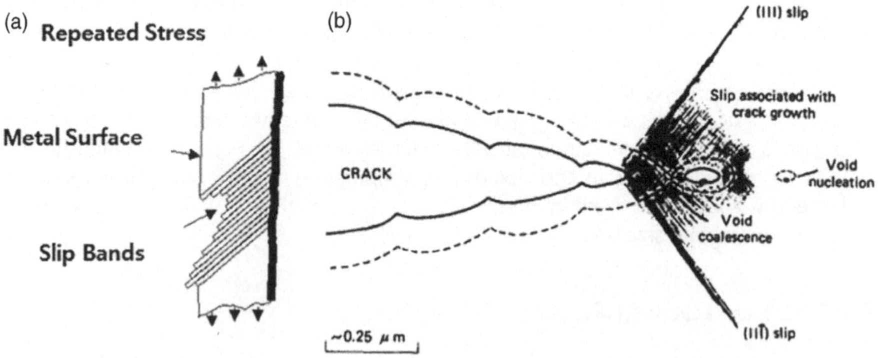

FIGURE 7.16 A schematic diagram of general slip producing nucleation and growth of voids: (a) fatigue due to repeated loads and (b) nucleation and growth of voids.

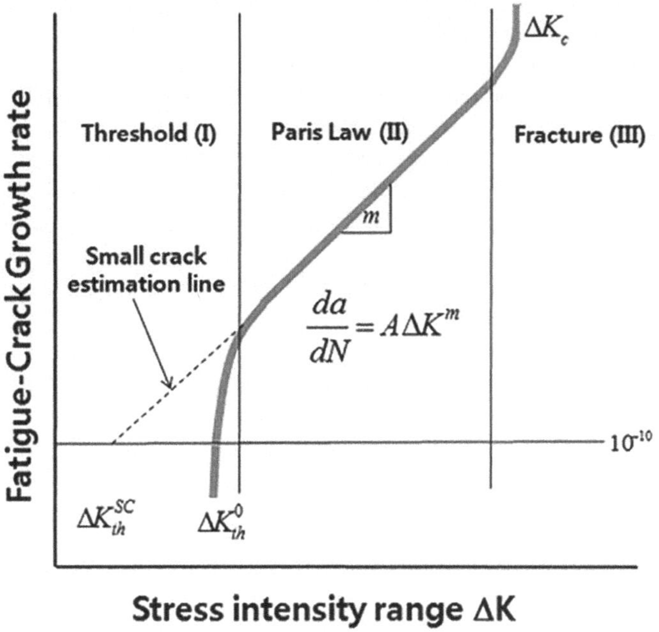

FIGURE 7.17 Paris law for fatigue crack growth rates.

TABLE 7.2
Numerical Parameters in the Paris Equation

Alloy	*m*	*A*
Steel	3	10^{-11}
Aluminum	3	10^{-12}
Nickel	3.3	4×10^{-12}
Titanium	5	10^{-11}

Some specific values of constants m and A for various alloys are given in Table 7.2. The exponent m is often near 4 for metallic systems, which might be rationalized as the damage accumulation being related to the volume V_p of the plastic zone: since the volume V_p of the zone scales with r_p^2 and $r_p \propto K_I^2$, then $da/dn \propto \Delta K^4$.

7.5 FATIGUE FAILURE

7.5.1 Introduction

As previously mentioned, approximately 90% of all structural failures happen through a fatigue mechanism. Fatigue occurs after a member is subjected to repeated cyclic loadings and repetitive deformations. It depends on the shape, material, and how close to the elastic limit it deforms. It also represents itself in the formation of cracks, which are developing at specific locations in the diverse types of mechanical structures that might include mechanisms such as airplanes, boats, cranes, overhead cranes, turbines, machine parts, reactor vessels, bridges, offshore platforms, canal lock doors, transmission towers, chimneys, and masts.

The crack tip travels a very small distance in each loading cycle, if supplied that the stress is high enough, but not too high to give rise to unexpected comprehensive fracture. It is not nearly possible to discover any growing variations in material behavior during the fatigue process, so brittle failures usually happen without caution. Rest time, with the fatigue stress eliminated, does not cause any noticeable recovery or healing. With the bare eye, we can see a 'clam shell' structure in the crack plane. Under a microscope, 'striations' can be seen, which mark the locations of the crack tip after each individual loading cycle. Because crack growth is very small in each single load cycle, a big number of cycles are required before the total failure happens. The fatigue failure was earliest discovered in the nineteenth century by watching the destitute service life of railroad axles designed based on static design limits (Figure 7.18).

The fatigue life of a structure subjected to repetitive cyclic loadings is expressed as the number of stress cycles it can rise before failure. The physical result of a repetitive load on a material is dissimilar from that of a static load. Failure every time may be brittle fracture regardless of whether the material is ductile or brittle. Mainly fatigue failure happens at stress well below the static elastic strength of the material. Depending upon the structural particular geometry, its fabrication or the material used, four crucial parameters can affect the fatigue strength: (1) the stress difference called stress range, (2) the material, (3) the structural geometry, and (4) the environment.

The fatigue failure in ductile metals represents itself in the formation of cracks having high stress concentration, such as holes and grooves, and propagates it [6,7].

FIGURE 7.18 Facture of train wreck due to metal fatigue failure of rail (Wikipedia).

Fatigue stresses in mechanical product take the form of a sinusoidal pattern. In periodic patterns, the peaks on both the low side (minimum) and the high side (maximum) are crucial. The modified Goodman diagram describes the fatigue of part with alternating stress on the *y*-axis and average stress on the *x*-axis. Thus, it can be represented as follows:

$$\frac{S_a}{S_e} + \frac{S_m}{S_{ut}} = 1 \tag{7.31}$$

where S_a is the alternating stress, S_m is the mean stress, S_e is the fatigue limit for completely reversed loading, and S_{ut} is the ultimate tensile strength of the material.

The cycles to failure are represented as a function of mean stress and range along the lines of constant *R*-values [8]. The modified Goodman relation defines a failure envelope such that any alternating stress that drops in the diagram will not give rise to failure (Figure 7.19).

However, fatigue failure criteria including the representative Goodman diagram [10–13] are hard to estimate the lifetime cycles of multi-module products because small samples of part, not module, are tested and part failures due to design faults rarely occur in field. These methodologies fail to reproduce the field failure and have been controversial.

If there are design flaws where a product is subjected to repeated loads, the product will fail in its lifetime. In other words, we can say that fatigue failure occurs. Fatigue is the enfeebling of a material caused by cyclic loading. Especially, low-cycle fatigue (LCF) regime is distinguished by high cyclic stress levels in excess of the endurance limit of the material and is usually comprehended to be in the area of 10^4–10^5 cycles. A great deal of attention is currently being given to the LCF performance of super-alloys, especially in the area of turbine-engine designs [14–18].

Lots of papers have been published concerning the LCF of nickel-base polycrystalline [19–21] and mono-crystalline alloys [22]. There are a variety of fatigue tests on different patterns of samples according to stated fatigue testing standards [23–25]. However, it is difficult to obtain experimental data of multi-module product available

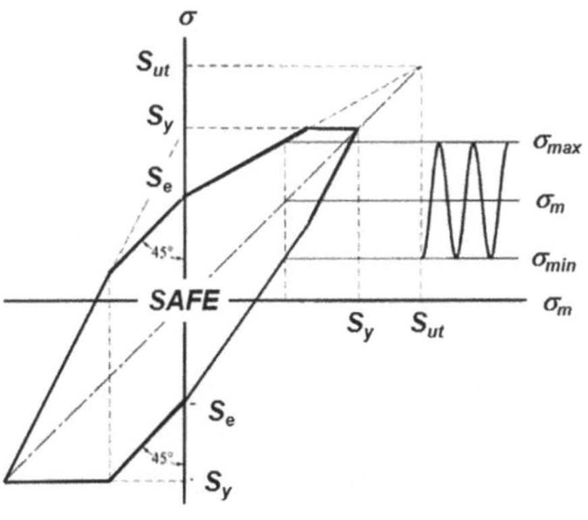

FIGURE 7.19 The 'complete' modified Goodman diagram [9].

to straightly build S–N curve for LCF because an extensive testing would have been too costly due to time and sample size.

In Chapter 8, we will discuss a methodology for reliability tests for improving the LCF failure of a mechanical system subjected to repetitive loading. It includes (1) a parametric ALT plan based on product BX lifetime, (2) a load analysis for accelerated lifetime test, (3) a tailored sample of parametric ALT with the design changes, and (4) an assessment of whether the last designs of the mechanical system achieve the target BX lifetime.

7.5.2 Fluctuating Load

Most of the mechanical structures – especially, mechanism – are subjected to repeatedly variable, fluctuating loading. This type of repeated loading is very threatening not only because limiting stresses are greatly lower than those accepted for static loading, but because of the essence of fatigue failure, products fail abruptly without any signs of yielding in their lifetime. There are three dissimilar fluctuating stress–time patterns in fatigue loading: (1) symmetrical about zero stress, (2) asymmetrical about zero stress, and (3) random stress cycle. For asymmetrical about zero stress, a fluctuating load (stress) that results in fatigue is distinguished by mean stress σ_m, the range of stress $\Delta\sigma$, the amplitude stress σ_a, ratio of the mean stress over the amplitude stress χ, and the stress ratio R (Figure 7.20).

The fatigue cycle begins when the nominal stress is at point 'a'. As the stress increases in tension through point 'b', the crack tip is open to give rise to confined plastic deformation, while it expands into the virginal metal at point 'c'. As the tensile stress just now decreases through 'd', the crack tip closed and the lasting plastic deformation generates a distinguishing saw tooth profile familiar as a 'striation'. On accomplishment of the cycle at point 'e', the crack has now proceeded through length Δa, and has formed an extra striation.

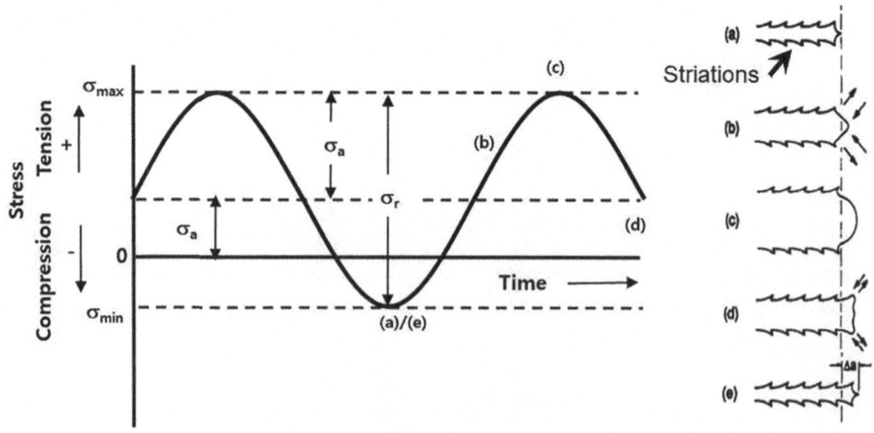

FIGURE 7.20 Fatigue with striations under fluctuating and cyclic stresses asymmetrical about zero stress.

As seen in Figures 7.21 and 7.22, additional cyclic stresses that result in fatigue are (1) periodic and symmetrical about zero stress and (2) random stress fluctuations. In mechanical/civil systems such as aircraft, automobiles, refrigerators, machine components, and bridges, fatigue failures under fluctuating/cyclic stresses are required:

- A maximum tensile stress of sufficiently high value
- A large enough variation or fluctuation in the applied stress
- A sufficiently large number of applied stress cycles.

They are represented as the following equations:

$$\sigma_m = \frac{(\sigma_{\max} + \sigma_{\min})}{2} \tag{7.32}$$

$$\Delta\sigma = (\sigma_{\max} - \sigma_{\min}) \tag{7.33}$$

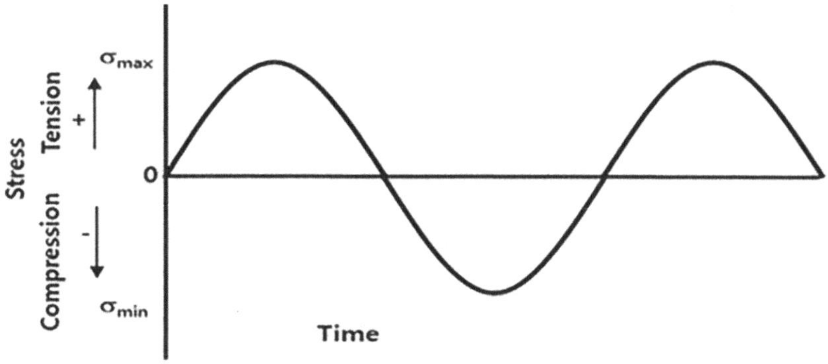

FIGURE 7.21 Periodic and symmetrical about zero stress in fatigue.

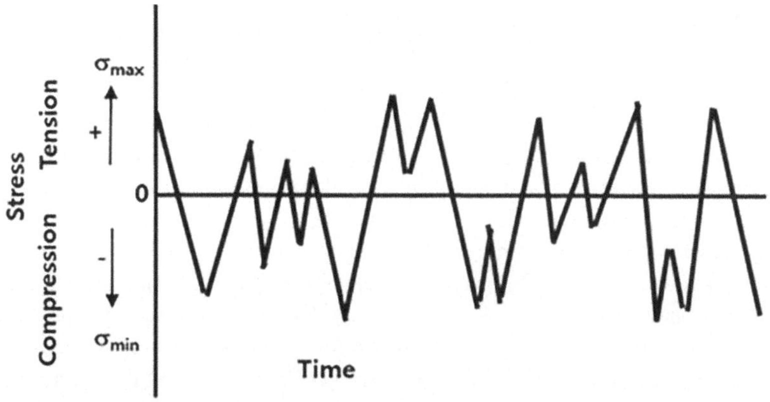

FIGURE 7.22 Random stress fluctuations in fatigue.

$$\sigma_a = \frac{(\sigma_{\max} - \sigma_{\min})}{2} \tag{7.34}$$

$$\chi = \frac{\sigma_m}{\sigma_a} \tag{7.35}$$

$$R = \sigma_{\min}/\sigma_{\max} \tag{7.36}$$

There are two typical load patterns: one with $\chi = 1$, the load is pulsing, and the other with $\chi = 0$, the load is reversed. The latter one is the most menacing shape of load variation. The stress bound for fluctuating loading (the endurance limit) is expressed as a value of stress that is secure for a given specimen regardless of the number of load repetitions. It stays usually in a compact relationship to the ultimate stress limit.

There are two methods to decide when a part is in danger of metal fatigue: by predicting when a failure will happen due to the shape/iteration/material/force mix and substituting the unsafe materials before this happens, or carrying out examinations to identify the microscopic cracks and performing replacements once they happen. Choosing materials that are not likely to be hurt by metal fatigue during the life of the product is the best solution, but it is not always possible. Avoiding shapes with sharp corners restricts metal fatigue by lessening stress concentrations, but it does not eradicate it.

Fatigue in product designs is a crucial failure mechanism to be contemplated. It is a process in which damage cumulates due to the repeated loads below the yield point, which is brittle-like even in usually ductile materials. Fatigue cracks start at very small and initially grow very slowly until the crack length approaches the crucial length. So, it is dangerous because it is difficult to originally identify accumulative fatigue damage with the naked eye until the crack has grown to near-crucial length. General fracture surface is perpendicular to the direction of acted stress. Fatigue failure has three distinct phases: (1) crack starting in the regions of stress concentration (near stress raisers), (2) incremental crack propagation, and (3) last catastrophic failure.

An instance of 'fatigue' for a multitude of causes has been studied as the Versailles rail accident, and the disaster of Comet aircraft happened when they became big enough to propagate catastrophically (see Chapter 2). A fatigue failure occurs in both metallic and non-metallic materials, and it is accountable for approximately estimated 80%–90% of all structural failures – aircraft landing, motor shaft, gear machine parts, automobile crank-shaft, bridges, etc. Consequently, designing for a maximum stress will not make sure proper product lifetime. Most fracture produced associates to this classification.

Engineering stress is asymmetrical around stress raisers such as notches, fillets, or holes that concentrate on the stress. In compound drawings, engineers often omit these design flaws that may cause the reliability disasters. For example, the vibration of aircraft wing during a long flight can result in tens of thousands of load cycles. If not properly designed, these structures will collapse. It is significant to find the design faults and correct them. In Chapter 8, we will deal with how to search the design faults by using the parametric ALT.

The main difficulty in designing against fracture in high-strength materials is that the pre-existence of cracks can modify the local stresses to such an extent that the elastic stress analyses done so carefully by the designers are inadequate. When a crack extends to a definite crucial length, it can propagate catastrophically through the structure, even though the gross stress is much less than would usually cause yield or failure in a tensile specimen. The term 'fracture mechanics' refers to an essential specialization in solid mechanics in which the existence of a crack is presumed, and we wish to search out quantitative relationships between the crack length, the material's intrinsic resistance to crack growth, and the stress at which the crack propagates at high speed to give rise to a structural failure.

A fast fracture can happen within a few loading cycles. For instance, fatigue failures in 1,200 rpm motor shafts took less than 12 hours from installation to last fracture, about 830,000 cycles. On the other hand, crack growth in steadily rotating process equipment shafts has taken many months and more than 10,000,000 cycles to fail.

7.6 FACTURE FAILURE

7.6.1 Introduction

Fracture is the breakdown of a body into pieces when subjected to a (repetitive) stress. It occurs whenever the acted loads (or stresses) are more than the resisting strength of the body. It begins with a crack that breaks with no making completely apart. Fracture due to overstress is probably the most widespread failure mechanism in mechanical/civil system and might be classified as brittle (cleavage) fracture, ductile (shear) fracture, fatigue fracture, de-adhesion, and crazing.

When a crack propagates, a new free surface is produced, possessing a specific surface energy γ. This energy is supplied by the exterior load and is also obtainable as accumulated elastic energy. Not all available energy, however, is utilized for the production of new crack surfaces. It is also transformed into other energies, like dissipative heat or kinetic energy. When lots of available energy is utilized for crack growth, the fracture is said to be brittle. When lots of energy is transformed into other energies, mostly due to dissipative mechanisms, the fracture is designated to be ductile.

As seen in Figure 7.23, brittle fracture propagates quickly on a crack with minimum energy absorption and plastic deformation. Brittle fracture happens along characteristic crystallographic planes defined as cleavage planes. The mechanism

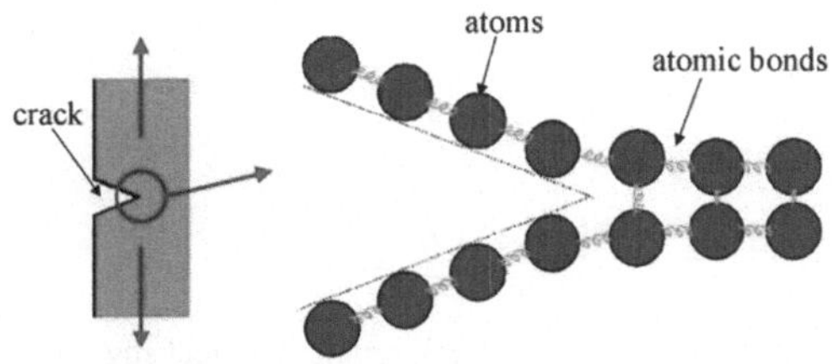

FIGURE 7.23 Brittle (cleavage) fracture mechanism and its example (glass).

of brittle fracture was originally explained by the Griffith theory [26]. Griffith suggested that there are micro-cracks in a brittle material which act to the concentrated stress at their tips. The crack could come from a number of sources as flow happened during a surface scratch or solidification. When plastic deformation at the crack tip is forbidden, the crack can travel through grains by breaking atom bonds in lattice planes. This cleavage fracture will exist in materials with little or no close-packed planes, having BCC or HCP crystal structure.

Brittle materials are ceramics, glasses, metals, and some polymers. They have the following attributes:

- No appreciable plastic deformation so that crack propagation is very fast.
- Crack propagates almost perpendicular to the direction of the acted stress.
- Crack frequently propagates by cleavage – breaking atomic bonds along specific crystallographic planes (cleavage planes).

For ductile fracture, as a crystalline material is filled, dislocations will start to move through the lattice due to local shear stresses. Also lots dislocations will increase. Because the interior structure is changed irreversibly, the macroscopic deformation is permanent (plastic). The dislocations will unite at grain boundaries and accumulate to make a void. These voids will grow, and one or more of them will move in a macroscopic crack.

Because the start and growth of cracks are aroused by shear stresses, this mechanism is mentioned as shearing. Plastic deformation is crucial, so this mechanism will usually be seen in FCC crystals, which have lots of close-packed planes. The fracture surface has a 'dough-like' structure with dimples, the shape of which designates the loading of the crack (Figures 7.24 and 7.25).

- Brittle fracture: Separation along crystallographic planes due to breakdown of atomic bonds (V-shaped chevron, cleavage, inter-granular)
- Ductile fracture: Initiation, growth, and coalescence of micro-voids (cup-and-cone, dimple).

However, material in sale is composed of polycrystalline, whose crystal axes are situated at random. When a polycrystalline material is subjected to stress, slip begins first in those grains in which the slip system is most positively situated with respect to

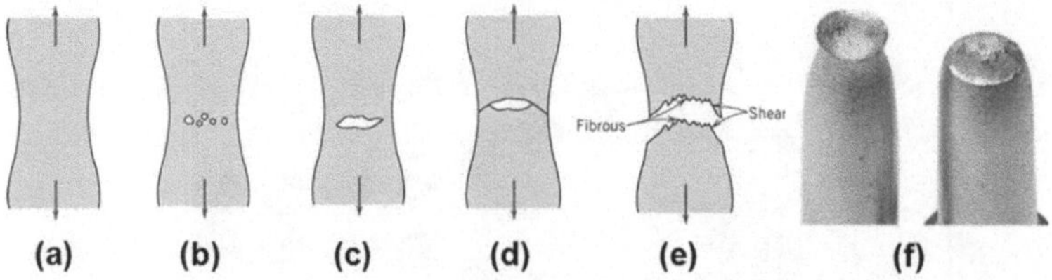

FIGURE 7.24 Ductile fracture failure mechanism: (a) necking, (b) formation of micro-voids, (c) coalescence of micro-voids, (d) crack propagation by shear, (e) fracture, and (f) cup and cone.

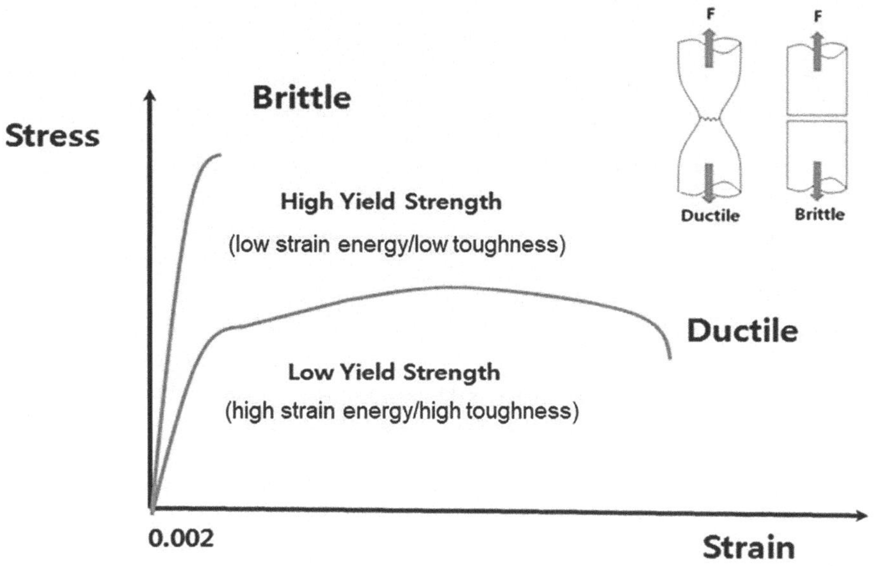

FIGURE 7.25 Brittle versus ductile fracture in material.

the acted stress. Since contact at the grain boundaries is continued, it may be requisite for more than one slip system to run. The rotation into the axis of tension causes other grains, initially less positively situated, into a position where they can now deform. As deformation and rotation proceed, the individual grains tend to elongate in the direction of flow.

When a crystal deforms, there is some distortion of the lattice structure. This deformation is greatest on the slip planes and grain boundaries, and increases with deformation. This is obvious by an increase in resistance to further deformation. The material is going through strain hardening or work hardening. Since dislocations pile up at grain boundaries, metals can be hardened by lessening the size of the grains.

7.6.2 Ductile–Brittle Transition Temperature (DBTT)

The ductile-to-brittle transition temperature (DBTT) is widely noticed in metals that are dependent on its composition. For some steels, the transition temperature can be around 0°C, and in winter, the temperature in some areas of the world might be below this. Consequently, some steel structures are very likely to fail in winter. That is, at high temperatures, where the impact energy absorbed before failure is big, ductile failure occurs with extensive elastic and plastic deformation before failure. At low temperatures, where the impact energy absorbed before failure is low, brittle fracture happens with little deformation before failure.

The governing mechanism of this transition still remains uncertain despite the lots of efforts made in theoretical and experimental research. All ferrous materials (except the austenitic grades) show a transition from ductile to brittle when tested above and below a definite temperature, defined as DBTT. FCC metals such as Cu and Ni remain ductile down to very low temperatures. For ceramics, this type of transition happens at much higher temperatures than for metals (Figure 7.26).

Steels were utilized possessing DBTTs just below room temperature. At low temperatures, steels may become seriously brittle. At higher temperatures, the impact energy is big, being consistent with a ductile mode of fracture. As the temperature lowers, the impact energy drops abruptly over a comparatively narrow temperature range, which is consistent with the mode of brittle fracture. Fatigue cracks form a nucleus at the corners of square hatches and propagate rapidly by brittle fracture.

As the well-known weld fractures in some US army ships (Liberty Ships, tankers) during World War II are explored, the ductile-to-brittle transition can be computed by impact testing such as Charpy V-notch testing (Figure 7.27). Although the tensile stress–strain curve already supplies a sign for ductile/brittle failure, the archetype experiment to look over this is the Charpy V-notch test. The principal benefit of this test is that it supplies a straightforward measure for the dissipated energy during fast crack propagation.

The sample is a beam with a 2mm deep V-shaped notch, which possesses a 90° angle and a 0.25mm root radius. It is carried and loaded as in a three-point bending test. The load is supplied by the impact of a weight at the end of a pendulum. A crack will begin at the tip of the V-notch and run through the sample. The material deforms at a strain rate of generally 103 s^{-1}. The energy that is dissipated during fracture can be calculated easily from the height of the pendulum weight, before and after impact. The dissipated energy is the impact toughness C_v.

The impact toughness can be decided for diverse specimen temperatures *T*. For inherent brittle materials like high-strength steel, the dissipated energy will be low for all *T*. For inherent ductile materials like FCC metals, C_v will be high for all *T*. Lots of materials represent a transition from brittle to ductile fracture with increasing temperature.

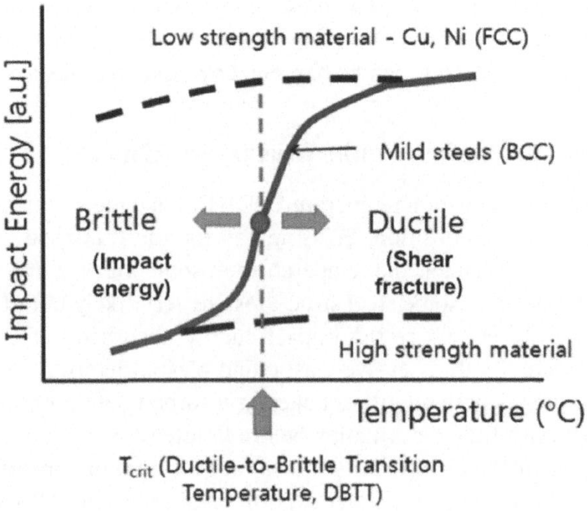

FIGURE 7.26 Ductile-to-brittle transition temperature.

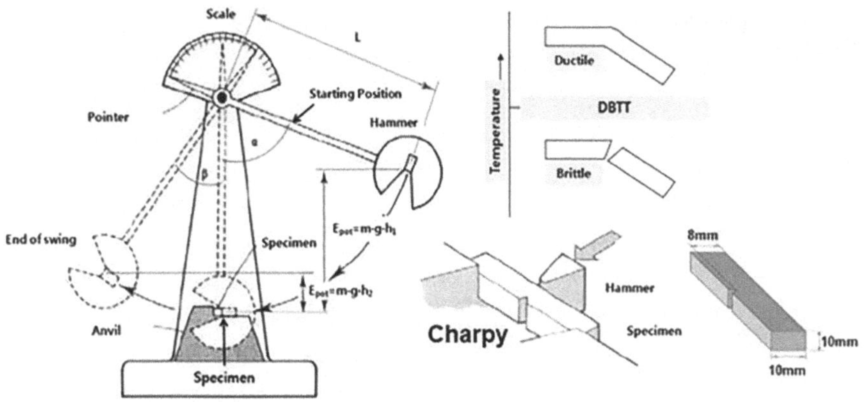

FIGURE 7.27 Schematic of a conventional Charpy V-notch testing.

7.6.3 Case Study: Fracture Faces of Products Subjected to Loads

Fatigue failure can be recognized from the sample fracture surface with different growth areas and crucial physical characteristics: (1) area of slow crack growth is generally obvious in the form of a 'clamshell' concentric around the location of the initial defect such as void, (2) clamshell region usually possesses concentric 'beach marks' at which the crack may become big enough to fulfill the energy or stress intensity criteria for quick propagation, and (3) the last phase generates the granular rough surface before the last brittle fracture.

For instance, the suction reed valves open and close to allow refrigerant to flow into the compressor during the intake cycle of the piston. Due to repeated stresses, the suction reed valves of domestic refrigerator compressors utilized in the field were cracking and fracturing, leading to the failure of refrigerator. As judging from scanning electron microscopic (SEM) image, the fracture begins in the void of the suction reed valve, propagates, and collapses at the end (Figure 7.28).

The fracture face of a fatigue failure represents both the load types (tension, bending, torsion, or combination of these) and the quantity of the load. To figure out the pattern of load, look at the direction of crack propagation. It is always going to be perpendicular to the plane of maximum stress. The following instances reflect the fracture paths in accordance with a variety of loads.

Figure 7.29 describes the reversed torsional fatigue failure of splined shaft from a differential drive gear. The mating halves of the fracture show how two unconnected cracks started in a circumferential cavity adjoining the ends of the splines and started to propagate into the cross section succeeding helical paths. Because the cycles of twisting forces are applied oppositely, each crack follows opposing helices which progressively lessens the effective cross-sectional area and, thus, increases the quantities of cyclic stresses from the same acted loads. Immediately before the shaft lastly broke down, bending forces began a third crack at the opposite side of the shaft and this had begun to propagate as a plane fracture at 90° to the shaft axis until the splined ending lastly broke away.

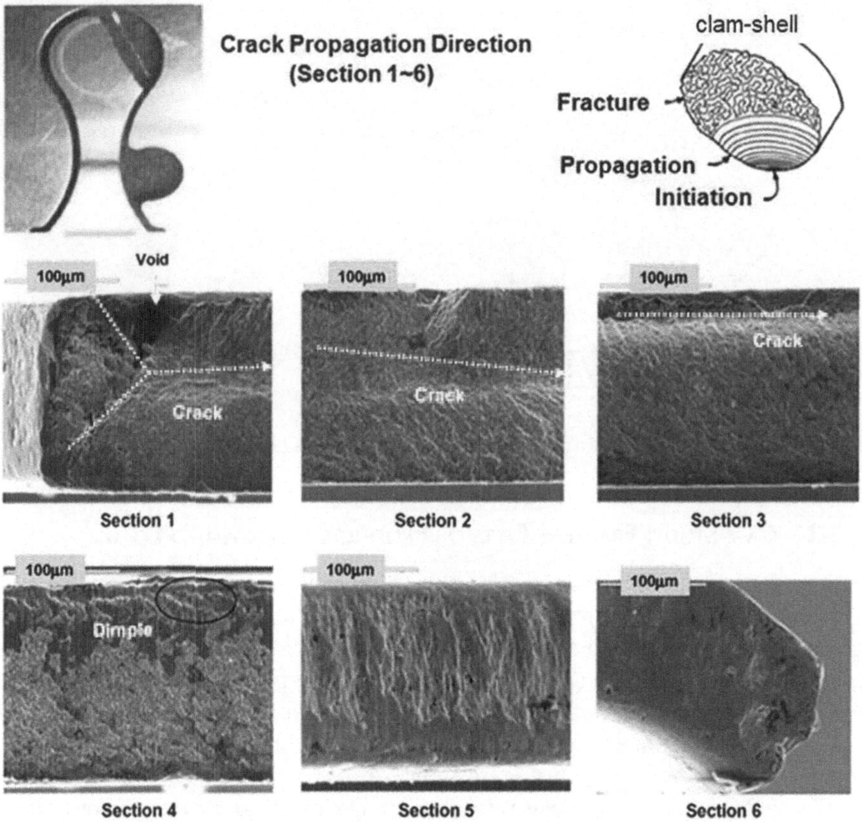

FIGURE 7.28 Fatigue fracture surface of compressor suction reed valve due to impact.

FIGURE 7.29 Fatigue failure of splined shaft due to reversed torsional loads (example).

Torsional fatigue is related to 10%–25% of rotating apparatus failures. Torsion fatigue failures can find them as the fracture situated 45° to the shaft centerline. The fracture look generally to have one or more originations, a fatigue zone with progression lines and an immediate zone. A big fatigue zone and a small immediate zone mean that the fatigue load is small. A small fatigue zone and large immediate zone mean that the fatigue load is high.

Torsional fatigue fractures occasionally happen inside a hub or coupling of a shaft. These fractures usually start at the bottom of a keyway and progress around the shaft's circumference. The fracture travels around the shaft, climbing toward the surface, so the outer part of the shaft looks like it was peeled away. The fracture surface has characteristics of a fatigue fracture: one or more originations, ratchet marks, and a fatigue zone with progression lines. The shaft fragment is generally held in place by the coupling or hub, so there is typically a very small or no instantaneous zone.

A shaft fracture may have both torsion and bending fatigue forces. When this happens, the positioning of the fracture face may vary from 45° to 90° with respect to the shaft centerline. As the fracture is closer to 90°, the shaft combines ruling bending with torsion. The fracture angle, therefore, supplies crucial evidence as follows:

- Closer to 90°, it is a major bending force.
- Midway between 45° and 90°, it is a combination of torsion and bending forces.
- Closer to 45°, it is a dominant torsion force.

Verification of torsional fatigue may also be found on gear and coupling teeth. Most apparatus operates in one direction, so wear is expected on one side of a gear or coupling teeth. Wear on both sides of a gear or coupling teeth that rotates in one direction is a sign of changing torsional force. When coupling alignment is good and wear happens uniformly on both sides of all coupling teeth, it usually indicates torsional vibration. Alignment quality can be verified from vibration spectra and phase readings. The absence of 2X running speed spectral peaks and uniform phase across the coupling occurs when the alignment is good.

7.6.4 Stress–Strength Analysis and Its Limitation

Stress–strength interference analysis in reliability engineering is the analysis of the strength of the materials and the interference of the stresses placed on the materials. It is competent to figure out the failure of mechanical parts in airplanes, automobiles, refrigerators, and the others when there are no major design defects. A mechanical product's probability of failure is identical to the probability that the stress underwent by that product will surpass its strength. If given one probability distribution function for a product's stress and strength, the probability of failure can be estimated by calculating the area of the overlap between the two distributions. This partly covering region may also be mentioned as stress–strength interference (Figure 7.30).

To express the reliability function from the standpoint of load–strength interference, the strength of a product is modeled as a random variable S. The product is exposed to a load L that is also modeled as a random variable. A failure will happen

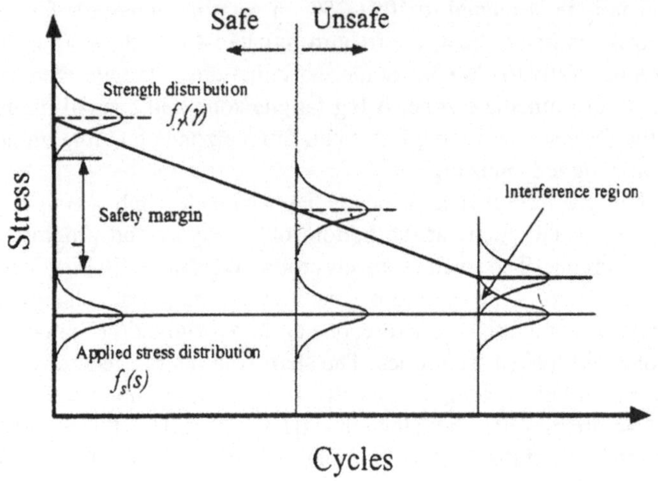

FIGURE 7.30 Applied fatigue stress–strength interference model.

as soon as the load is higher than the strength. The reliability R of the item is defined as the probability that the strength is greater than the load:

$$R(t) = P(S > L) \tag{7.37}$$

where $P(\cdot)$ is the probability.

The load will generally vary with time and may be modeled as a time-dependent variable $L(t)$. The product will deteriorate with time, due to failure mechanisms like fatigue, fracture, and corrosion. The strength of the product will, therefore, also be a function of time, $S(t)$. The time to failure T of the product is the (shortest) time until $S(t) < L(t)$:

$$T = \min\{t : S(t) \prec L(t)\} \tag{7.38}$$

If the distributions for both the stress and the strength pursuit a normal distribution, the expected probability of failure, F, can be expressed as follows:

$$F = P[\text{stress} \geq \text{strength}] = \int_0^\infty f_{\text{strength}}(x) \cdot R_{\text{stress}}(x)\,dx \tag{7.39}$$

The expected reliability, R, is calculated as follows:

$$R = P\left[\text{stress} \leq \text{strength}\right] = \int_0^\infty f_{\text{stress}}(x) \cdot R_{\text{strength}}(x)\,dx \tag{7.40}$$

There are two methods to increase reliability: (1) increase the difference (or safety margin) between the mean stress and strength values and (2) decrease the standard deviations of the distributions of stress and strength. The approximates of stresses and strengths for all parts of a product would be perfectly accurate, but this is too costly to accomplish. And the stress conditions depend on the way the product is used – the customer profiles and environmental conditions.

Environmental stresses have a distribution with an average μ_x and a standard deviation S_x, and part strengths have a distribution with a mean μ_y and a standard deviation S_y. The overlap of these distributions is the probability of failure Z. This overlap is also referred to as stress–strength interference.

If stress and strength are normally distributed random variables independent of each other, the standard normal distribution and Z tables can be used to quantitatively determine the probability of failure. First, the Z-statistic is calculated as follows:

$$Z = -\frac{\mu_x - \mu_y}{\sqrt{S_x^2 + S_y^2}} \tag{7.41}$$

Using the Z value table for a standard normal distribution, the area above the calculated Z-statistic is the probability of failure. $P(Z)$ can be determined from a Z table or a statistical software package.

The stress-strength has been utilized in mechanical component designs to find the reasons of product failures. It expresses the product failure as the probability of stress exceeding strength. However, it neglects the fact that the product failure occurs randomly from the weakest parts in the product. Because of that, it requires the concepts of life-stress model and fracture mechanics.

However, if there are design flaws such as stress raiser in a product's structure, stress–strength interference analysis is not a good expression for the root cause of reliability disasters. That is, the weakest part firstly will fracture before the product reaches to the lifetime. The strength distribution mainly depends on the product material and its shape. To improve the product reliability, strength (or stiffness) of the product in the design phase should be increased by utilizing the optimal design and reliability testing. Otherwise, if there are design faults that cause inadequacy of strength (or stiffness) when a product is subjected to load, the product suddenly fails in its lifetime. The leading method to correct the product failure is the parametric ALT that will be explained in Chapters 8 and 9.

Example 7.3

If μ_x is 2,500 kPa, μ_y is 4,500 kPa, S_x is 500 kPa, and S_y is 400 kPa, the probability of failure can be calculated as follows:

$$Z = -\frac{\mu_x - \mu_y}{\sqrt{S_x^2 + S_y^2}} = -\frac{2{,}500 - 4{,}500}{\sqrt{500^2 + 400^2}} = 2.34$$

Using the Z-value table for a standard normal distribution, the area above a Z value of 2.34 (2.34 standard deviations) is 0.0096. Therefore, the probability of failure is 0.96%. Likewise, reliability is $1 - 0.0096 = 0.9904$ or 99.04%.

REFERENCES

1. Duga, J.J., Fisher, W.H., Buxaum, R.W., Rosenfield, A.R., Buhr, A.R., Honton, E.J. and McMillan, S.C., "The economic effects of fracture in the United States," Final Report, September 30, 1982, Battelle Laboratories, Columbus, OH. Available as NBS Special Publication 647-2.
2. Wöhler, A., 1870, Über die Festigkeitsversuche mit Eisen und Stahl, *Zeitschrift für Bauwesen*, 20, 73–106.
3. Inglis, C.E., 1913, Stresses in a plate due to the presence of cracks and sharp corners, *Transactions of the Royal Institute of Naval Architectes*, 60, 219–241.
4. Neugebauer, G.H., 1943, Stress concentration factors and their effect in design, *Product Engineering*, 14, 82–87.
5. Paris, P.C., Gomez, M.P., Anderson, W.E., 1961, A rational analytic theory of fatigue, *The Trend in Engineering*, 13, 9–14.
6. Kim, W.H., Laird, C., 1978, Crack nucleation and stage I propagation in high strain fatigue—II. Mechanism, *Acta Metallurgica*, 26 (5), 789–799.
7. Woo, S., O'Neal, D., Pecht, M., 2010, Reliability design of a reciprocating compressor suction reed valve in a common refrigerator subjected to repetitive pressure loads, *Engineering Failure Analysis*, 17 (4), 979–991.
8. Sutherland, H.J., 1999, *On the fatigue analysis of wind turbines*, SAND99-0089, Albuquerque, NM: Sandia National Laboratories, p. 132.
9. Nisbett, R.G. Budynas, J.K., 2014, *Shigley's mechanical engineering design*, 10th ed., Boston, MA: McGraw-Hill Higher Education, pp. 308–324.
10. Miner, M.A., 1945, Cumulative damage in fatigue, *Journal of Applied Mechanics*, 12, 149–164.
11. Palmgren, A.G., 1924, Die Lebensdauer von Kugellagern Zeitschrift des Vereines Deutscher Ingenieure, *Scientific Research*, 68 (14), 339–341.
12. Burhan, I., Kim, H., 2018, S-N curve models for composite materials characterization—An evaluative review, *Journal of Composites Science*, 2 (3), 38–66.
13. Sutherland, H.J., Mandell, J.F., 2005, Optimized Goodman diagram for the analysis of fiberglass composites used in wind turbines blades. *43rd AIAA Aerospace Sciences Meeting and Exhibit*, Reno, Nevada.
14. Murakami, Y., Miller, K.J., 2005, What is fatigue damage? A view point from the observation of low cycle fatigue process, *International Journal of Fatigue*, 27 (8), 991–1005.
15. Qiu, B., Kan, Q., Kang, G., Yu, C., Xie X., 2019, Rate-dependent transformation ratcheting-fatigue interaction of super-elastic NiTi alloy under uniaxial and torsional loadings—Experimental observation, *International Journal of Fatigue*, 127, 470–478.
16. Sun, J., Su, A., Wang, T., Chen, W., Guo, W., 2019, Effect of laser shock processing with post-machining and deep cryogenic treatment on fatigue life of GH4169 super alloy, *International Journal of Fatigue*, 119, 261–267.
17. Kumar, N., Goel, S., Jayaganthan, R., Owolabi, G.M., 2018, The influence of metallurgical factors on low cycle fatigue behavior of ultra-fine grained 6082 Al alloy, *International Journal of Fatigue*, 110, 130–143.
18. Maier, G., Riedel, H., Somsen, C., 2013, Cyclic deformation and lifetime of Alloy617B during isothermal low cycle fatigue, *International Journal of Fatigue*, 55, 126–135.
19. Sola, J.F., Kelton, R., Meletis, E.I., Huang, H., 2019, Predicting crack initiation site in polycrystalline nickel through surface topography changes, *International Journal of Fatigue*, 124, 70–81.
20. Liu, Y., Kang, M., Wu, Y., Wang, M., Wang, J., 2018, Crack formation and microstructure-sensitive propagation in low cycle fatigue of a polycrystalline nickel-based superalloy with different heat treatments, *International Journal of Fatigue*, 108, 79–89.

21. Wahi, R.P., Auerswald, J., 1997, Damage mechanisms of single and polycrystalline nickel base superalloys SC16 and IN738LC under high temperature LCF loading, *International Journal of Fatigue*, 19 (93), 89–94.
22. Wilhelm, F., Affeldt, E., Fleischmann, E., Glatzel, U., Hammer, J., 2017, Modeling of the deformation behavior of single crystalline Nickel-based superalloys under thermal mechanical loading, *International Journal of Fatigue*, 97, 1–8.
23. American Society for Testing and Materials, 1999a, Standard practice for strain-controlled fatigue testing, In—Annual Book of ASTM Standard, Vol. 03.01," American Society for Testing and Materials, West Conshohocken, PA, no. E606.
24. American Society for Testing and Materials, 1999b, Standard practice for cycle counting in fatigue analysis, In—Annual Book of ASTM Standard, Vol. 03.01," American Society for Testing and Materials, West Conshohocken, PA, no. E1049.
25. American Welding Society, 1996, *Structural welding code—steel*, 15th ed., Miami, FL: American Welding Society, prepared by AWS Structural Welding Committee.
26. Griffith, A.A., 1921, The phenomena of rupture and flow in solids, *Philosophical Transactions of the Royal Society of London, A*, 221, 163–198.

8 Statistical Inference
Parametric Accelerated Life Testing

8.1 INTRODUCTION

Customers use modern products including mechanical systems because they are of good quality with high performance and usability, without any trouble or technological complexity. If the product price of a company is similar to that of its competitor, the product should be manufactured with good quality, as proposed by D. Garvin [1]. From the standpoint of a customer, the reliability of products is one of the acceptable attributes, regardless of both the type of products and the number of parts integrated into the products. A mechanical product is required to have a variety of functions that satisfy a customer's desires. As specifications for basic features are assigned in the developing process, a newly designed module in the product will properly have the structure including the mechanism. To satisfy these requirements, manufacturers try to seek new design plans that cost less but are of high quality. We can say that reliability can be expressed as the ability of a system or module to function under stated operational/environmental conditions for a described period of time or total elapsed distance. Products should endure the operating or environmental conditions subjected to them by the customers who purchase and use the mechanical product. Global companies, therefore, should have a sharp eye for technology to develop a new product with high performance. Simultaneously, manufacturers need a new reliability methodology to evaluate their designs.

To keep away from a product failure – especially fatigue – in the field due to design faults under customer usages [2], any design faults should be detected and corrected through statistical methodology or reliability testing before a product is released. Robust design methods, including statistical design of experiments (DOE) [3] and the Taguchi approach [4], were proposed to perform optimal design for mechanical products. DOE is a structured method to decide the relationship between factors affecting a process and its output. DOE is performed for a variety of factors that make an impact on the design (or lifetime) for new product functionality. The effectiveness of design factors can be assessed through analysis of variance. There is one missing factor in DOE. That is, if there is a design fault in the mechanical components, the design will not endure the number of repetitive loads it is subjected to in its expected lifetime. Neglecting the critical factors of failure mechanisms such as load and design, DOE is used to identify optimal designs without understanding failure mechanisms such as fatigue. Consequently, it might require substantial testing and time.

In particular, Taguchi's robust design method utilizes parametric design to avoid random "noise" from affecting the outcome. Thus, mechanical systems can be utilized to make out the proper design parameters and their levels [5–9]. By utilizing interactions between control factors and noise factors, the parametric design of a mechanical system can be used to decide the right control factors that make the design robust regardless of the change of noise factors. In an orthogonal array, the control factors are assigned to an inner array, and the noise factors are assigned to an outer array. However, because many of the design parameters for a mechanical structure need to be considered, the Taguchi product array needs enormous experimental computations, and crucial design factors would be missed. As new products are frequently introduced with missing design parameters in the mechanical structure, the product may result in recalls and loss of brand name value.

To express failures, failure modes, and failure effects in a product, there are typical methodologies of reliability design, which are stress–strength interference analysis [10], failure modes, effects, and criticality analysis (FMEA/FMECA) [11,12], and fault-tree analysis (FTA) [13,14]. In the product design process, these analyses are typically documented and performed by the company's department experts. Because the critical design factors may be missed in reviewing a new product, the product may experience failures and then have to be recalled from the marketplace. Especially, the stress–strength interference model has been utilized in the mechanical part design to determine why the product fails. It defines product failure as the probability of stress exceeding strength. However, it neglects the fact that product failure occurs randomly from the fragile parts of the product. Because of that, it requires the concepts of fracture mechanics [15,16] and the life-stress (LS) model [17].

Based on the reliability block diagrams (RBDs) [18,19], experimental methods such as accelerated life testing (ALT) have been carried out. When these methods are carried out, several debating key issues arise – test, planning for the product, failure mechanics, accelerated method, sample size equation, and so on. Elsayed (1996) classified the models into three categories: statistics-based models, physics-statistics models, and physics-experimental-based models [20]. Meeker (1985) suggested some practical guidelines to plan an accelerated life test [21]. However, these methodologies fail to reproduce the design faults and have been disputable. That is, as failure mechanisms – fatigue – in a mechanical system are not fully understood, small samples among products are tested so that it is hard to reproduce product failures that rarely occur in the field [22–26].

Because of daily consumer usage and problematic designs, a product may have structural failure and its lifetime may be reduced. To design mechanical structure by adapting mechanism and properly choosing material, conventional product design approaches the strength of materials [27,28]. On the other hand, recent fracture mechanics suggests critical factors such as fracture toughness instead of strength as the relevant material property [29]. As the electronic technology advances, engineers recognize product failures from micro-void coalescence [30]. To better understand product failures, we recognize the necessities of the better LS model that is applicable in electronic parts [31].

Another engineering approach includes the finite element method (FEM) [32,33]. Because product recalls are infrequent, many engineers think that they can be assessed by (1) mathematical modeling using Newtonian methods, (2) assessing the time response of the system for dynamic loads and finding the product stress, (3) utilizing the rain-flow counting method for Von Mises stress [34,35], and (4) estimating system damage using the Palmgren-Miner rule [36]. However, utilizing an analytical methodology that can generate a closed-form exact result often requires invoking many suppositions that are not capable of recognizing multi-module product failures due to design flaws such as inadequate stiffness and a small crack or pre-existing defect that is clearly unfeasible to model using FEM.

Based on literature reviews, we know that parametric ALT as a reliability methodology should have the following features: (1) product BX lifetime test plans of the multi-module product based on its reliability definition and block diagram, (2) general LS model and acceleration method based on the design concepts of a mechanical system including the mechanism, (3) derivation of sample size equation combined with product lifetime target and AF besides the fundamental concepts of statistics including probability, (4) ALT with reproduction of design problems and action plans, and (5) evaluation of the last product design to ensure that the lifetime criteria are fulfilled.

8.2 RELIABILITY DIAGRAM FOR ANALYZING THE STRUCTURE OF MECHANICAL PRODUCTS

8.2.1 Introduction

Examples of representative mechanical systems among modern products include airplane, automobile, refrigerator, domestic appliance, agricultural machinery, machine tools, heavy construction equipment, and so on. The mechanical systems use power to accomplish a task that involves forces and movement, which produce mechanical advantages by adapting product mechanisms. To evaluate the design of new modules in the product, we need to understand how the product can be broken down to its single parts.

For example, an airplane moves forward by thrust from a jet engine, which is known as the Brayton thermodynamic cycle. Airplanes are composed of approximately 200,000 components, including engine system, aviation control system, power delivery system, machine system, door system, and other parts. An automobile is a wheeled motor vehicle utilized for transportation. Its power is generated by the engine and transmitted to each wheel through mechanisms such as the transmission and drive train. An automobile consists of several different modules – engine, transmission, drive train, electrical systems, and body parts. The components in an automobile may count as many as 20,000 totally. On the other hand, to store food fresh, a refrigerator is designed to supply chilled air from the evaporator to the freezer (or refrigerator) compartment. It consists of several different modules – cabinet, door, internal fixture (shelves and drawers), controls and instruments, generating parts (motor or compressor), heat exchanger, water supply device, and other miscellaneous parts. Total components have approximately 2,000 pieces.

Lifetimes of an airplane, automobile, or a refrigerator are targeted to be at least B20 life 10 years. Because a refrigerator consists of 20 units, where each unit has 100 components, the lifetime of each unit is strictly targeted to be B1 life 10 years. An automobile also consists of 200 units and the lifetime of each unit is targeted to be B0.1 life 10 years. Airplanes have almost 2,000 units (or 100 modules), where each module is equivalent to the total components of a refrigerator. The lifetime of each unit is strictly targeted to be B0.01 life 10 years. As seen in Figure 8.4, the product lifetime depends on module #3, which might be expected to have design faults and no longer meet the established specifications for its proper functionality (Figures 8.1–8.4 and Table 8.1).

The reliability of a product has a direct effect on the patterns of modules, their quantities, their qualities, and how they are arranged in the system. To assess whether the mechanical system has proper acceptable performance and reliability, firstly, we need to figure out how the product arranges a collection of modules that can break down to the individual parts. In addition to the reliability of the modules, the relationship between these product modules is also considered, and proper decisions as to the selection of modules can improve or optimize the general system reliability, maintainability, and/or availability.

The reliability relationship, expressed as RBDs, usually uses logic diagrams, which can be widely utilized in engineering and science and exist in many different forms. For product reliability analysis, it describes the interrelationship between a system and its modules. For example, each subsystem in an automobile is broken down further into multiple lower-level subsystems. The RBD of a representative automobile possesses over 20,000 blocks, which include the parts. The reliability design of automobiles might focus on the modules. They can effortlessly compute the module reliability due to the connection of the serial system (see Figure 8.2).

However, we know that it may vary how the modules are physically connected. Think about a two-wheel back drive car. The layout of power transmission in which the engine (Figure 8.5), transmission, driveshaft, differential, axles, and wheels are physically connected is quite different from their reliability wise configuration like RBD. So we need to comprehend the design of the mechanical system so that we can develop a time-to-failure product through another approach like mechanism and load analysis (Chapter 5), system design (Chapter 6), and mechanical failure (Chapter 7).

8.2.2 Reliability Block Diagram

RBD is a graphical presentation of the reliability relationship between the system and its modules (or parts). The diagram shows the reliability of the system (or module) in terms of the working states of its parts. RBD is a success-oriented system that the logical connections of (functioning) parts in product will satisfy an intended function. So RBDs are used in reliability evaluations to help the analyst who wants to figure out the structure of the product and improve it.

Based on the motivation of system analysis, a block may show a lowest-level component, module, subsystem, and system. It is treated as a block box for which the physical details may not need to be known. Consider an automobile. The configurations of a complex system can be generated from the series, parallel, and other connections between modules (or parts). In an RBD, system parts are represented by rectangular blocks, which are connected by straight lines according to their logic relations (Figures 8.6 and 8.7).

Because the system has more than one (intended) function (or purpose), each function can begin a separate RBD. RBD also is satisfactory for systems in which the

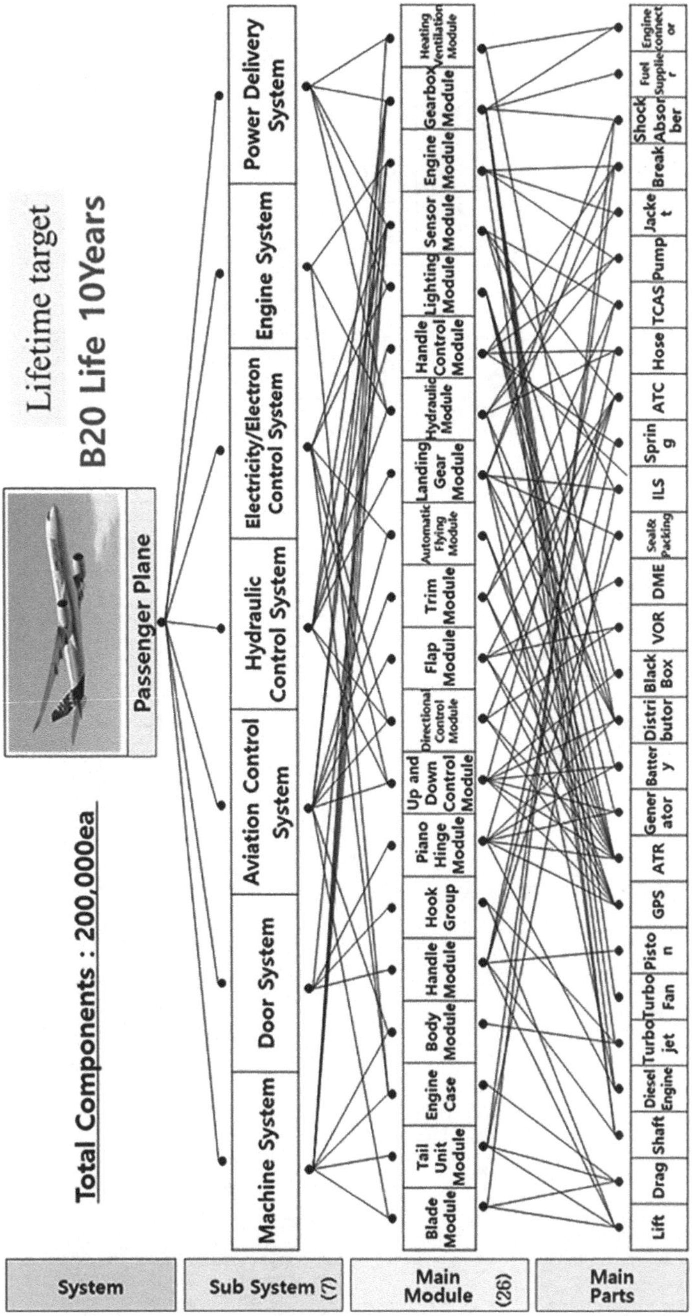

FIGURE 8.1 Breakdown of an airplane with multi-modules.

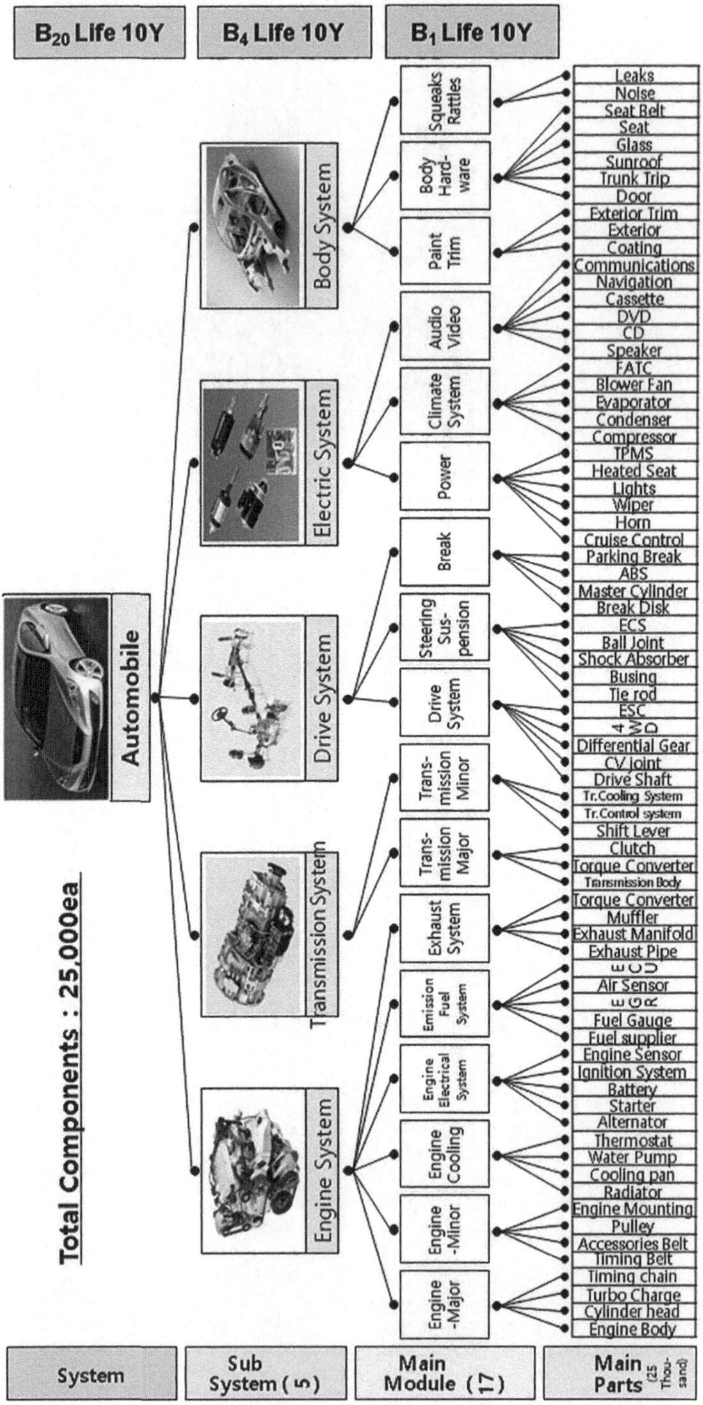

FIGURE 8.2 Breakdown of an automobile with multi-modules.

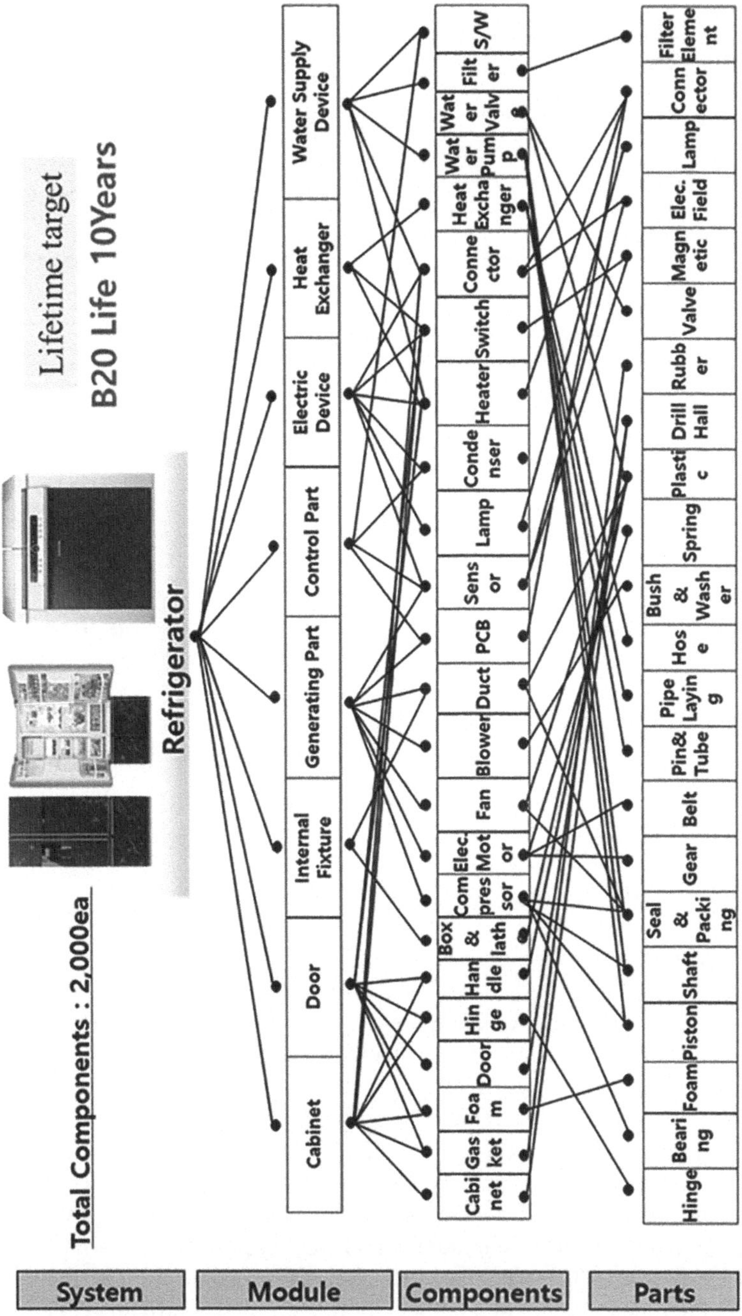

FIGURE 8.3 Breakdown of an automobile with multi-modules.

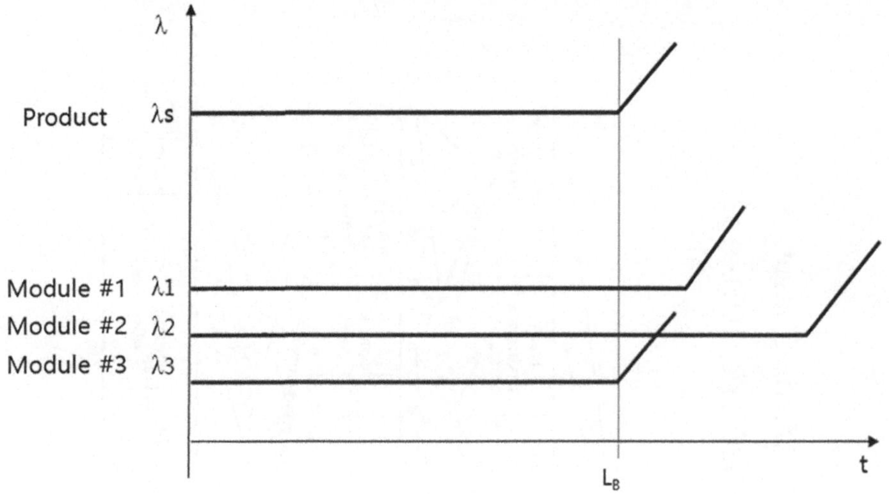

FIGURE 8.4 Product lifetime L_B and failure rate λ_s from the standpoint of a multi-module product.

TABLE 8.1
Lifetime Target of Mechanical Product, Units, and Components (L_{BX})

Mechanical Product	Refrigerator	Automobile	Aircraft
Lifetime target of the product	B20 life 10Y	B20 life 10Y	B20 life 10Y
Number of components in the product	2,000	20,000	200,000
Accumulative failure rate	0.01%	0.001%	0.0001%
BX life of component	B0.01 life 10Y	B0.001 life 10Y	B0.0001 life 10Y
Number of units in the product (each unit has 100 components)	20	200	2,000
Accumulative failure rate	1%	0.1%	0.01%
BX life of units	B1 life 10Y	B0.1 life 10Y	B0.01 life 10Y

failure order does not matter. In automobiles, power is generated in the engine and then transferred to transmission, drive system, and wheels. Each block shows critical modules such as transmission. The reliability diagrams represent the serial, parallel, and other models. Each module reliability in a product can be computed from failure rate and lifetime, if the exponential distribution is supposed to follow.

For any life data analysis, a mechanical engineer selects one product module at which no more particular information about the object of analysis is known to be examined. In that module, the engineer regards the purpose of analysis as a "black box or block diagram". The reliability data analysis is to get a life distribution that can explain the time to failure of a product (or module) and approximate the parameters in Weibull. It is based on time-to-failure data of the product (or module) in a block diagram that composes serial-connected, parallel-connected parts, and so on.

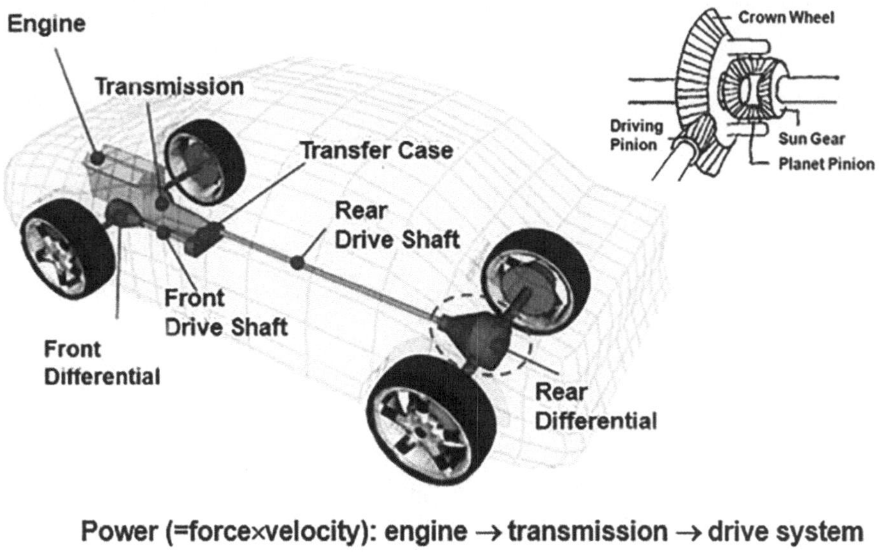

FIGURE 8.5 Typical power transmission of an automobile (example).

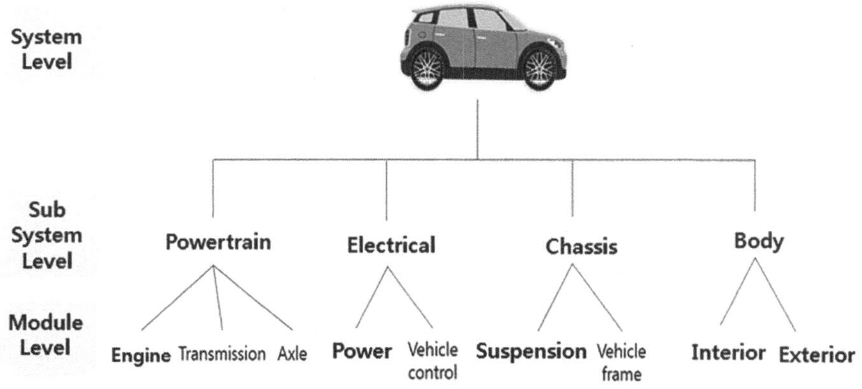

FIGURE 8.6 Automobile that consists of multi-modules connected serially.

Think about a system (or module) that n parts are described by a block. If there is a connection between the ending points a and b, component i in the product is working. It means that the stated function of the component is achieved. Therefore, it can represent reliability that is expressed as failure rate and lifetime. It is also possible to put more information into the block that shortly describes the necessitated function of the component (Figure 8.8).

Reliability is the chance that a device will work successfully for time t (or number of cycles). Reliability is frequently denoted as $R(t)$. Generally, modules in mechanical products are connected independently in a series. $R_i(t)$ will be defined as the

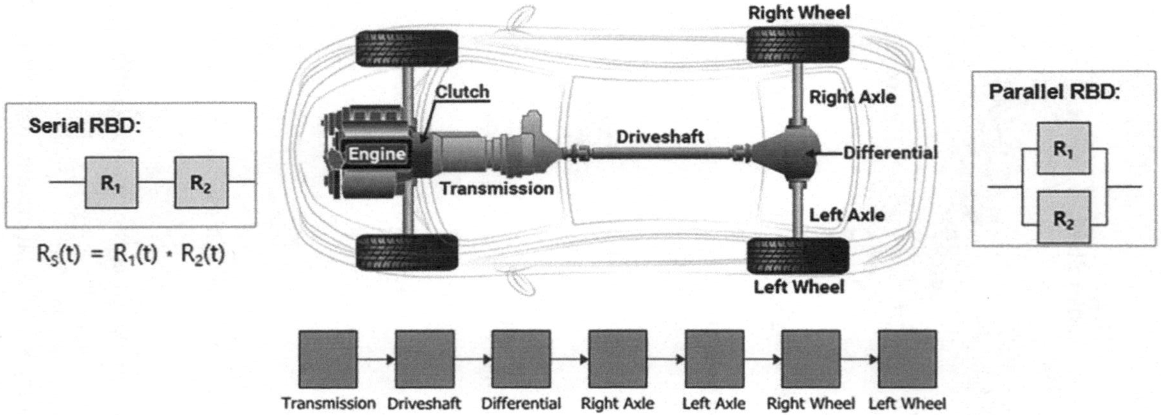

FIGURE 8.7 Automobile and its module for a reliability block diagram.

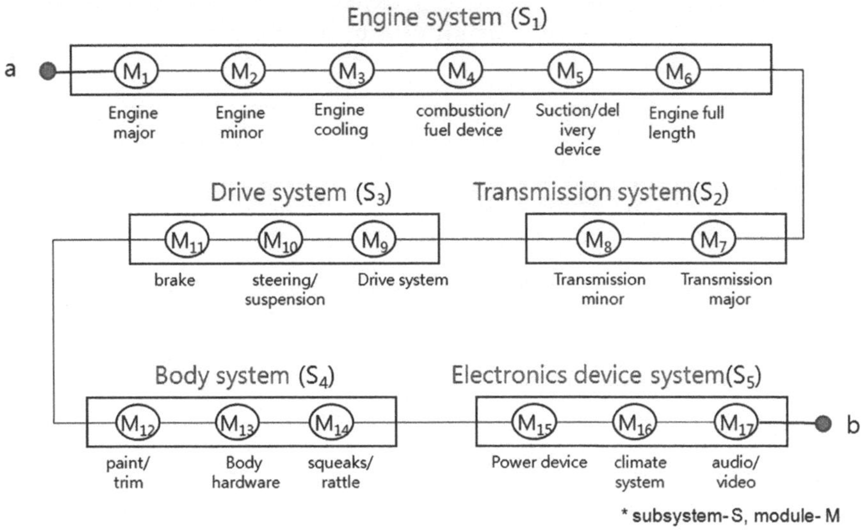

FIGURE 8.8 Reliability block diagram for an automobile (example).

reliability of module i at time t, and $F_i(t) = 1 - R_i(t)$ will be called the unreliability of module i at time t, for $i = 1, 2, \ldots, n$.

Similarly, if $R(t)$ is called the system (or module) reliability, $F(t) = 1 - R(t)$ will be defined as the system unreliability at a time t.

8.2.2.1 Serial Model

A series of configuration in Figure 8.9 indicates that all product parts must operate. Product has a connection between the ending points a and b (the system is functioning) if and only if it has a connection through all the n blocks showing the parts (or module). For a serial structure, if the product consists of n modules with each module reliability being $R_1, R_2, R_3, \ldots, R_n$, the system reliability can be expressed as follows:

$$R = R_1 \cdot R_2 \cdot R_3 \ldots R_n = \prod_{i=1}^{n} R_i \tag{8.1}$$

where R is the total reliability of the assembly and $R_1, R_2, R_3, \ldots, R_n$ are the individual reliabilities of components.

The reliability of components can be represented in several different ways. However, the most widely accepted method of product reliability that can be described is the exponential distribution (i.e. constant instantaneous failure rate),

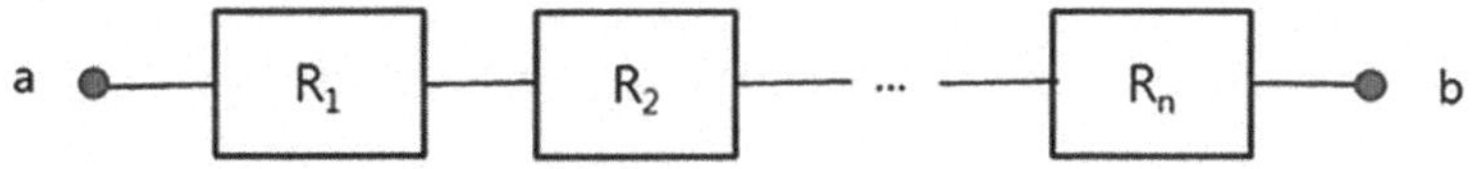

FIGURE 8.9 Serial model configuration.

which is discussed in Chapter 3. That is, if reliability function for a component is defined as $R_i = e^{-\lambda_i t}$, the system reliability of a series model from Equation (8.1) can be represented as follows:

$$R = e^{-\lambda t} = R_1 \times R_2 \times \ldots \times R_n = e^{-\lambda_1 t} \times e^{-\lambda_2 t} \times \ldots \times e^{-\lambda_n t} = \prod_{i=1}^{n} e^{-\lambda_i t} \quad (8.2a)$$

$$R = e^{-(\lambda_1 + \lambda_2 + \ldots + \lambda_n)t} = e^{-\left(\sum_{i=1}^{n} \lambda_i\right)t} \quad (8.2b)$$

where λ_i is the failure rate of the i component and λ is the failure rate of the system.

Example 8.1

Think about a serially connected mechanical product of three independent modules. At a stated point of time t, the module reliabilities are $R_1 = 0.95$, $R_2 = 0.97$, and $R_3 = 0.94$. The system reliability at time t is $R = R_1 \cdot R_2 \cdot R_3 = 0.95 \cdot 0.97 \cdot 0.94 \approx 0.866$.

Example 8.2

A mechanical product consists of three items. The failure rate for item A is 0.0002 failures/hour, item B is 0.00001 failures/hour, and item C is 0.0001 failures/hour. What is the product reliability for 75 hours?

Solution: Since the failure rates are additive:

$$\sum \lambda_i = \lambda_A + \lambda_B + \lambda_C = 0.0002 + 0.0001 + 0.00001 = 0.00031 \text{ failures/hour}$$

$$\sum \lambda_i t = (0.00031) \cdot (75) = 0.02325$$

$$R_{(75)} = e^{-0.02325} = 0.977$$

8.2.2.2 Parallel Model

A system that is working if at least one of its n parts (or modules) is working is defined as a parallel structure (Figure 8.10). The product has a connection between the ending points a and b (the system is working) if and only if it has a connection through all the n blocks showing the parts (or module).

A parallel configuration shows that at least one of the modules must work – redundancy. Thus, RBD supplies a framework for seeing system configuration. For parallel structure, if the system consists of n modules with each part reliability being R_1, R_2, R_3, …, R_n, the system reliability can be represented as follows:

$$R = R_1 \times R_1 \times \ldots \times R_n = 1 - (1 - R_1) \cdot (1 - R_2) \cdots (1 - R_n) = 1 - \prod_{i=1}^{n} (1 - R_i) \quad (8.3)$$

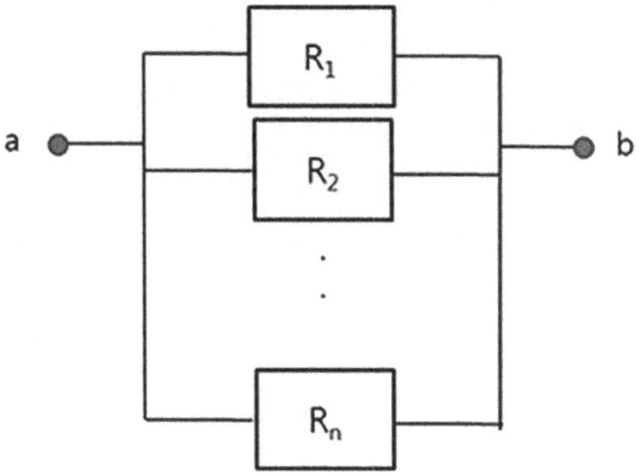

FIGURE 8.10 Parallel model configuration.

Example 8.3

The reliability of item *A* is 0.95, and the reliability of item *B* is 0.90. If these two parts are in parallel, what is the system reliability?

Solution: Using the previous formula, we get

$$R = 1 - (1 - 0.95) \cdot (1 - 0.9) = 0.995$$

8.2.2.3 Standby System

When there is parallel redundancy and both systems are not active, but one is in a standby mode waiting to be switched only when required, reliability can be enhanced, especially if switching is very reliable. In the most simplistic of cases, the switch is considered to have a reliability or probability of success of 1 (Figure 8.11). While there is no real reliability of 1, there are instances where human intervention is the switching mechanism and the probability of switching approaches 1.

The formula for computing the reliability of a system, with one unit in standby and a perfect switch, is

$$R = e^{-\lambda t}(1 + \lambda t) \tag{8.4}$$

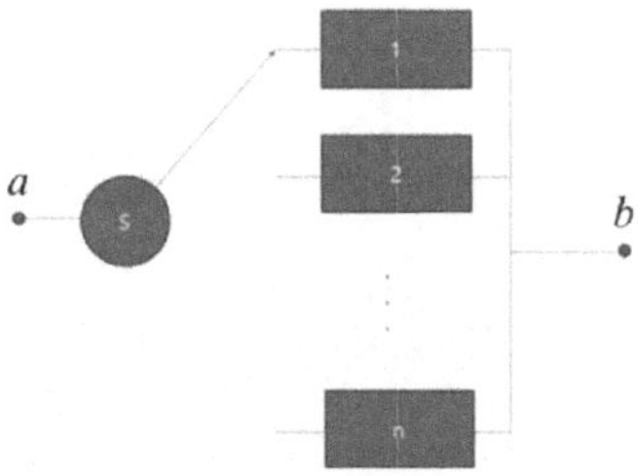

FIGURE 8.11 Standby model configuration.

Example 8.4

Two parts are similar in a parallel configuration where the second unit is in standby and will be activated only upon failure of the primary unit. The switching to the backup unit is considered certain. The failure rate for each part is 0.002 failures/hour. What is the reliability for 200 hours?

Solution: $R(200) = e^{-(0.002)(200)}\left(1+(0.002)(200)\right) = e^{-0.4}(1.4) = 0.93845$

8.2.3 Comparison between Reliability Block Diagram and Fault Tree

In some practical applications, an engineer may decide whether the product is modeled by selecting a structure by a fault tree (FT) or by an RBD. Because the FT is limited to only OR gates and AND gates, the FT can be changed to an RBD (or vice versa).

RBDs and FTs supply similar information. In an RBD, block connections signify that the part (or modules) in the product is working. The FT takes a failure standpoint, while the RBD takes a success standpoint in terms of reliability. A series structure is equivalent to an FT where all the basic events are connected through an OR gate. The TOP event happens if either component 1 or component 2 or component 3 or…component n fails. Therefore, a serial configuration in an FT is similar to a parallel configuration in an RBD.

On the other hand, a parallel structure might be shown as an FT where all the primary events are connected through an AND gate. The TOP event happens if component 1 and component 2 and component 3, …, and component n fail. The parallel configuration in an FT is similar to the serial configuration in an RBD. The following examples explain how the same analysis scenario using either RBDs or FTs is modeled.

Example 8.5

If a product that composes three parts can be drawn in an RBD as the following figure, find the equivalent FT.

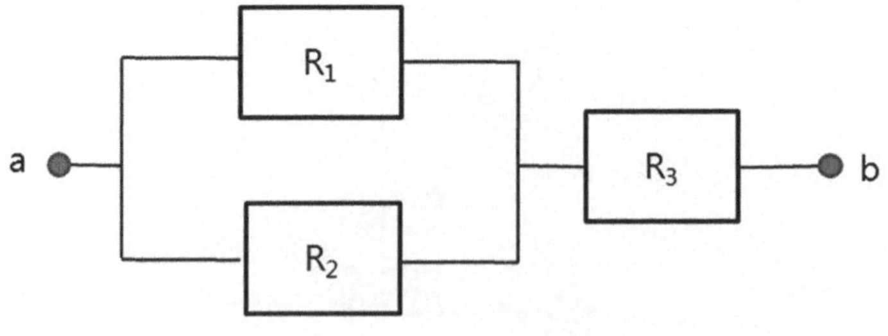

Since part 1 and part 2 in the RBD are connected in parallel, the equivalent FT can be represented as an AND gate. Because part 3 and the equivalent of part 1 and part 2 in the RBD are connected in series, the total product can be represented as an OR gate. If summarized in process, the equivalent FT can be drawn as the following figure.

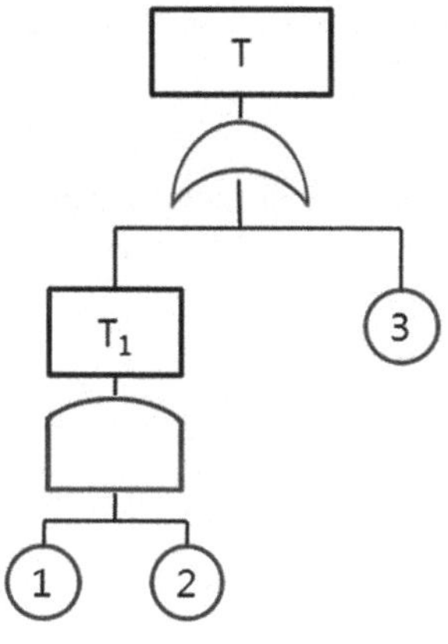

Example 8.6

Assume that the probability of occurrence of five fault events is 0.1, equally. Calculate the probability of occurrence of the top event in the following FT.

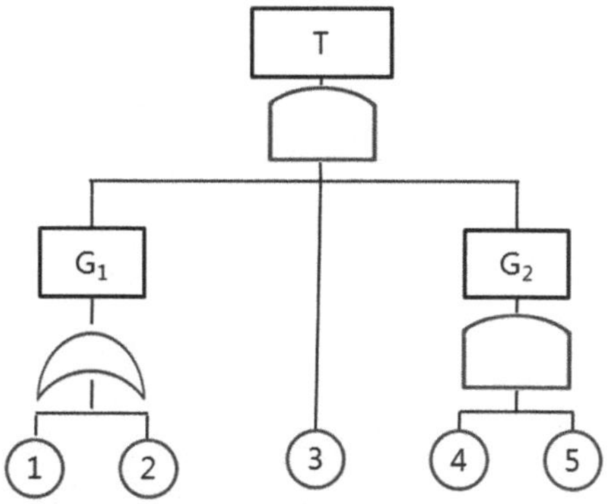

Because the chance of occurrence of G_1 is connected as an OR gate, it can be computed as follows:

$$G_1 = 1 - (1 - 0.1) \times (1 - 0.1) = 0.19$$

Because the chance of occurrence of G_2 is connected as an AND gate, the chance of occurrence of G_2 can be computed as follows:

$$G_1 = 0.1 \times 0.1 = 0.01$$

Thus, the chance of occurrence of the top event can be obtained as follows:

$$T = G_1 \times 3 \times G_2 = 0.19 \times 0.1 \times 0.01 = 0.00019$$

8.2.4 Allocation of Reliability Target for a Product Module

We see that some parts are assigned to very high-reliability goals, which may be unachievable at all. On the other hand, although there are major components whose failure causes safety, environmental, or legal issues, they will be allocated to low-reliability targets. Therefore, it is critical to start some criteria in reliability allocation. The job of reliability allocation is to choose part reliability targets, $R_1^*, R_2^*, \ldots, R_n^*$ which fulfill the following equality equation:

$$R_S^* \leq g\left(R_1^*, R_2^* R_2, \ldots\ldots, R_n^*\right) \tag{8.5}$$

Mathematically, there is an unlimited number of such sets. Distinctly, these sets are not equally good, and even some of them are not feasible. The usual criteria are expressed as follows:

1. Failure possibility. Parts that have a high likelihood of failure formerly should be given a low-reliability target because of the rigorous attempt required to make the reliability better. Reversely, for reliable parts, it is sensible to allocate a high-reliability goal.
2. Complexity. The number of basic parts (or modules) in a subsystem indicates the complication of the subsystem. A higher complexity leads to a lower reliability. It is similar to the motive of failure possibility.
3. Criticality. The failure of some parts may cause serious effects, including, for instance, loss of life and lasting environmental harm. The circumstances will be acute when such parts have a high likelihood of failure. Seemingly, criticality is a product of seriousness and failure probability, as defined in the FMEA technique described in Chapter 4. If a design cannot remove severe failure modes, the parts should have the lowest likelihood of failure. As a result, high-reliability targets should be allocated to them.
4. Cost. Cost is a crucial criterion that is a target subject to minimization in the commercial industry. The cost results for attaining reliability rely on parts. Some parts persuade a high cost to improve reliability a little because of the difficulty in design, verification, and production. Therefore, it may be advantageous to assign a higher-reliability target to the parts that have less cost result to increase reliability.

Although some ways for reliability assignment have been developed, the easiest method suggested here is the equal allocation method. This method can only be applied when the system reliability configuration is in series. The system reliability is computed by

$$R_S^* = \prod_{i=1}^{n} R_i \tag{8.6}$$

The assigned reliability for each subsystem is

$$R_i = \left(R_S^*\right)^{1/k} \tag{8.7}$$

8.3 RELIABILITY DESIGN IN A MECHANICAL SYSTEM

8.3.1 Introduction

Reliability is explained as the ability of a system or module to carry out the required function under stated environmental (or operational) conditions for an identified period of time. As a customer uses a product during its lifetime, the product's reliability can be represented using a bathtub curve. If a product suits the bathtub curve, the manufacturer might be required to eliminate the product from the marketplace because of the higher failure rates during the initial introduction of the product or an overall shorter lifetime. In such an instance, the product is required to have a reliability design that can be evaluated by lifetime L_B and failure rate λ from the product's final design.

Reliability design for a new product can be summarized as (1) reliability targeting for product, (2) reliability testing and data analysis in a Weibull chart for test or field, (3) searching out the design problems of the suspicious parts and correcting them, and (4) demonstrating the effectiveness of test through the field data. While achieving the targeted reliability, the mechanical product has a desired quality – enough strength and stiffness.

If reliability of a product module is targeted, the design faults of the mechanical system missed in the design stage might be found by reliability methodologies such as parametric ALTs embedded in the design process. Therefore, these reliability methodologies require the following: (1) product reliability test plans based on their reliability definition, (2) LS model combined with design concepts of a mechanical system including the mechanism and structure, and (3) the derivation of sample size equation with an acceleration factor (AF).

8.3.2 Lifetime of a Mechanical System on Reliability Test Plans

Most mechanical products such as automobiles, appliances, or airplanes are composed of multi-module structures, which can be expressed as a serial model. If one module has a critical problem, all systems will stop. For instance, an automobile is a wheeled motor vehicle utilized for transportation. Its power is produced by the engine and transmitted to each wheel through mechanisms such as transmission and drive system. An automobile consists of several different modules – engine, transmission, drive train, electrical systems, and body parts. If the engine has a problem, everything will be broken (Figure 8.12).

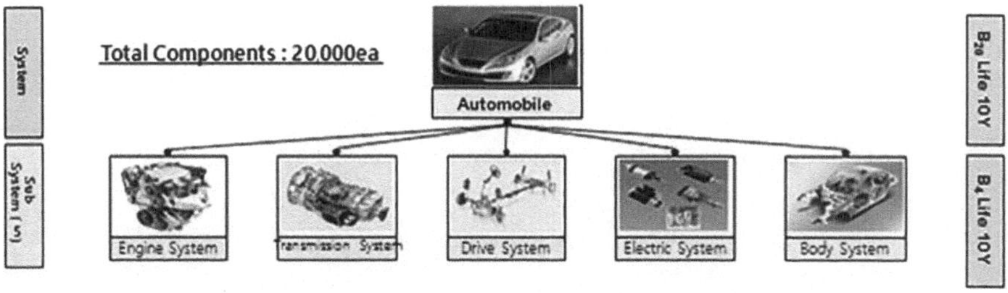

FIGURE 8.12 Automobile with several different modules.

The total components in an automobile may be as many as 20,000. A proper index of product lifetime such as BX lifetime can be defined for performing parametric ALT. BX lifetime is the time at which *X* percent of the items in a population will not have succeeded in achieving their goal. For instance, if a product has B20 life 10 years, 20% of the population will have failed during 10 years of working time. 'BX life Y years' appropriately decides the cumulative failure rate of a product and its lifetime.

The automobile lifetime is targeted to have B20 life 10 years. Because the automobile is composed of five modules, we can expect that its modules have B4 life 10 years. If modules are assembled, the product can work properly and perform its own intended functions with proper reliability that can be explained by the product of failure rate and its lifetime. The cumulative failure rate of an automobile over its lifetime would be the summation of the failure rate of each module. On the other hand, assume that there were no starting failures in a product, the product lifetime could be determined by the problematic module that there are design faults. Thus, an engineer needs the reliability test plans to enlarge the lifetime of a problematic module, which can be extended by finding the design faults and modifying it.

8.4 PARAMETRIC ACCELERATED LIFE TESTING

8.4.1 Introduction

Most mechanical products such as automobiles, appliances, or airplanes are composed of multi-module structures. If they are assembled, a product can work properly and perform its own intended functions. For example, as seen Figure 8.13a, a refrigerator

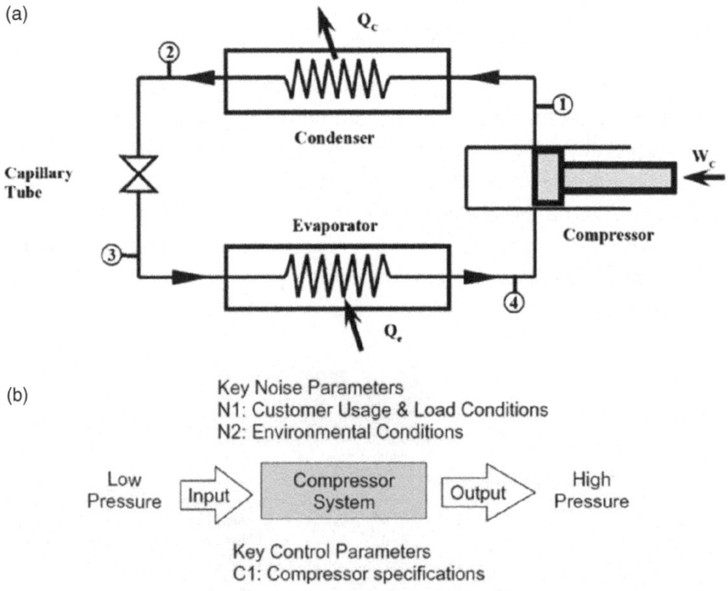

FIGURE 8.13 (a) Vapor-compression refrigeration cycle and (b) robust design schematic of the compressor.

usually supplies chilled air from the evaporator to both the freezer and refrigerator compartments where fresh food is stocked. Refrigerators use the vapor-compression refrigeration cycle where the refrigerant experiences phase changes between vapor and liquid.

The refrigerant at a low-side pressure goes into the compressor. It then departs from the compressor and goes into the condenser at a high-side pressure; the refrigerant is condensed as heat is moved to the surroundings. The refrigerant then departs from the condenser as a high-pressure liquid. The pressure of the liquid is decreased as it runs through the expansion valves, and thus, some of the liquid flashes into chilled vapor. The residual liquid at a low pressure and temperature is vaporized in the evaporator as heat is moved from the fresh/freezer compartment. This vapor then again goes into the compressor.

As seen in Figure 8.13b, a refrigerator compressor has input and output units that have their own (intended) functions that increase pressure in the vapor-compression cycle. If there is a void (design fault) in the system that causes an inadequacy of strength (or stiffness) when subjected to loads, the structure will fall in its lifetime. The representative failure mechanisms of the mechanical system are fatigue and fracture. That is, repeated displacement, strain, and stress due to loads induce voids in weak parts that have design faults. The fracture started at the void and propagated to the end. Eventually, a mechanical failure such as fatigue occurs. Especially, fatigue failure appears, not due to conceptual stresses in an ideal part, but rather due to the existence of a small crack or pre-existing flaws on the exterior of a component that the stress and strain become plastic (Figures 8.14 and 8.15).

If a structure has the desired quality – enough strength and stiffness – it will withstand loads near the targeted product lifetime. On the other hand, if the repetitive load is acted at the stress raisers such as a shoulder fillet, the structure in which damage is piled up will crack. The more, the system will fracture abruptly after repetitive stresses in its lifetime. The product engineer would want to reshape the design faults where the stress is acted. The design of a mechanical product might be expressed as two factors: (1) the load (or stress) on the structure and (2) materials and the structure (shape). That is, after changing the material type of a product structure for repeated loads, find its optimal design shape.

The failure location of the product structure could be discovered when the failed products are disassembled in the field or after the samples of a parametric ALT failed. Because the electric products in mechanical systems are typically housed, both electronic

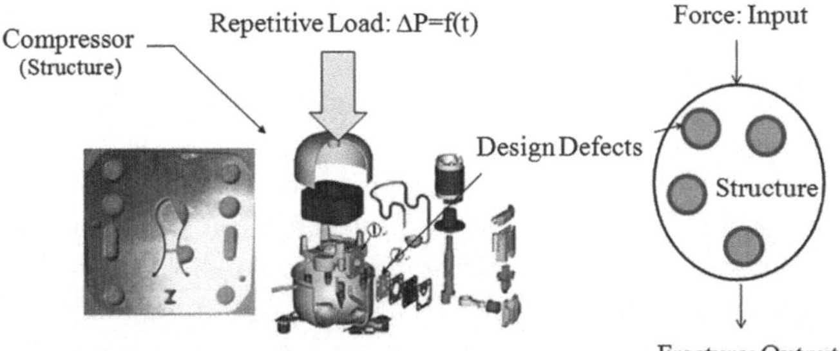

FIGURE 8.14 Fatigue failure made by repetitive load and design defects.

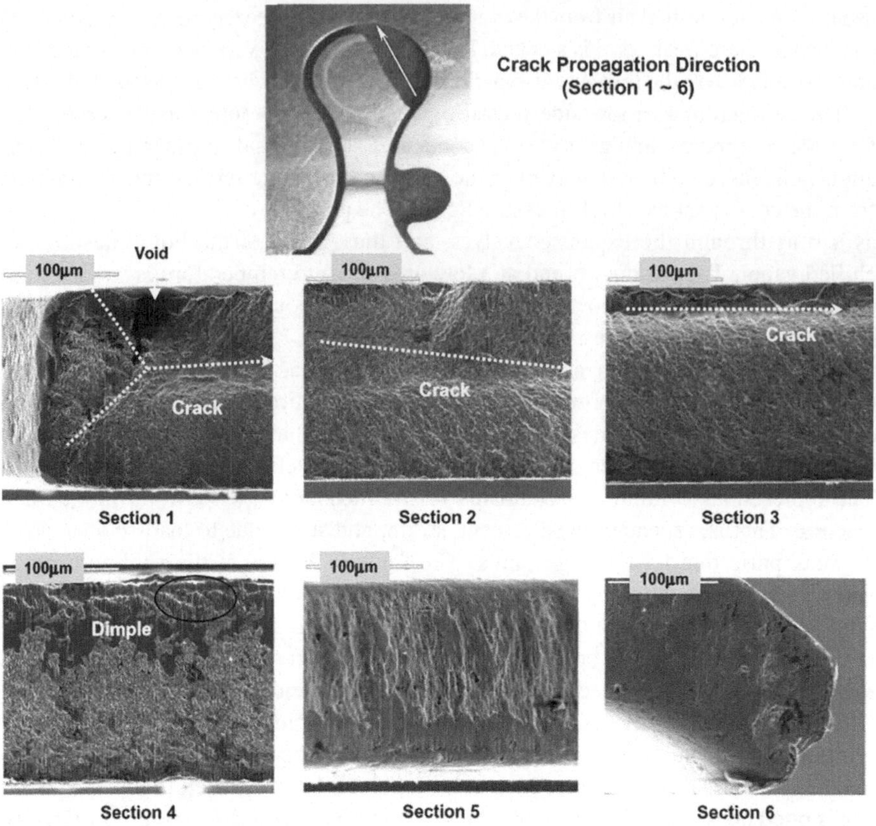

FIGURE 8.15 Fractography of the compressor suction reed valve on SEM.

and mechanical products might apply to the design and reliability testing. It, therefore, is critical to search out the design faults by the reliability testing like parametric ALT in the design process and modify them before new product launches (Figure 8.16).

With the emergence of Finite Element Analysis (FEA) tools, design failures such as fatigue can be evaluated in an implicit environment. Though fatigue assessments using FEA do not totally replace with fatigue testing, an engineer will discover the detailed or optimal design in the structure of the new product. However, because crack starts from micro-voids due to design defects, the product might have failure shown in the field. If modules have a problem due to an inappropriate design, the engineer will affirm if the new design endures for loads by reliability testing. Otherwise, in the field, if a mechanical module works inappropriately, customers would request the problematic product to be replaced. Because the field usage conditions are unknown, engineers should confirm if newly designed functions work robustly by proper testing.

Thus, modules with specific intended functions need to be robustly designed to withstand a variety of loads. Such design targeting might be known to be conventionally achieved through the statistical design of the experiment and the Taguchi methods (SDE). Especially, Taguchi methods, familiar to robust designs, utilize the loss

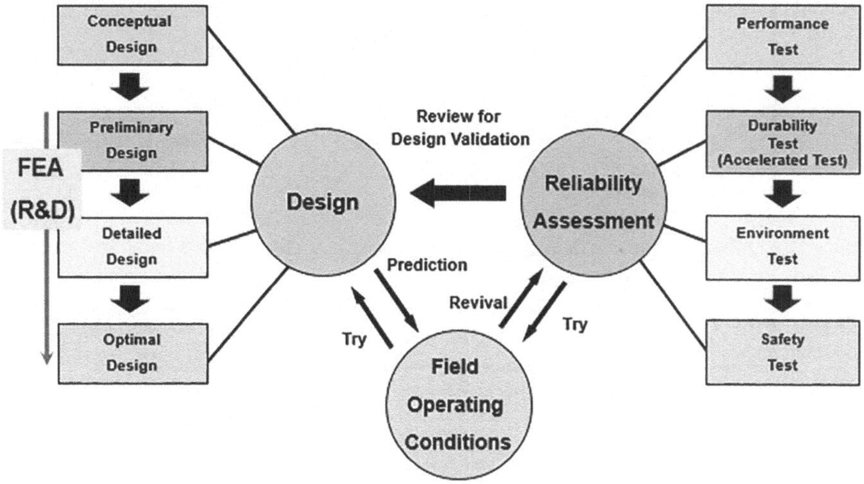

FIGURE 8.16 Optimal design and reliability assessment.

function, which quantifies the amount of loss based on departure from the desired performance. It puts a design factor in an optimal place where utilizing cost function random "noise" factors are less likely to hurt the design and it helps decide the design parameters (or best control factors). However, for a not complex mechanical/civil structure, such as a beam, Taguchi methods should examine many design parameters. In the design process, it is impossible to think about the total range of the chemical, physical, and mathematical conditions that could influence the design.

Parametric ALT as another experimental methodology might be suggested. Parametric ALT utilizes the sample size equation with *AF*. As for finding the faulty designs, engineers help to experimentally find the optimal designs that can endure several repetitive loads. Parametric ALT can also be utilized to estimate product reliability – lifetime, L_B, and failure rate, λ. The new parametric ALT discussed in the next section requires an LS model and sample size formulation on the basis of a test plan [37–61].

8.4.2 Setting an Overall Parametric Accelerated Life Testing Plan

Reliability can be expressed as the ability of a system or module to work under stated conditions for a described period of time or total distance [62]. It might be explained using a diagram called a "bathtub curve" that comprises three parts [63]. First, there is a decreasing failure rate in the premature life of the product ($\beta < 1$). Then, there is a constant failure rate ($\beta = 1$). Lastly, there is an increasing failure rate toward the ending of the product's life ($\beta > 1$).

As seen in Figure 8.17, if a product pursuits the bathtub curve, it may have difficulties prospering from the marketplace because of the higher failure rates and short lifetimes due to its design fault. Companies can firmly make better the design of a product by targeting reliability for new products to (1) lessen premature failures, (2) decrease random failures during the product working time, and (3) lengthen product lifetime. As the designs of a mechanical product are improved, the failure rate of a product in

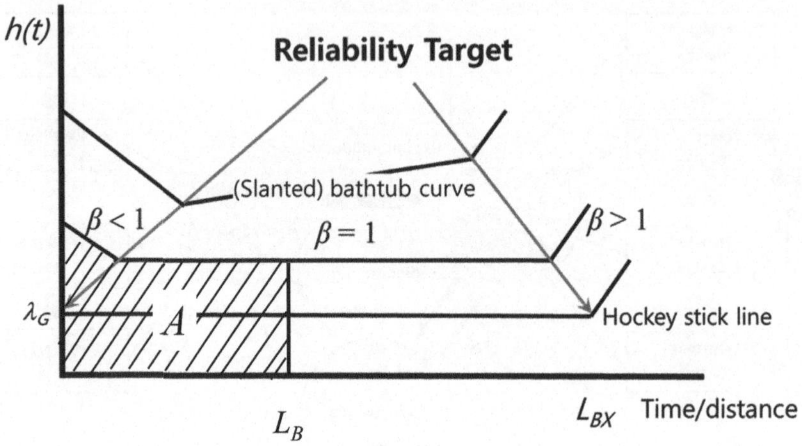

FIGURE 8.17 Bathtub curve and straight line with slope β.

the field declines and its lifetime is extended. For such a situation, the convention bathtub curve can be changed to an unswerving line with the shape parameter β.

The failure rate on the bathtub can be defined as

$$\lambda = \frac{f}{R} = \frac{dF/dt}{R} = \frac{(1-R)'}{R} = \frac{-R'}{R} \tag{8.8}$$

If Equation (8.8) is integrated to time, we can obtain the life of $X\%$ cumulative failure. That is,

$$\int \lambda \, dt = -\ln R \tag{8.9a}$$

It can be expressed as follows:

$$A = \langle \lambda \rangle \cdot L_B = \int_0^{L_B} \lambda(t) \cdot dt = -\ln R(L_B) = -\ln(1-F) \cong F(L_B) \tag{8.9b}$$

Because of the straight line that follows an exponential distribution with a low failure rate, the reliability of a mechanical product might be estimated from the product lifetime L_{BX} and failure rate λ as follows:

$$R(L_B) = 1 - F(L_B) = e^{-\lambda L_{BX}} \cong 1 - \lambda L_{BX} \tag{8.10}$$

where R is the reliability function, F is the cumulative distribution function (CDF), λ is the failure rate, and L_{BX} is the BX lifetime.

This equation is relevant to less than 20% of accumulative failure rates [64]. The reliability of the mechanical system therefore might be achieved without problematic designs. After the product lifetime L_{BX} is targeted, we can find any design faults and modify them through parametric ALT. Consequently, a product can achieve product optimization and targeted reliability (Table 8.2).

TABLE 8.2
Overall Parametric ALT Plan for Targeting the Lifetime of a Mechanical Product (BX Life Y Years) Including Airplane and Refrigerator, Based on the Market/Expected Data

		Market Data, F ($BX = 1.8$)				Expected, F ($BX = 1.8$)		Targeted, F ($BX = 1.0$)	
No.	Module Name	Yearly Failure Rate, %/year	L_{BX} Life year	Design	Conversion	Yearly Failure Rate, %/year	L_{BX} Life year	Yearly Failure Rate, λ_G, %/year	L_{BX} Life year
1	Module A	0.34	5.3	New	x5	1.70	1.1	0.10	10
2	Module B	0.35	5.1	Given	x1	0.35	5.1	0.10	10
3	Module C	0.25	7.2	Modified	x2	0.50	3.6	0.10	10
4	Module D	0.20	9.0	Modified	x2	0.40	4.5	0.10	10
5	Module E	0.15	12.0	Given	x1	0.15	12.0	0.15	10 ($BX = 1.5$)
6	Others	0.50	12.0	Given	x1	0.50	12.0	0.45	10 ($BX = 4.5$)
Total	Product (R-set)	1.79	5.1	-	-	3.60	3.6	1.00	10 ($BX = 10$)

If the product lifetime was given by Y and the total failure rate was X, the yearly failure rate can be computed by dividing total failure rate X by product lifetime Y. The product lifetime may be expressed as L_{BX} life Y years with a yearly failure rate of X/Y. Based on failure data from the field, we can target the lifetime of a module in a product. For example, if the life of an automobile is targeted to have B20 life 10 years, we know that the yearly failure rate is 2% that can be obtained by dividing 20% by 10 years. Appliance lifetime is targeted to have B10–20 life 8–12 years. On the other hand, an automobile is targeted to have B40–120 life 12 years.

The testing plan of the product for parametric ALT can be categorized as a newly designed module, modified module, and no changing module. In targeting the lifetime of a new module where there was no market reliability data, the data for similar modules are frequently utilized as a source. If there has been a crucial redesign of the module, the failure rate from the marketplace may be expected to be higher and the product lifetime will decrease. Thus, the reliability of the product will rely on the following factors:

1. How well the new design maintains a similar structure to the prior design,
2. For each new module, new manufacturers are assumed to supply parts for the product,
3. Magnitude of the loads compared to the prior design, and
4. How much technological change and additional functions are incorporated into the new design.

For Module A, based on market data, the failure rate was 0.34%/year and its lifetime was 5.3 years. However, the expected failure rate was 1.7%/year and its expected lifetime was 1.1 years because there was no market data on the reliability of the new design. The lifetime of the new design was targeted to be over an L_{BX} ($BX = 1.0$) life 10 years with a yearly failure rate of 0.10%. To satisfy the anticipated product lifetime, the parametric ALT should help determine design parameters that could influence the product lifetime.

For module B, a machine system from the field had yearly failure rates of 0.35% per year and a lifetime of B1.8 life 5.1 years. To reply to customer requirements, a new lifetime for the machine system is targeted to B1 life 10 years. For module D, a modified module, the yearly failure rate was 0.2%/year and L_{BX} ($BX = 1.8$) life was 9 years from the field data. Because this was a modified design, the anticipated failure rate was 0.4%/year and the anticipated L_{BX} ($BX = 1.8$) life was 4.5. To lengthen the targeted product life, the lifetime of the new design was targeted to be an L_{BX} ($BX = 1.0$) life 10 years with a yearly failure rate of 0.1%. The product reliability might be decided by summing up the failure rates of each module and the lifetimes of each module. The product lifetime is targeted to be over L_{BX} ($BX = 10.0$) life 10 years.

8.4.3 BX Life and Mean Time to Failure

A proper index of product lifetime such as BX life can be chosen for performing parametric ALT. BX life is defined as *the time at which X percent of the items in a population will have failed.* For example, if a product lifetime has B20 life 10 years, then 20% of the population will be unsuccessful in achieving their goal during 10 years of working time. 'BX life Y years' helps satisfactorily decide the accumulative failure rate of a product and its lifetime responding to market requirements. The mean time to failure (MTTF) as the inverse of the failure rate cannot be utilized for the lifetime because it is equivalent to a B60 life, which is too big. BX life indicates a more proper standard of lifetime compared to MTTF.

8.4.4 Parametric Accelerated Life Testing of Mechanical Systems

Product failures such as fatigue occur when a problematic member in a module is subjected to repetitive cyclic loadings. That is, design flaws such as stress raisers produce deformation. After repetitive deformations, cracks can start to appear and propagate. Fatigue failure appears, not due to pure stresses in an ideal part, but rather due to the existence of a small crack or pre-existing flaw on the exterior of a part that the stress and strain become plastic. Relying on the material, shape, and proximity to the elastic limit when it deforms, fatigue failure may require many deformation cycles.

Mechanical failure – especially fatigue – can be distinguished by two factors: (1) the stress due to loads on the structure and (2) the type of materials (or shape) used in the product. If there are design faults in the product where the repetitive loads are acted, the structure can fracture in its expected lifetime (Figure 8.18).

The product engineer would like to find the problematic designs in the structure and modify them by experiment using parametric ALT. It is essential to make sure of the soundness of the product design, including predicted design life, by assessing the fatigue behavior of the material. The major matter in the reliability test is how premature the possible failure mode can be obtained. To do this, it is obligatory to work out a failure model and determine the related coefficients. First of all, we can configure the LS model, which incorporates stresses and reaction parameters. This equation can explain several failures such as fatigue in the mechanical structure.

To better understand it, an engineer recognizes how small crack or pre-exited material defects in a material generate. That is, because system failure starts from the presence of material defects formed on a microscopic level when repeatedly subjected to a variable tensile and compression load randomly, we might define the LS model from such a standpoint. For example, we can figure out the following processes utilized for solid-state diffusion of impurities in silicon that is popularly used

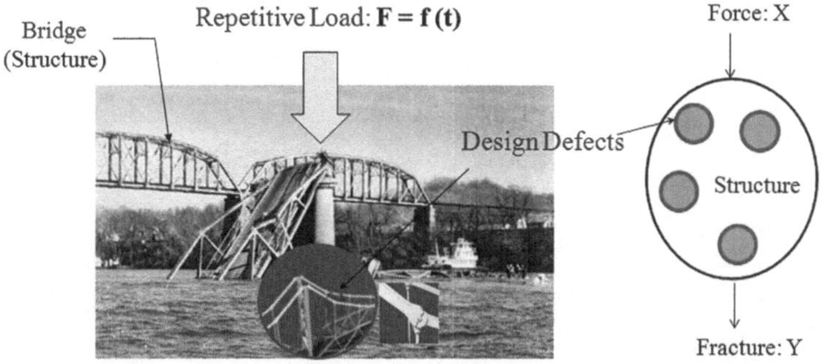

FIGURE 8.18 Fatigue failure made by repetitive load and design defects.

as a semi-conduct material: (1) electro-migration-induced voiding; (2) build-up of chloride ions; and (3) trapping of electrons or holes.

When the electric field, ξ, is applied, we know that the impurities in silicon are easily migrated because the energy barrier lowers. For solid-state diffusion of impurities in silicon, the junction equation J might be defined as [65,66] (Figure 8.19):

$$J = [aC(x-a)] \cdot \exp\left[-\frac{q}{kT}\left[w - \frac{1}{2}a\xi\right]\right] \cdot v$$

$$[\text{Density/Area}] \cdot [\text{Jump Probability}] \cdot [\text{Jump Frequency}]$$

$$= -\left[a^2 v e^{-qw/kT}\right] \cosh\frac{qa\xi}{2kT}\frac{\partial C}{\partial x} + \left[2ave^{-qw/kT}\right] C \sinh\frac{qa\xi}{2kT}$$

$$\cong \Phi(x,t,T)\sinh(a\xi)\exp\left(-\frac{Q}{kT}\right)$$

$$= B\sinh(a\xi)\exp\left(-\frac{Q}{kT}\right) \tag{8.11}$$

where B is constant, a is the distance between (silicon) atoms, ξ is the applied field, k is Boltzmann's constant, Q is the energy, and T is the temperature.

On the other hand, a reaction process that is dependent on speed might be defined as follows:

$$K = K^+ - K^- = a\frac{kT}{h}e^{-\frac{\Delta E - \alpha S}{kT}} - a\frac{kT}{h}e^{-\frac{\Delta E + \alpha S}{kT}} = 2\frac{kT}{h}e^{-\frac{\Delta E}{kT}} \cdot \sinh\left(\frac{\alpha S}{kT}\right)$$

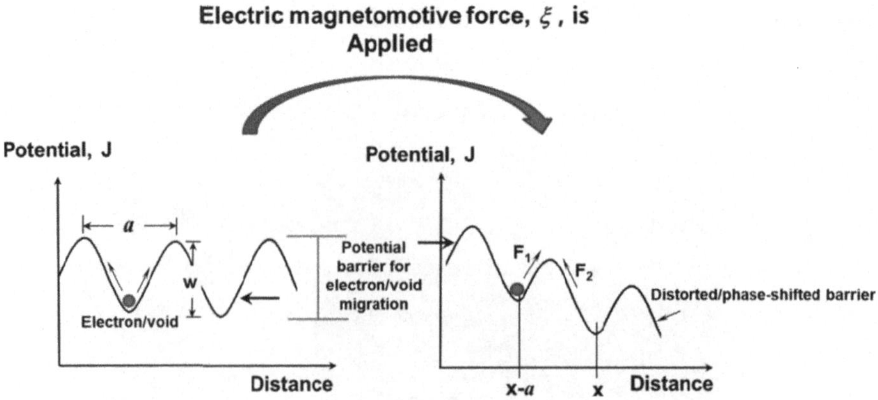

FIGURE 8.19 Potential (barrier) change in material for electron/void migration after an electric magneto-motive force is applied.

$$= B\sinh(aS)\exp\left(-\frac{\Delta E}{kT}\right) \tag{8.12}$$

Thus, the reaction rate, K, from Equations (8.11) and (8.12) can be abridged as follows:

$$K = B\sinh(aS)\exp\left(-\frac{E_a}{kT}\right) \tag{8.13}$$

where B is constant.

If the reaction rate in Equation (8.13) takes an inverse function, the generalized stress model can be defined as

$$TF = A\left[\sinh(aS)\right]^{-1}\exp\left(\frac{E_a}{kT}\right) \tag{8.14}$$

As seen in Figure 8.20, the hyperbolic sine stress term can be explained as follows: (1) initially $(S)^{-1}$ in low stress effect, (2) $(S)^{-n}$ in medium stress effect, and (3) $\left(e^{aS}\right)^{-1}$ in high stress effect. From the S–N curve, we can also find similar trends. That is, we account that the transition point (approximately 1,000 cycles) of the S–N curve is not sustainable because the nominal stresses are elastic-plastic (high stress region). We will know that fatigue is worked by the release of plastic shear strain energy; thus above yield, stress does not obey the linear relationship with strain and cannot be used. Between the transition and the endurance limit (about 10^7 cycles), the S–N based analysis is sustainable (medium-stress region). Above the endurance limit, the slope of the curve reduces strikingly, and as such, this is usually defined as the 'infinite life' region (low stress region).

Ductile metals in the stress–strain curve surpass the specification limits, operating limit, yield point, and ultimate stress point into the fracture (Figure 8.21). Because

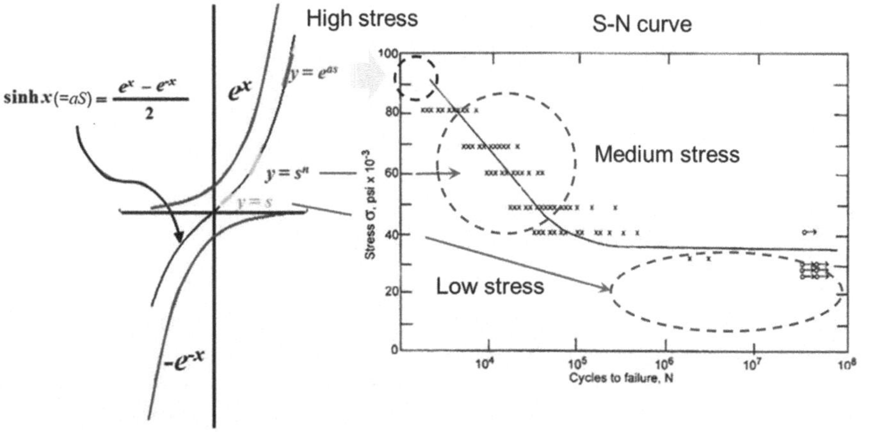

FIGURE 8.20 Hyperbolic sine stress term.

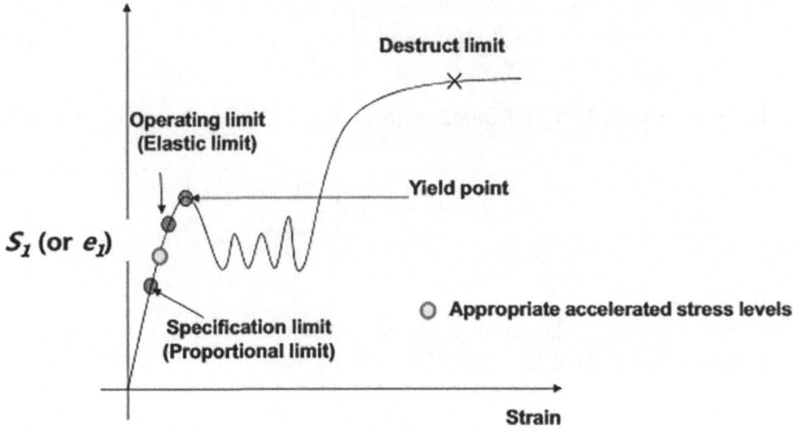

FIGURE 8.21 Typical stress–strain curves for ductile materials.

accelerated testing will be performed in the medium range between specification limits and the operating limits, Equation (8.14) is restated as follows:

$$TF = A(S)^{-n} \exp\left(\frac{E_a}{kT}\right) \tag{8.15}$$

8.4.5 Acceleration Factor

The internal (or external) stress in a product is not easy to quantify and use in accelerated testing because stress can be acquired by the analytic method. It is necessary to alter Equation (8.15) into a more relevant form. The power (or energy flow) in a physical system including the structure and mechanism that was discussed in Chapter 5 can generally be demonstrated as efforts and flows. Thus, stresses in mechanical systems may come from the efforts (or loads) like force, torque, pressure, or voltage (Table 8.3).

For a mechanical/civil system, when substituting stress into effort, the time to failure can be restated as follows:

$$TF = A(S)^{-n} \exp\left(\frac{E_a}{kT}\right) = A(e)^{-\lambda} \exp\left(\frac{E_a}{kT}\right) \tag{8.16}$$

TABLE 8.3
Energy Flow in the Multi-Port Physical System [67]

Modules	Effort, e(t)	Flow, f(t)
Mechanical translation	Force, $F(t)$	Velocity, $V(t)$
Mechanical rotation	Torque, $\tau(t)$	Angular velocity, $\omega(t)$
Compressor, pump	Pressure difference, $\Delta P(t)$	Volume flow rate, $Q(t)$
Electric	Voltage, $V(t)$	Current, $i(t)$
Magnetic	Magneto-motive force, e_m	Magnetic flux, φ

where λ is the power index or damage coefficient.

Because the material strength depraves steadily, it may require longer times to test a module until mechanical failures such as fatigue happen. The major difficulties in discovering wear-induced failures and overstressed failures are the testing time and cost when carrying out the test without ignoring failure mechanisms such as fatigue. To resolve these problems, the reliability engineer usually favors testing under serious conditions. Due to overstress, the module failures can be effortlessly found with parametric ALT.

The more the accelerated conditions, the shorter the testing time. Thus, the sample size will decrease in the accelerated areas that will be discussed with the reduction factor in the latter section. This concept is crucial to carry out the accelerated life tests, but the scope of the accelerated life tests will be decided by whether the results in the accelerated tests are the same as that normally found in the field.

The stress–strain curve is a way to visualize the behavior of a material when it is subjected to load. A result of stresses in the vertical axis has the corresponding strains along the horizontal axis. Mild steel subjected to loads proceeds specification limits (proportional limit), operating limits (elastic limit), yield point, and ultimate stress point into the fracture (destruct limit). In accelerated testing, the proper accelerated stress levels (S_1 or e_1) will typically fall outside the specification limits but inside the operating limits (Figure 8.22).

When a module has been tested for several hours under the accelerated stressed condition, one would like to know the identical working time at the normal stress condition. After the multiple accelerated testing under diverse conditions, the identical working time at normal (or actual) operation time can be attained.

From the time to failure in Equation (8.14), the AF can be expressed as the ratio between the appropriate accelerated stress levels and normal stress levels. The AF can also be restated to incorporate the effort concepts:

$$AF = \left(\frac{S_1}{S_0}\right)^n \left[\frac{E_a}{k}\left(\frac{1}{T_0} - \frac{1}{T_1}\right)\right] = \left(\frac{e_1}{e_0}\right)^\lambda \left[\frac{E_a}{k}\left(\frac{1}{T_0} - \frac{1}{T_1}\right)\right] \tag{8.17}$$

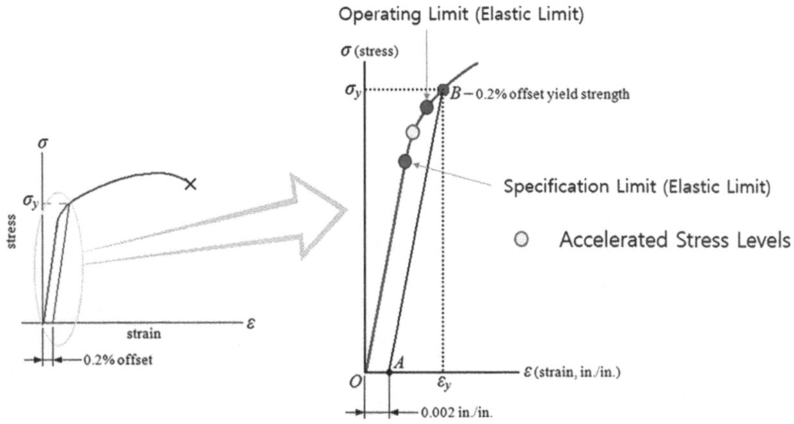

FIGURE 8.22 Strain–stress curve in mild steel.

where n is the stress dependence and λ is the cumulative damage exponent that depends on materials.

In Equation (8.17), the first expression is the outside effort (or load) and the second is the internal energy. Under acute conditions, the outside higher load releases the energy barrier, and the accelerated (or high) temperature activates the material elements. Finally, the material degrades and fails. The equation has two parameters, which are effort and temperature. Utilizing a three-level test under accelerated conditions, these parameters can be attained. In addition, the quantified value, *activation energy* E_a, is called the reaction rate due to temperature rises.

Under acute conditions, the duty effect with on/off cycles involves the repetitive stress (or load), which shortens the module lifetime [68]. The equation needed to determine the sample size for the parametric ALT is discussed in the next section.

8.4.6 Derivation of General Sample Size Equation

Because of the cost and time limit, it is not easy to perform a test with big samples for reliability testing of a product. If fewer components are tested, the reliability testing in statistical analysis will become more unknown. For a more exact result, enough samples should be tested. However, this testing will require lots of time and cost. Thus, utilizing the sample size equation with AF in Equation (8.17) and combining the fundamentals of statistics such as confidence level, it is important to develop a method for the accelerated testing.

In statistical testing, the first step of parametric ALT requires to decide how the sample size should be drawn from the population. The test samples are chosen randomly for a representative test sample. It is also connected with the confidence levels and the statistical range of the measured failure values. Thus, the confidence levels for the lifetime assessment are necessary because it is impossible to gather the lifetimes of limited several sample sizes. In statistics, the failure behavior of the limited sample may strongly deviate from the actual failure behavior of the population itself. The central concept supplies further help through the confidence levels, which can specify the confidence of the test results and estimate the failure behavior of the population.

Various methods have been developed to decide the sample size equation. The Weibull analysis is a famous and extensively accepted way of examining reliability data. The Weibayes model based on the Weibull analysis can be utilized for deciding sample size. However, because it is difficult to directly apply because of its mathematical complication, we have to differentiate the case of failures ($r \geq 1$) from the case of no failures ($r = 0$). A clarified equation therefore must be developed.

In choosing the set of values of the model parameters that make as large as possible the likelihood function, maximum-likelihood estimation (MLE) in statistics is a way of approximating the parameters of a statistical model from a given data set. The characteristic life η_{MLE} can be expressed as follows:

$$\eta_{\text{MLE}}^{\beta} = \sum_{i=1}^{n} \frac{t_i^{\beta}}{r} \tag{8.18}$$

If the number of failures is $r \geq 1$ and the confidence level is 100(1 − α), the characteristic life, η_α, would be approximated from Equation (8.16):

$$\eta_\alpha^\beta = \frac{2r}{\chi_\alpha^2(2r+2)} \cdot \eta_{\text{MLE}}^\beta = \frac{2}{\chi_\alpha^2(2r+2)} \cdot \sum_{i=1}^{n} t_i^\beta \text{ for } r \geq 1 \tag{8.19}$$

Assuming there are no failures, the p-value is α and ln(1/α) is mathematically identical to the Chi-squared value, $\frac{\chi_\alpha^2(2)}{2}$ [69]. That is,

$$p-\text{value}: \alpha = \int_{\chi_\alpha^2(2)}^{\infty} \left(\frac{e^{-\frac{x}{2}} x^{\frac{\nu}{2}-1}}{2^{\frac{\nu}{2}} \Gamma\left(\frac{\nu}{2}\right)} \right) dx = \int_{2\ln\alpha^{-1}}^{\infty} \left(\frac{e^{-\frac{x}{2}} x^{\frac{\nu}{2}-1}}{2^{\frac{\nu}{2}} \Gamma\left(\frac{\nu}{2}\right)} \right) dx \quad \text{for } x \geq 0 \tag{8.20}$$

where ν is the shape parameter and Γ is the gamma function.

For $r = 0$, the characteristic life η_α from Equation (8.19) would be expressed as follows:

$$\eta_\alpha^\beta = \frac{2}{\chi_\alpha^2(2)} \cdot \sum_{i=1}^{n} t_i^\beta = \frac{1}{\ln\frac{1}{\alpha}} \cdot \sum_{i=1}^{n} t_i^\beta, \quad \text{for } r = 0 \tag{8.21}$$

Because Equation (8.19) is established for all cases $r \geq 0$, characteristic life, η_α, can be redefined as follows:

$$\eta_\alpha^\beta = \frac{2}{\chi_\alpha^2(2r+2)} \cdot \sum_{i=1}^{n} t_i^\beta \quad \text{for } r \geq 0 \tag{8.22}$$

From CDF in Weibull, the relationship between BX life and characteristic life can be expressed as follows:

$$L_{BX}^\beta = \left(\ln\frac{1}{1-x} \right) \cdot \eta^\beta \tag{8.23}$$

If the approximated characteristic life of the p-value α, η_α, in Equation (8.22), is substituted into Equation (8.23), we obtain the BX life equation:

$$L_{BX}^\beta = \left(\ln\frac{1}{1-x} \right) \cdot \frac{2}{\chi_\alpha^2(2r+2)} \cdot \sum_{i=1}^{n} t_i^\beta \tag{8.24}$$

Because most life testing usually has inadequate samples to approximate lifetime and the allowed number of failures would be less than that of the sample size, the planned testing time will proceed as follows:

$$n \cdot h^{\beta} \geq \sum t_i^{\beta} \geq (n-r) \cdot h^{\beta} \tag{8.25}$$

If Equation (8.25) is substituted into Equation (8.24), the BX life equation can be expressed as follows:

$$L_{BX}^{\beta} \cong \left(\ln\frac{1}{1-x}\right) \cdot \frac{2}{\chi_{\alpha}^2(2r+2)} \cdot nh^{\beta} \geq \left(\ln\frac{1}{1-x}\right) \cdot \frac{2}{\chi_{\alpha}^2(2r+2)} \cdot (n-r)h^{\beta} \geq L_{BX}^{*\beta} \tag{8.26}$$

The sample size equation with the number of failures can also be obtained as follows:

$$n \geq \frac{\chi_{\alpha}^2(2r+2)}{2} \cdot \frac{1}{\left(\ln\frac{1}{1-x}\right)} \cdot \left(\frac{L_{BX}^{*}}{h}\right)^{\beta} + r \tag{8.27}$$

For a 60% confidence level, the first term $\frac{\chi_{\alpha}^2(2r+2)}{2}$ in Equation (8.27) can be approximated to $(r + 1)$ [70]. In addition, if the cumulative failure rate, x, is below about 20%, the denominator of the second term $\ln\frac{1}{1-X}$ approximates to x by the Taylor expansion (Table 8.4).

So the sample size equation (8.27) can be approximated as follows:

$$n \geq (r+1) \cdot \frac{1}{x} \cdot \left(\frac{L_{BX}^{*}}{h}\right)^{\beta} + r \tag{8.28}$$

So we know that sample size equation $n \approx$ (number of failures + 1)·(1/cumulative failure rate) ((target lifetime/(test hours))^β + r.

TABLE 8.4
Characteristics of $\frac{\chi_{0.0}^2(2r+2)}{2}$ at α = 60% Confidence Level

R	$1-\alpha$	$\frac{\chi_{0.4}^2(2r+2)}{2}$	$\frac{\chi_{\alpha}^2(2r+2)}{2} \approx r+1$	$1-\alpha$
0	0.4	0.92	1	0.63
1	0.4	2.02	2	0.59
2	0.4	3.11	3	0.58
3	0.4	4.18	4	0.57

If the *AF*s in Equation (8.17) are added into the planned testing time *h*, Equation (8.28) becomes

$$n \geq (r+1) \cdot \frac{1}{x} \cdot \left(\frac{L_{BX}^{*}}{AF \cdot h_a} \right)^{\beta} + r \tag{8.29}$$

If the lifetime of a refrigerator compressor was targeted to have B1 life 10 years, the mission cycles can be obtained for a given set of samples subjected to the accelerated conditions. Because $\frac{\chi_{\alpha}^{2}(2r+2)}{2} \cong (r+1)$, $\ln(1-x)^{-1} \cong x$, and $h = h_a{\cdot}AF$ in Equation (8.26), the estimated lifetime L_{BX} in each ALT is also approximated as follows:

$$L_{BX}^{\beta} \cong x \cdot \frac{n \cdot (h_a \cdot AF)^{\beta}}{r+1} \tag{8.30}$$

Let $x = \lambda{\cdot}L_{BX}$. The estimated failure rate of the design samples λ can be described as

$$\lambda \cong \frac{1}{L_{BX}} \cdot (r+1) \cdot \frac{L_{BX}^{\beta}}{n \cdot (h_a \cdot AF)^{\beta}} \tag{8.31}$$

In each ALT, we can quantify the reliability from the multiplication of the estimated L_{BX} life and failure rate λ. The usual operating cycles of a product in its lifetime are computed under the anticipated customer usage conditions. If the failed number, targeted lifetime, *AF*, and cumulative failure rate are decided, the required actual testing cycles under the accelerated conditions can be obtained from Equation (8.29).

To apply the accelerated effort to samples, ALT equipment will be inventively designed based on the load analysis of the product and its working mechanism. Utilizing parametric ALT with an approximated sample size of an *AF*, we can get the actual mission cycles, h_a from Equation (8.29). After obtaining the failed samples in mission cycles, we can determine whether the reliability target is achieved. To prove the accuracy of the approximated sample size in Equation (8.28), we present in Table 8.5 without considering the *AF*.

TABLE 8.5
The Calculated Sample Size with a Testing Time (*h*) of 1,080 Hours

		Sample Size	
β	Failure Number	Equation (8.27) by Minitab	Equation (8.28)
2	0	3	3
2	1	7	7
3	0	1	1
3	1	3	3

If the approximated failure rate from the reliability testing is not larger than the targeted failure rate (λ^*), the number of sample size (n) might also be obtained. The estimated failure rate with a commonsense level of confidence (λ) can be expressed as follows:

$$\lambda^* \geq \lambda \cong \frac{r+1}{n \cdot (AF \cdot h_a)} \tag{8.32}$$

By resolving Equation (8.32), we can also get the sample size

$$n \geq (r+1) \cdot \frac{1}{\lambda^*} \cdot \frac{1}{AF \cdot h_a} \tag{8.33}$$

By multiplying the targeted lifetime (L^*_{BX}) into the numerator and denominator of Equation (8.33), we can get another sample size equation:

$$n \geq (r+1) \cdot \frac{1}{\lambda^* \cdot L^*_{BX}} \cdot \frac{L^*_{BX}}{AF \cdot h_a} = (r+1) \cdot \frac{1}{x} \cdot \left(\frac{L^*_{BX}}{AF \cdot h_a} \right)^1 \tag{8.34}$$

Here, we know that $\lambda^* \cdot L^*_{BX}$ is transformed into the cumulative failure rate x.

We can find two equations for sample size that have a close form – Equations (8.28) and (8.34). For wear-out failure, it is fascinating that the exponent of the third term for two equations is 1 or β, which is greater than 1. Because the sample size equation for the failure rate is included and the allowed failed number r is 0, the sample size equation (8.28) for the lifetime might be a generalized equation to attain the reliability target.

If the testing time of an item (h) is more than the targeted lifetime (L^*_{BX}), the reduction factor R is close to 1. The generalized equation for sample size in Equation (8.28) might be restated as follows:

$$n \geq (r+1) \cdot \frac{1}{x} \tag{8.35}$$

If the targeted reliability for a module is assigned to have B1 life 10 years, the targeted lifetime (L^*_{BX}) is effortlessly obtained from the calculation by hand. For a refrigerator, the number of operating cycles for one day was five; the worst case was nine. Thus, the targeted lifetime for ten years might be 32,850 cycles.

The other type of sample size equation that is derived by Wasserman [64] can be expressed as follows:

$$n = -\frac{\chi^2_\alpha(2r+2)}{2m^\beta \ln R_L} = \frac{\chi^2_\alpha(2r+2)}{2m^\beta \ln R_L^{-1}} = \frac{\chi^2_\alpha(2r+2)}{2m^\beta \ln\left(1 - F_L^{-1}\right)} = \frac{\chi^2_\alpha(2r+2)}{2} \cdot \frac{1}{\ln\left(1 - F_L^{-1}\right)} \cdot \left(\frac{L_{BX}}{h} \right)^\beta \tag{8.36}$$

where $m \cong h/L_{BX}$ and $n \gg r$.

When $r = 0$, the sample size equation can be obtained as follows:

$$n = \frac{\ln(1-C)}{m^{\beta}\ln R_L} = \frac{-\ln(1-C)}{-m^{\beta}\ln R_L} = \frac{\ln(1-C)^{-1}}{m^{\beta}\ln R_L^{-1}} = \frac{\ln\alpha^{-1}}{m^{\beta}\ln R_L^{-1}} = \frac{\chi_{\alpha}^{2}(2)}{2}\cdot\frac{1}{\ln\left(1-F_L^{-1}\right)}\cdot\left(\frac{L_{BX}}{h}\right)^{\beta} \tag{8.37}$$

Thus, we know that Wasserman's sample size equation (8.37) is identical to Equation (8.28). Especially, the ratio between product lifetime versus the testing time in Equation (8.26) can be defined as a reduction factor. It can be used to determine whether ALT is proper. That is,

$$R = \left(\frac{L_{BX}^{*}}{h}\right)^{\beta} = \left(\frac{L_{BX}^{*}}{AF\cdot h_a}\right)^{\beta} \tag{8.38}$$

To successfully begin the parametric ALT, we have to obtain the acute conditions that will increase the AF and the shape factor β. At that time, the location and shape of the failed product in both market and parameter ALT results are similar. If the actual testing time h_a is longer than the testing time that is stated in the reliability target, the reduction fraction will be less than one. Thus, we can obtain the accelerated conditions that can decrease the testing time and sample size number.

8.4.7 Simplified Sample Size Equation

Typically, the characteristic life is expressed as follows [71]:

$$\eta^{\beta} = \frac{\sum t_i^{\beta}}{r} \cong \frac{n\cdot h^{\beta}}{r} \tag{8.39}$$

where β is the shape parameter in a Weibull distribution.

As product (or part) reliability becomes better, failure of the part becomes less recurrent in laboratory tests. It becomes more difficult to assess the characteristic life in Equation (8.36). We can give a practical definition for characteristic life as follows:

$$\eta^{\beta} \cong \frac{n\cdot h^{\beta}}{r+1} \tag{8.40}$$

If a random variable T is represented to the lifetime of a product, the Weibull CDF of T, denoted as $F(t)$, may be expressed as all individuals having a $T \leq t$, divided by the total number of individuals. This function also gives the probability P of randomly selecting an individual having a value of T equal to or less than t, and thus we have

$$P(T \leq t) = F(t). \tag{8.41}$$

If the product follows Weibull distribution, the accumulated failure rate, $F(t)$, is expressed as follows:

$$F(t) = 1 - e^{-\left(\frac{t}{\eta}\right)^{\beta}} \tag{8.42}$$

When $t = L_{BX}$ in Equation (8.42), the relationship between BX life (L_{BX}) and characteristic life η can be defined as follows:

$$L_{BX}^{\beta} \cong x \cdot \eta^{\beta} \tag{8.43}$$

where $x = 0.01F(t)$.

If Equation (8.40) is inserted into (8.43), the BX life can be redefined as follows:

$$L_{BX}^{\beta} \cong x \cdot \eta^{\beta} \cong x \cdot \frac{n \cdot h^{\beta}}{r+1} \geq L_{BX}^{*\beta} \tag{8.44}$$

The simplified sample size equation can be found from Equation (8.44):

$$n \geq (r+1) \cdot \frac{1}{x} \cdot \left(\frac{L_{BX}}{h}\right)^{\beta} \tag{8.45}$$

If the AFs in Equation (8.17) are added into the planned testing time h, Equation (8.45) becomes

$$n \geq (r+1) \frac{1}{x} \left(\frac{L_{BX}}{AF \cdot h_a}\right)^{\beta} \tag{8.46}$$

Example 8.7

Parametric ALT for a mechanical system – Geared Motor.

Parametric ALT is an accelerated life test for using the following parameters of the approximated sample size Equation (8.26) – reliability target, AF equation, shape parameter (β), allowed number of failures (r), cumulative damage exponent in Palmgren-Miner's rule (λ), and activation energy (E_a). To apply parametric ALT, the reliability for a mechanical system such as a robot is targeted to have B20 life years. Because a robot consists of a fifth module, the reliability for an actuator module such as a geared motor might have B4 life 5 years by a part count method. So, the calculated lifetime of the product would be 3,600 hours for 5 years:

$$\text{Product lifetime}(L_{B4}) = (2\ \text{hours/day}) \cdot (360\ \text{days/year}) \cdot (5\ \text{years}) = 3{,}600\ \text{hours} \tag{8.47}$$

Now consider a test plan for evaluating the mechatronic actuator (geared motor). As the normal conditions are at 0.10 kg·cm (100 mA) and 20°C, the accelerated conditions are elevated to 0.16 kg·cm (140 mA) and 40°C. The *AF* can be calculated as follows:

$$AF = \left(\frac{0.10\ \text{kg}\cdot\text{cm}}{0.16\ \text{kg}\cdot\text{cm}}\right)^2 \cdot \exp\left[\frac{0.55\ \text{eV}\cdot 1.60\times 10^{-19}}{1.38\times 10^{-23}}\left(\frac{1}{273+20}-\frac{1}{273+40}\right)\right] = 10.3 \cong 10 \tag{8.48}$$

where the activation energy is 0.55 eV (typical reaction rate), and exponent, λ, is 2 because the loss of electrical input power dissipates in a conductor by I^2R [72].

By applying the *AF*, the actual test time will be reduced to 360 hours (1/10 times). When the product lifetime for the geared motor (module) is targeted as B4 life 5 years and the accelerated test is carried out for 360 hours (15-day testing), the required sample size equation in the case of no failure allowed ($r = 0$) for 60% confidence level is calculated from Equation (8.27). That is,

$$n = (r+1)\left(\frac{1}{x}\right)\left(\frac{L_{BX}}{AF\cdot h}\right)^{\beta} = (1)\cdot\left(\frac{1}{0.04}\right)\left(\frac{3{,}600}{10\times 360}\right)^3 = 25 \tag{8.49}$$

We know that 25 samples might be tested in the case of no failure allowed ($r = 0$) because the last term of Equation (8.27) becomes one. This sample size still is required to have too many numbers for the reliability testing of a mechanical product. As the test time under accelerated conditions increases, we can find a suitable solution. If parametric ALT is carried out for 830 hours (35-day testing), the required sample size equation under no failure ($r = 0$) for a 60% confidence level is calculated as follows:

$$n = (r+1)\left(\frac{1}{x}\right)\left(\frac{L_{BX}}{AF\cdot h}\right)^{\beta} = (1)\cdot\left(\frac{1}{0.04}\right)\left(\frac{3{,}600}{10\times 360\times 2.3}\right)^3 = 2.0 \tag{8.50}$$

And if the accelerated test is carried out for 1,080 hours (360 hours × 3 times = 45-day testing) and the shape parameter is three, the required sample size equation under no failure ($r = 0$) for 60% confidence level is calculated as follows:

$$n = (r+1)\left(\frac{1}{x}\right)\left(\frac{L_{BX}}{AF\cdot h}\right)^{\beta} = (1)\cdot\left(\frac{1}{0.04}\right)\left(\frac{3{,}600}{10\times 360\times 3}\right)^3 = 1.0 \tag{8.51}$$

That is, the sample size will decrease by the cube of one-third ($=(1/3)^3$). Consequently, the required sample size will decrease to one.

Example 8.8

Printed Circuit Assembly

To apply parametric ALT, the reliability for a mechanical system such as a robot is targeted to have B20 life years. Because a robot consists of a fifth module, the reliability for a printed circuit assembly might be B4 life 5 years by a part count method. So, the calculated lifetime of the product would be 1,800 cycles for 5 years:

$$\text{Product lifetime}(L_{B4}) = (1\ \text{cycle/day})\cdot(360\ \text{days/year})\cdot(5\ \text{years}) = 1{,}800\ \text{cycles} \quad (8.52)$$

Now consider a test plan for evaluating the printed circuit assembly. As thermal shock under the normal conditions, ΔT_1, is 10°C, thermal shock under the accelerated conditions, ΔT_2, is 50°C. The *AF* can be calculated as follows:

$$AF = \left(\frac{\Delta T_2}{\Delta T_1}\right)^2 = \left(\frac{50°\text{C}}{10°\text{C}}\right)^2 \cong 25 \quad (8.53)$$

By applying the *AF*, the actual test time will be reduced to 72 cycles (1/25 times). When the product lifetime for the geared motor (module) is targeted as B4 life 5 years and the accelerated test is carried out for 144 hours (6-day testing), the required sample size equation in the case of no failure allowed ($r = 0$) for 60% confidence level is calculated from Equation (8.27). That is,

$$n = (r+1)\left(\frac{1}{x}\right)\left(\frac{L_{BX}}{AF\cdot h}\right)^{\beta} = (1)\cdot\left(\frac{1}{0.04}\right)\left(\frac{1{,}800\ \text{cycle}}{25\times 72\ \text{cycle}}\right)^2 = 25 \quad (8.54)$$

We know that 25 samples might be tested in the case of no failure allowed ($r = 0$) because the last term of Equation (8.27) becomes one. This sample size still is required to have too many numbers for the reliability testing of a mechanical product. As the test time under accelerated conditions increases, we can find a suitable solution. If parametric ALT is carried out for 520 hours (22-day testing), the required sample size equation under no failure ($r = 0$) for a 60% confidence level is calculated as follows:

$$n = (r+1)\left(\frac{1}{x}\right)\left(\frac{L_{BX}}{AF\cdot h}\right)^{\beta} = (1)\cdot\left(\frac{1}{0.04}\right)\left(\frac{1{,}800\ \text{cycle}}{25\times 260\ \text{cycle}}\right)^2 \cong 2.0 \quad (8.55)$$

REFERENCES

1. Garvin, D.A., Nov.–Dec. 1987, Competing on the eight dimensions of quality, *Harvard Business Review*, 65 (6), 101–109.
2. Magaziner, I.C., Patinkin, M., March–April 1989, Cold competition: GE wages the refrigerator war, *Harvard Business Review*, 89 (2), 114–124.
3. Montgomery, D., 2013. *Design and analysis of experiments*, 8th ed., Hoboken, NJ: John Wiley & Sons, Inc.
4. Taguchi, G., 1978, Off-line and on-line quality control systems. *Proceedings of the International Conference on Quality Control*, Tokyo, Japan.
5. Taguchi, G., Shih-Chung, T., 1992, *Introduction to quality engineering: Bringing quality engineering upstream*. New York: American Society of Mechanical Engineering.
6. Ashley, S., 1992, Applying Taguchi's quality engineering to technology development. *Mechanical Engineering*, 114 (7), 58.
7. Wilkins, J., 2000, Putting Taguchi methods to work to solve design flaws, *Quality Progress*, 33 (5), 55–59.

8. Phadke, M., 1989, *Quality engineering using robust design*. Englewood Cliffs, NJ: Prentice Hall.
9. Byrne, D., Taguchi, S., 1987, The Taguchi approach to parameter design. *Quality Progress*, 20 (12), 19–26.
10. Tersmette, T., 2013, *Mechanical stress/strength interference theory*. New York: Quanterion.
11. Procedures for Performing a Failure Mode, Effects and Criticality Analysis, U.S. Department of Defense, 1949, MIL-P-1629.
12. Neal, R.A., 1962, *Modes of failure analysis summary for the Nerva B-2 reactor*. Pittsburgh, PA: Westinghouse Electric Corporation Astronuclear Laboratory.
13. Goldberg, B.E., Everhart, K., Stevens, R., Babbitt, N., Clemens, P., Stout, L., 1994, *System engineering toolbox for design-oriented engineers*, Marshall Space Flight Center.
14. Center for Chemical Process Safety, 2010, *Guidelines for hazard evaluation procedures*, 3rd ed., Hoboken, NJ: Wiley-AIChE.
15. Griffith, A.A., 1921, The phenomena of rupture and flow in solids, *Philosophical Transactions of the Royal Society of London, A*, 221, 163–198.
16. Irwin, G., 1957, Analysis of stresses and strains near the end of a crack traversing a plate, *Journal of Applied Mechanics*, 24, 361–364.
17. McPherson, J., 1989, Accelerated Testing, Volume 1: Packaging. In: M. L. Minges, ed., *Electronic materials handbook*, Materials Park, OH: ASM International Publishing, 887–894.
18. Modarres, M., Kaminskiy, M., Krivtsov, V., 1999, *Reliability engineering and risk analysis: A practical guide*, Boca Raton, FL: CRC Press.
19. Distefano, S., Puliafito, A., 2009, Dependability evaluation with dynamic reliability block diagrams and dynamic fault trees, *IEEE Transactions on Dependable and Secure Computing*, 6(1), 4–17.
20. Elsayed, E.A., 1996, *Reliability engineering*, Reading, MA: Addison Wesley Longman.
21. Meeker, W.Q., Hahn, G.J., 1985, *Volume 10: How to plan an accelerated life test: Some practical guidelines*, Milwaukee: ASQ Basic References in Quality Control: Statistical Techniques.
22. Bertsche, B., 2010, *Reliability in automotive and mechanical engineering: Determination of component and system reliability*, Berlin: Springer.
23. O'Connor, P., Kleyner, A., 2012, *Practical reliability engineering*, 5th ed., Hoboken, NJ: Wiley & Sons.
24. Yang, G., 2007, *Life cycle reliability engineering*, Hoboken, NJ: Wiley.
25. Lewi, E., 1995, *Introduction to reliability engineering*, Hoboken, NJ: Wiley.
26. Saunders, S, 2007, *Reliability, life testing and the prediction of service lives*, New York: Springer.
27. Timoshenko, S.P., Young, D.H., 1968, *Elements of strength of materials*, 5th ed., Princeton, NJ: Van Nostrand Co.
28. Gere, J.M., 2004, *Mechanics of materials*, 6th ed., Boston, MA: Thomson Brooks Cole Learning, Inc.
29. Anderson, T., 2005, *Fracture mechanics – Fundamentals and applications*, 3rd ed., Boca Raton, FL: CRC Press.
30. Hertzberg, R.W., 1996, *Deformation and fracture mechanics of engineering materials*, 4th ed., Hoboken, NJ: John Wiley and Sons, Inc.
31. McPherson, J., 2010, *Reliability physics and engineering: Time-to-failure modeling*, New York: Springer.
32. Hrennikoff, A., 1941, Solution of problems of elasticity by the framework method, *Journal of Applied Mechanics*, 8, 169–175.
33. Courant, R., 1943, Variational methods for the solution of problems of equilibrium and vibrations, *Bulletin of the American Mathematical Society*, 49, 1–23.

34. Matsuishi, M., Endo, T., 1968, Fatigue of metals subjected to varying stress. *Japan Society of Mechanical Engineering*, 68 (2), 37–40.
35. Mott, R.L., 2004, *Machine elements in mechanical design*, 4th ed., Upper Saddle River, NJ: Pearson Prentice Hall, 190–192.
36. Palmgren, A.G., 1924, Die Lebensdauer von Kugellagern Zeitschrift des VereinesDeutscherIngenieure, *Scientific Research*, 68 (14), 339–341.
37. Woo, S., Pecht, M., 2008, Failure analysis and redesign of a helix upper dispenser, *Engineering Failure Analysis*, 15 (4), 642–653.
38. Woo, S., O'Neal, D., Pecht, M., 2009, Improving the reliability of a water dispenser lever in a refrigerator subjected to repetitive stresses, *Engineering Failure Analysis*, 16 (5), 1597–1606.
39. Woo, S., O'Neal, D., Pecht, M., 2009, Design of a hinge kit system in a kimchi refrigerator receiving repetitive stresses, *Engineering Failure Analysis*, 16 (5), 1655–1665.
40. Woo, S., Ryu, D., Pecht, M., 2009, Design evaluation of a French refrigerator drawer system subjected to repeated food storage loads, *Engineering Failure Analysis*, 16 (7), 2224–2234.
41. Woo, S., O'Neal, D., Pecht, M., 2010, Failure analysis and redesign of the evaporator tubing in a kimchi refrigerator, *Engineering Failure Analysis*, 17 (2), 369–379.
42. Woo, S., O'Neal, D., Pecht, M., 2010, Reliability design of a reciprocating compressor suction reed valve in a common refrigerator subjected to repetitive pressure loads, *Engineering Failure Analysis*, 7 (4), 979–991.
43. Woo, S., Pecht, M., O'Neal, D., 2009, Reliability design and case study of a refrigerator compressor subjected to repetitive loads, *International Journal of Refrigeration*, 32 (3), 478–486.
44. Woo, S., O'Neal, D., Pecht, M., 2011, Reliability design of residential sized refrigerators subjected to repetitive random vibration loads during rail transport, *Engineering Failure Analysis*, 18 (5), 1322–1332.
45. Woo, S., Park, J., Pecht, M., 2011, Reliability design and case study of refrigerator parts subjected to repetitive loads under consumer usage conditions, *Engineering Failure Analysis*, 18 (7), 1818–1830.
46. Woo, S., Park, J., Yoon, J., Jeon, H., 2012, Chapter 11: The Reliability Design and Its Direct Effect on the Energy Efficiency, Energy Efficiency. In: M. Eissa ed., *The innovative ways for smart energy, the future towards modern utilities,* Winchester: InTech.
47. Woo, S., 2015, Chapter 11: The Reliability Design of Mechanical System and Its Parametric ALT. In: A. S. H. Makhlouf and M. Aliofkhazraei, eds., *Handbook of materials failure analysis with case studies from the chemicals, concrete and power industries,* Oxford-Waltham, Butterworth Heinemann: Elsevier, 259–276.
48. Woo, S., O'Neal, D., 2016, Improving the reliability of a domestic refrigerator compressor subjected to repetitive loading, *Engineering*, 8, 99–115.
49. Woo, S., O'Neal, D., 2016, Reliability design of the hinge kit system subjected to repetitive loading in a commercial refrigerator, *Challenge Journal of Structural Mechanics*, 2 (2), 75–84.
50. Woo, S., 2017, *Reliability design of mechanical systems: A guide for mechanical and civil engineers*, 1st ed., Springer.
51. Woo, S., 2017, Chapter 2: Reliability Design of Mechanical Systems Subject to Repetitive Stresses, System Reliability. In: C. Volosencu, ed., *System reliability*, London, UK: InTech.
52. Woo, S., O'Neal, D., 2017, Improving reliability of the cooling enclosure system in refrigerator, *Innovations in Corrosion and Materials Science*, 2 (2), 127–134.
53. Woo, S., 2017, Improving the reliability of the design of a freezer drawer in a French door refrigerator subjected to repetitive food loading, *Recent Patents on Mechanical Engineering*, 10(4), 279–286.

54. Woo, S., 2017, Improving reliability of product subjected to random vibration during transportation, *WIT Transactions on The Built Environment in Urban Transport 2017*, 23, 361–371.
55. Woo, S., 2018, Chapter 9: Reliability Design of Mechanical Systems Subject to Repetitive Stresses, Volume 6. In: A. S. H. Makhlouf and M. Aliofkhazraei, eds., *Handbook of materials failure analysis with case studies from the construction industries*, Oxford-Waltham, Butterworth Heinemann: Elsevier, 171–186.
56. Woo, S., O'Neal, D., 2018, Improving the noise of mechanical compressor subjected to repetitive pressure loading, *Journal of Maintenance Engineering*, 2, 210–225.
57. Woo, S., O'Neal, D., 2019, Improving the reliability of mechanical components that have failed in the field due to repetitive stress, *Metals*, 9 (1), 38.
58. Woo, S., O'Neal, D., 2019, Reliability design and case study of mechanical system like a hinge kit system in refrigerator subjected to repetitive stresses, *Engineering Failure Analysis*, 99, 319–329.
59. Woo, S., Pecht, M., O'Neal, D., 2020, Reliability design and case study of the domestic compressor subjected to repetitive internal stresses, *Reliability Engineering & System Safety*, 193, 106604.
60. Woo, S., 2020, Chapter 8: Sample Size Equation of Mechanical System for Parametric Accelerated Life Testing. In: M. A. Mellal, ed., *Soft computing methods for system dependability,* Hershey, PA: IGI Global.
61. Woo, S., Matvienko, Y.G., O'Neal, D., 2020, Improving fatigue of mechanical systems such as freezer drawer subject to repetitive stresses, *Engineering Failure Analysis*, 110, 104404.
62. IEEE Std 610.12-1990, Sep. 1990, *IEEE standard glossary of software engineering terminology*, New York: Standards Coordinating Committee of the Computer Society of IEEE.
63. Klutke, G., Kiessler, P.C., Wortman, M.A., 2015, A critical look at the bathtub curve, *IEEE Transactions on Reliability*, 52 (1), 125–129.
64. Kreyszig, E., 2006, *Advanced engineering mathematics*, 9th ed., Hoboken, NJ: John Wiley and Son, 683
65. Grove, A., 1967, *Physics and technology of semiconductor device*, 1st ed., Wiley International Edition, 37.
66. ASM International, 1989, *Electronic materials handbook Volume 1: Packaging*,888
67. Karnopp, D.C., Margolis, D.L., Rosenberg, R.C., 2012, *System dynamics: Modeling, simulation, and control of mechatronic systems*, 5th ed., New York: John Wiley & Sons.
68. Ajiki, T., Sugimoto, M., Higuchi, H., 1979, A new cyclic biased THB power dissipating ICs. *Proceedings of the 17th International Reliability Physics Symposium*, San Diego CA.
69. Wasserman, G., 2003, *Reliability verification, testing, and analysis in engineering design*. New York: Marcel Dekker, 228.
70. Ryu, D., Chang, S., 2005, Novel concept for reliability technology, *Microelectronics Reliability*, 45(3), 611–622.
71. Lee, S.Y., 2003, *Reliability engineering*, Seoul: Hyung Seol, 73.
72. Hughes, A., 1993, *Electric motors and drives*. Oxford, UK: Newness.

9 Case Studies of Parametric Accelerated Life Testing

9.1 RELIABILITY DESIGN OF THE HELIX UPPER DISPENSER IN AN ICEMAKER

The basic function of a refrigerator is to store fresh and/or frozen foods. Since customers need ice, an icemaker is designed to harvest ice. The primary parts in an icemaker consist of a bucket case, helix support, helix dispenser clamp, blade dispenser, helix upper dispenser, and blade, as shown in Figure 9.1.

In the field, these icemaker parts in a refrigerator cracked and fractured due to design failures under unknown customer usage conditions. Field data indicated that the damaged products may have had two structural design flaws: (1) a 2-mm gap between the blade dispenser and the helix upper dispenser and (2) a weld line around the impact area of the helix upper dispenser. Below −20°C, the rotating blade dispenser (stainless steel) impacts the fixed helix upper dispenser (plastic) while the crushed ice is being harvested. A crack may occur in the helix dispenser (Figure 9.2).

By utilizing the above failure analysis (and tests), fractures that started in voids have been shown to propagate to their ends. Cracks in the helix dispenser require the manufacturer to redesign the product to keep it functioning for its expected lifetime. If there are design faults when the product structure is subjected to repetitive loads, the structure will fail short of its desired lifetime. Therefore, an icemaker's actual lifetime depends on problematic parts like the helix upper dispenser. To reproduce the part(s) and modify them, an engineer is required to conduct parametric accelerated life testing (ALT) as reliability testing for a new design. The process consists of (1) a load analysis for the returned product, (2) the utilization of parametric ALTs with action plans, and (3) the verification of whether the lifetime target of final designs has been achieved.

Ice-making involves several repetitive mechanical processes: (1) filtered water supplies the tray; (2) water freezes into ice via cold air in the heat exchanger; (3) the ice harvests until the bucket is full; and (4) when the customer pushes the lever by force, cubed or crushed ice is then dispensed. In this ice-making process, the icemaker parts receive a variety of mechanical loads. In the United States, refrigerators are designed to produce 10 cubes per use and up to 200 cubes a day. Because the icemaker system is repetitively used in both cubed and crushed ice modes, it is continuously subjected to mechanical loads. Ice production may also be influenced by customer usage conditions such as water pressure, ice consumption, refrigerator notch settings, and the number of times the door is opened.

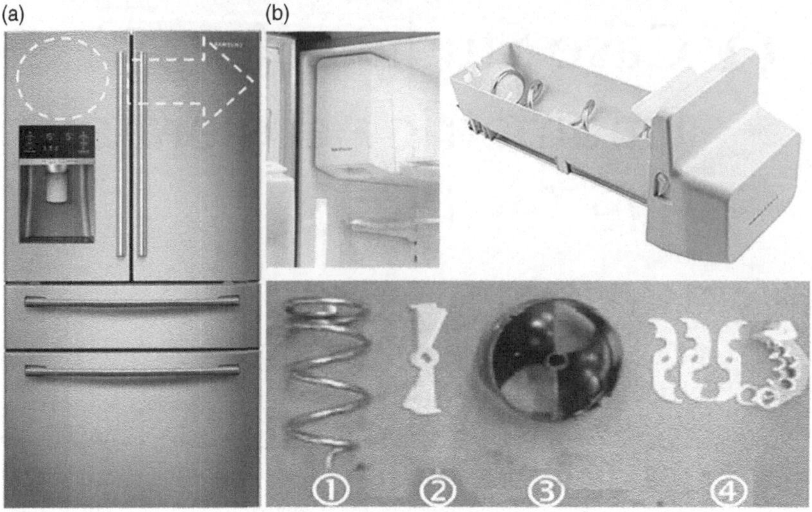

FIGURE 9.1 Refrigerator and icemaker assembly. (a) Refrigerator. (b) Mechanical parts of an icemaker assembly: helix support (1), blade dispenser (2), helix upper dispenser (3), and blade (4).

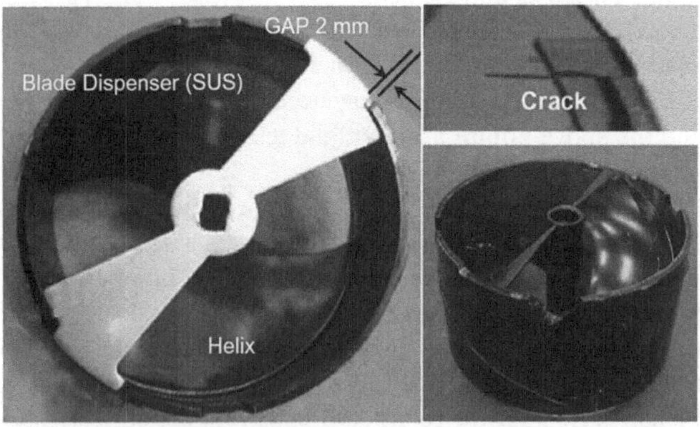

FIGURE 9.2 Damaged helix upper dispenser after use.

Figure 9.3 provides an overview of the schematic of an icemaker that represents the mechanical load transfer in the ice bucket assembly using a bond graph model. To generate enough torque to crush the ice, an AC motor provides power through the gear system, which is then transferred to the ice bucket assembly. Through the helix blade dispenser and the upper dispenser in the bucket, the ice is distributed by the blade. If subjected to different loads, the ice can also be crushed.

To derive the state equations, the bond graph model in Figure 9.3b can be solved at each node, that is,

$$df \times E_2/dt = 1/L_a \times eE_2 \tag{9.1}$$

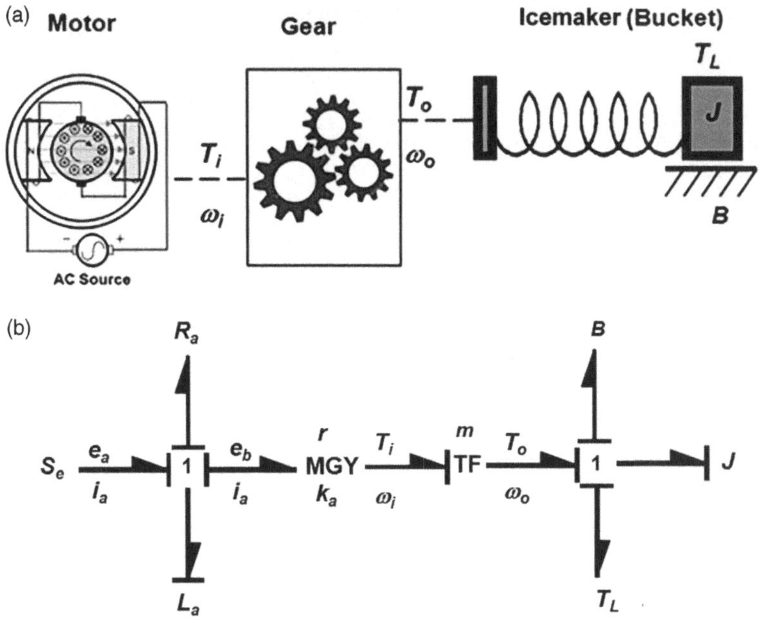

FIGURE 9.3 Design concept of an icemaker: (a) schematic diagram of the auger motor and ice bucket assembly and (b) bond graph modeling of an auger motor and ice bucket assembly.

$$dfM_2/dt = 1/J \times eM_2 \tag{9.2}$$

The junction from Equation (9.1) is

$$eE_2 = e_a - eE_3 \tag{9.3a}$$

$$eE_3 = R_a \times fE_3 \tag{9.3b}$$

The junction from Equation (9.2) is

$$eM_2 = eM_1 - eM_3 \tag{9.4a}$$

$$eM_1 = (K_a \times i) - T_{\text{Pulse}} \tag{9.4b}$$

$$eM_3 = B \times fM_3 \tag{9.4c}$$

Because $fM_1 = fM_2 = fM_3 = \omega$ and $i = fE_1 = fE_2 = fE_3 = i_a$ from Equation (9.3a and b),

$$eE_2 = e_a - R_a \times fE_3 \tag{9.5}$$

$$fE_2 = fE_3 = i_a \tag{9.6}$$

If substituting Equations (9.5) and (9.6) into Equation (9.1), then

$$di_a/dt = 1/L_a \times (e_a - R_a \times i_a) \tag{9.7}$$

In addition, from Equations (9.4), we can obtain

$$eM_2 = [(K_a \times i) - T_L] - B \times fM_3 \tag{9.8a}$$

$$i = i_a \tag{9.8b}$$

$$fM_3 = fM_2 = \omega \tag{9.8c}$$

If substituting Equations (9.8) into (9.2), then

$$d\omega/dt = 1/J \times [(K_a \times i) - T_L] - B \times \omega \tag{9.9}$$

We can obtain the state equation from Equations (9.8) and (9.9) as follows:

$$\begin{bmatrix} di_a/dt \\ d\omega/dt \end{bmatrix} = \begin{bmatrix} -R_a/L_a & 0 \\ mk_a & -B/J \end{bmatrix} \begin{bmatrix} i_a \\ \omega \end{bmatrix} + \begin{bmatrix} 1/L_a \\ 0 \end{bmatrix} e_a + \begin{bmatrix} 1 \\ -1/J \end{bmatrix} T_L \tag{9.10}$$

When Equation (9.10) is integrated, the output of the AC-motor and ice bucket assembly is obtained as

$$y_p = \begin{bmatrix} 0 & 1 \end{bmatrix} \begin{bmatrix} i_a \\ \omega \end{bmatrix} \tag{9.11}$$

From Equation (9.10), we know that the lifetime of the ice bucket assembly depends on the stress (or torque) due to forces required to crush the ice. The life-stress (LS) model in Equation (8.16) can then be modified as

$$TF = A(S)^{-n} = AT_L^{-\lambda} = A(F_c \times R)^{-\lambda} = B(F_c)^{-\lambda} \tag{9.12}$$

Therefore, the AF in Equation (8.17) can be modified as

$$AF = \left(\frac{S_1}{S_0}\right)^n = \left(\frac{T_1}{T_0}\right)^\lambda = \left(\frac{F_1 \times R}{F_0 \times R}\right)^\lambda = \left(\frac{F_1}{F_0}\right)^\lambda \tag{9.13}$$

We can carry out parametric ALT from Equation (8.29) until the required mission cycles that provide the reliability target of 10 years of B1 life are achieved.

The environmental operating conditions of the ice bucket assembly in a refrigerator icemaker can vary from approximately −15°C to −30°C with a relative humidity ranging from 0% to 20%. Depending on customer usage, an ice dispenser is used on an average of approximately 3–18 times per day. Under maximum use for 10 years, the dispenser incurs about 65,700 usage cycles. Data from the motor company specify that the normal torque is 0.69 kN-cm and the maximum torque is 1.47 kN-cm. Assuming the cumulative damage exponent $\lambda = 2$, the acceleration factor (AF) is approximately 5 in Equation (9.13).

For 10 years of B1 life, the test cycles for a sample of ten pieces (calculated using Equation 8.29) were approximately 42,000 cycles if the shape parameter was supposed to be 2.0. This parametric ALT is designed to ensure a B1 life of 10 years so that it would fail less than once during 42,000 cycles. Figure 9.4 shows the experimental setup of an ALT for the reproduction of the failed helix upper dispenser in the field. Figure 9.5 presents the duty cycles for the ice-crushing load T_L.

The equipment in the chamber was designed to operate down to a temperature of about −30°C. The controller outside can start or stop the equipment and can indicate the completed test cycles and the test periods, such as sample on/off time. To apply the maximum ice-crushing torque T_L, the helix upper dispenser and the blade dispenser were bolted together using a band clamper. As the controller gives the start signal, the auger motor rotates. At this point, the rotating blade dispenser impacts the fixed helix upper dispenser to the maximum mechanical ice-crushing torque (1.47 kN-cm).

In the first ALT, the helix upper dispenser fractured at 170, 5,200, 7,880, 8,800, and 11,600 cycles. Figure 9.6 shows a photograph comparing the failed product from

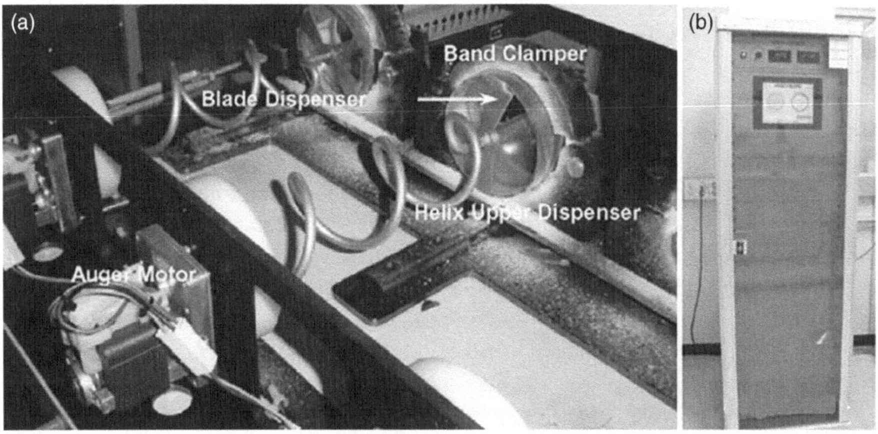

FIGURE 9.4 Equipment used in ALT and controller. (a) ALT equipment and (b) controller.

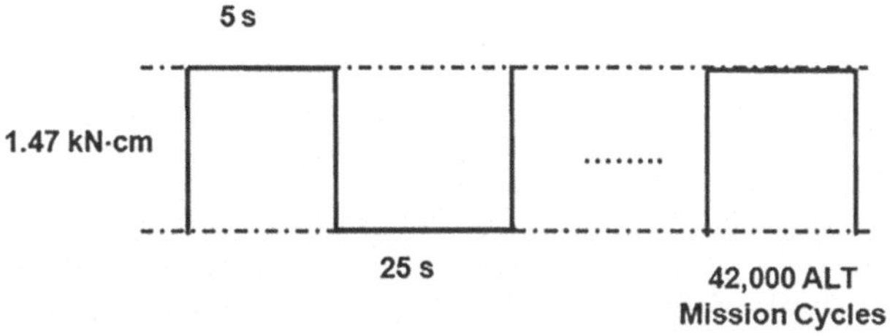

FIGURE 9.5 Duty cycles of disturbance load T_L on the band clamper.

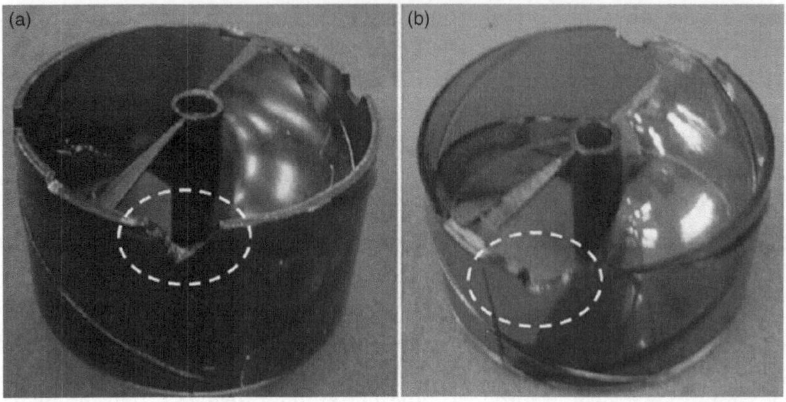

FIGURE 9.6 Failed helix upper dispensers in the field and during the first ALT: (a) failed product in the field and (b) product with crack after the first ALT.

the field and that from the first ALT, respectively. Because they are similar, by parametric ALT we were able to reproduce the fractured helix upper dispenser. There was a structural design flaw – a 2-mm gap between the blade dispenser and the helix upper dispenser. As the blade dispenser (stainless steel) struck the helix upper dispenser (plastic), it cracked and fractured. Figure 9.7 presents the graphical analysis of the ALT results and field data on a Weibull plot. The shape parameter in the first ALT was estimated to be 2.0. For the final design, the shape parameter from the Weibull plot was confirmed to be 4.78.

To withstand repetitive impact loads, the problematic helix upper dispenser used in the field was redesigned as follows: the 2 mm gap between the blade dispenser and helix upper dispenser was eliminated (Figure 9.8).

In the second ALT, the helix upper dispenser fractured at 17,000, 25,000, 28,200, and 38,000 cycles. When the gap between the blade dispenser and the helix upper dispenser was eliminated, the lifetime of the helix upper dispenser was extended. Because the helix upper dispenser did not have enough strength for stress, the target of 42,000 mission cycles in the second ALT was not met. As an action plan, a reinforced rib on the outside of the helix was added.

In the third ALT, there were no problems until 75,000 cycles. Throughout three ALTs with these design changes, the helix upper dispenser was guaranteed to have 10 years of B1 life. Table 9.1 shows a summary of the results of the ALTs.

9.2 RESIDENTIAL-SIZED REFRIGERATORS DURING TRANSPORTATION

To evaluate the ride quality of products mounted on a vehicle such as a train or an automobile, the most useful mathematical model of a vehicle suspension system is a quarter car model. Though it has two degrees of freedom and four state variables, it serves the purpose of figuring out the vehicle motion in transportation. The assumed

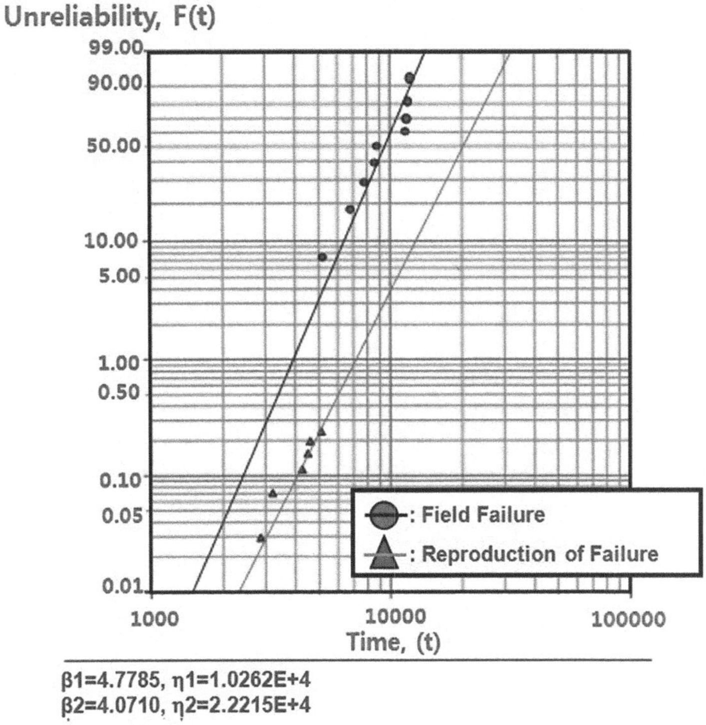

FIGURE 9.7 Field data and results of ALT on a Weibull chart.

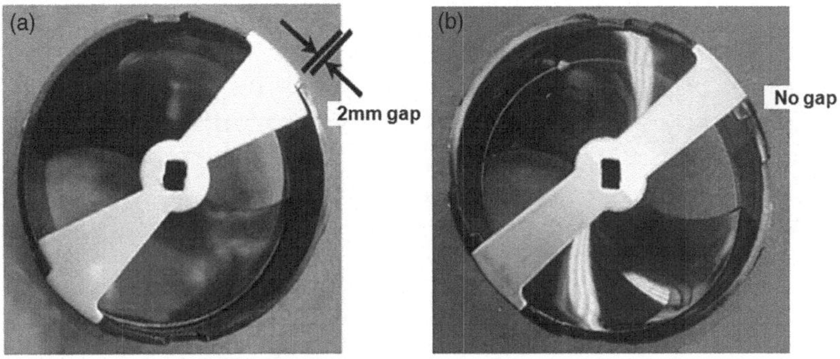

FIGURE 9.8 Redesigned helix upper dispenser: (a) old design and (b) new design.

model of the vehicle consists of the sprung mass and the unsprung mass, respectively. The sprung mass, m_s, represents 1/4 of the body of the vehicle, and the unsprung mass, m_{us}, represents one wheel of the vehicle. The main suspension is modeled as a spring k_s and a damper c_s in parallel, which connects the unsprung to the sprung mass. The tire (or rail) is modeled as a spring k_{us} and represents the transfer of the road force to the unsprung mass (Figure 9.9).

TABLE 9.1
Results of ALTs in Reliability Design of the Helix Upper Dispenser in an Ice-Maker

	First ALT	Second ALT	Third ALT
Parametric ALT	**Initial Design**	**Second Design**	**Final Design**
Over the course of 42,000 cycles, the helix upper dispenser has no problems	170 cycles: 1/10 fracture 5,200 cycles: 1/10 fracture 7,880 cycles: 2/10 fracture 8,880 cycles: 2/10 fracture 11,600 cycles: 4/10 fracture	17,000 cycles: 1/6 fracture 25,000 cycles: 3/6 fracture 28,000 cycles: 1/6 fracture 38,000 cycles: 1/6 fracture	42,000 cycles: 6/6 OK 75,000 cycles: 6/6 OK
Helix structure			–
Material and specification	C1: Gap of 2 mm → 0 mm	C2: Added rib on the outside of the helix	–

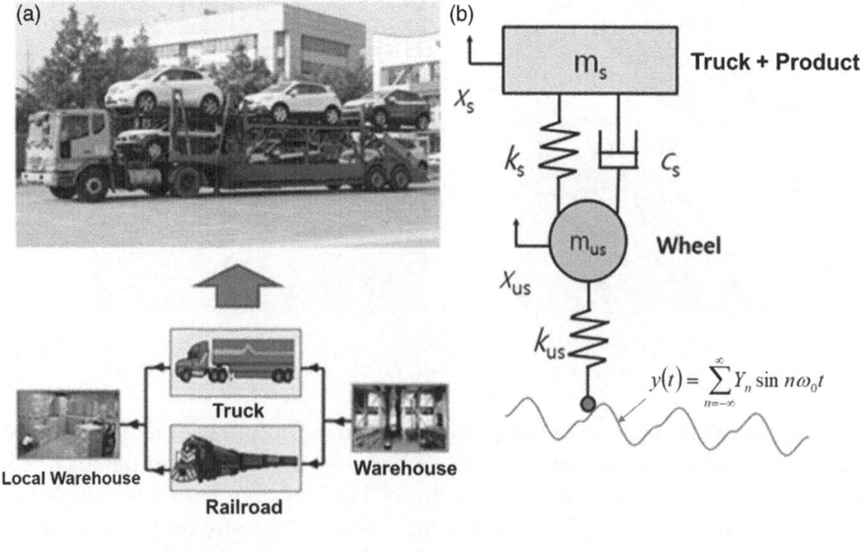

FIGURE 9.9 Product during transportation and its model: (a) a different mode of transportation and (b) a quarter car model.

The governing differential equations of motion for the quarter car model can be represented as follows:

$$m_s\ddot{x}_s + c_s(\dot{x}_s - \dot{x}_{us}) + k_s(x_s - x_{us}) = 0 \tag{9.14}$$

$$m_{us}\ddot{x}_{us} + c_s(\dot{x}_{us} - \dot{x}_s) + (k_{us} + k_s)x_{us} - k_s x_s = k_{us}y \tag{9.15}$$

So the above equations of motion can be concisely represented as follows:

$$\begin{bmatrix} m_s & 0 \\ 0 & m_{us} \end{bmatrix}\begin{bmatrix} \ddot{x}_s \\ \ddot{x}_{us} \end{bmatrix} + \begin{bmatrix} c_s & -c_s \\ -c_s & c_s \end{bmatrix}\begin{bmatrix} \dot{x}_s \\ \dot{x}_{us} \end{bmatrix}$$

$$+\begin{bmatrix} k_s & -k_s \\ -k_s & k_{us} + k_s \end{bmatrix}\begin{bmatrix} x_s \\ x_{us} \end{bmatrix} = \begin{bmatrix} 0 \\ k_{us}y \end{bmatrix} \tag{9.16}$$

As a result, Equation (9.16) can be expressed in a matrix form:

$$[M]\ddot{X} + [C]\dot{X} + [K]X = F \tag{9.17}$$

When Equation (9.17) is numerically integrated, we can find the time response of the state variables due to random vibration. By the Laplace transformation, we can obtain the power spectral density in the frequency domain (Figure 9.10).

A product in transportation would ride on a rough road or rail or wave height on the water or an airplane inducing wing load during flight. The random vibrations in the field are that the higher amplitude is the smaller frequency. On the other hand, the lower amplitude is the larger frequency. We can say that it follows the normal distribution (Figure 9.10a). If the response of the quarter car model displays in the frequency domain, we can obtain the plot of power spectral density as in Figure 9.10b that is the usual way to specify random vibration. We can judge the amplitude of the natural frequency of a product when random vibration is transmitted from the road in transportation.

The force transmitted to the product can be expressed as Q:

$$Q = \frac{F_T}{kY} = r^2\left[\frac{1 + (2\zeta r)^2}{(1 - r^2)^2 + (2\zeta r)^2}\right] \tag{9.18}$$

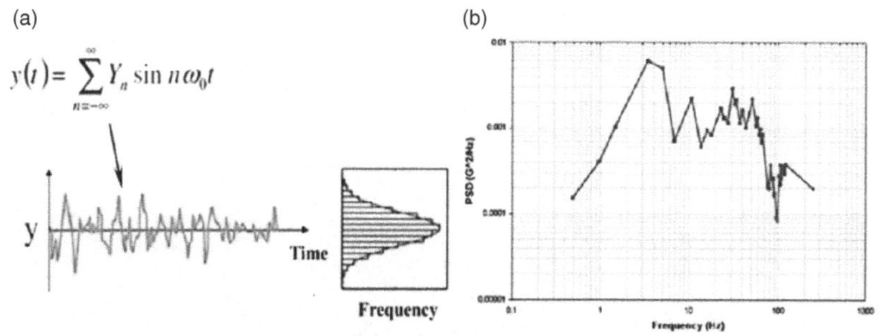

FIGURE 9.10 Base random vibrations and intermodal random vibration: (a) base random vibrations and (b) intermodal random vibration in the frequency domain.

Because the stress of the product in transportation comes from the transmitted vibration load (F_T) from the basis, Equation (8.16) can be modified as follows:

$$TF = A(S)^{-n} = A(F_T)^{-\lambda} \tag{9.19}$$

The AF in Equation (8.17) can be represented as the product of the amplitude ratio of gravitational acceleration R and force transmissibility Q. That is,

$$AF = \left(\frac{S_1}{S_0}\right)^n = \left(\frac{F_1}{F_0}\right)^\lambda = \left(\frac{g_1}{g_0}\frac{F_T}{kY}\right)^\lambda = (R \times Q)^\lambda \tag{9.20}$$

In the field, as the compressor rubber mounts in the mechanical compartments of refrigerators tear, the connecting tubes fracture under uninformed rail conditions. According to the market investigation, the distance of the first failure in rail transportation was approximately 2,500 km in 2 days. In Chicago, 27% of whole transported products were roughly failed. For 7 days, when the refrigerators were transported to 7,200 km that is the total distance from Los Angles to Boston, 67% of whole products were fractured (Figure 9.11).

As seen in Figure 9.12, a refrigerator is made up of a condenser, a capillary tube, an evaporator, and a compressor. As the refrigerating cycle transfers heat from the internal room of the refrigerator to the surrounding, the refrigerator can store fresh

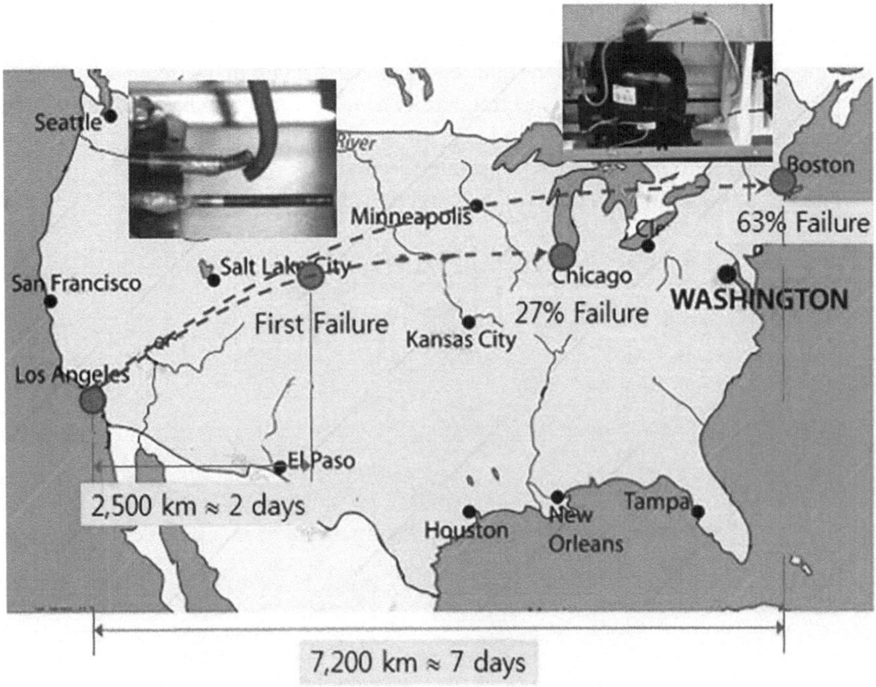

FIGURE 9.11 Failed locations in the field.

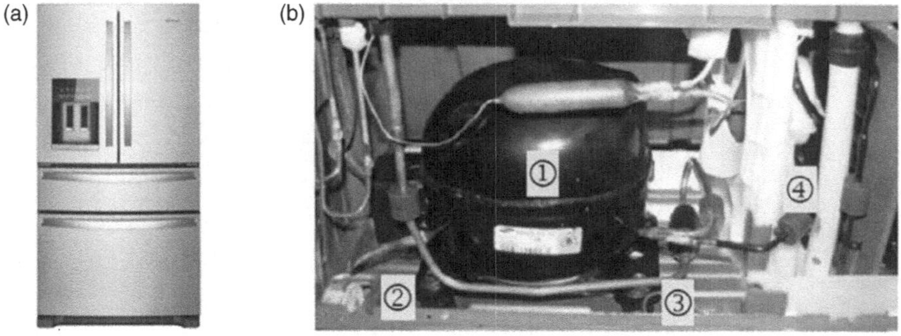

FIGURE 9.12 French door refrigerator and a mechanical compartment: (a) French door refrigerator and (b) mechanical compartment: (1) compressor, (2) rubber, (3) connecting tubes, and (4) fan and condenser.

or frozen food. To sell the refrigerator, after producing in a factory, the refrigerator is transported to the end-user by transportation means – air, rail, road, and water. When refrigerators are transported by a train, an automobile, or a ship, they are subjected to a different mode of random vibration loads. These random vibrations from the base in transportation are continually transmitted to the refrigerators. A product such as a refrigerator uses compressor rubber mounts to maintain proper stiffness and protect the system from the random vibration. Eventually, the refrigerators might be designed to outlive the damage that may occur during transport.

If improperly designed, the rubber and the tubes in the mechanical compartment subjected to the repeated random vibration in transportation fracture. The refrigerant, therefore, would leak out of the tubes and would not function anymore. Field data also showed that the failed refrigerators might have had design defects that prevented protection from random vibration during transportation. These design flaws could cause a crack to occur and thus result in failure. After reproducing the failed refrigerator in a laboratory test, the root causes of the problematic design should be exactly found.

The lifetime of a refrigerator in laboratory testing was targeted to be less than B1 life while transporting from Los Angles to Boston. Based on the field data, rail transportation was expected to move a refrigerator 2,500 km in 2 days. The refrigerator was transported 7,200 km from Los Angeles to Boston for a 7-day total travel transportation distance. From Equation (9.20), the AF is the product of the amplitude ratio of acceleration R and force transmissibility Q.

To obtain the accelerated factor, when the refrigerator was given an acceleration of 1 g on a shaker table, we could find that the natural frequency of horizontal vibration (left ↔ right) was 5 Hz. On the other hand, the natural frequency of vertical vibration (up ↔ down) was 9 Hz in the package test (Figure 9.13). Suppose that a small damping ratio ($\zeta = 0.1$) at a natural frequency ($r = 1.0$) was given, the force transmissibility had a value of approximately 5.1. The amplitude ratio of acceleration was 4.17 because the refrigerator was given an acceleration of 1 g on a shaker table. Using an accumulative damage exponent of 2.0, we knew that the AF was found to be 452.0 from Equation (9.18) (Table 9.2).

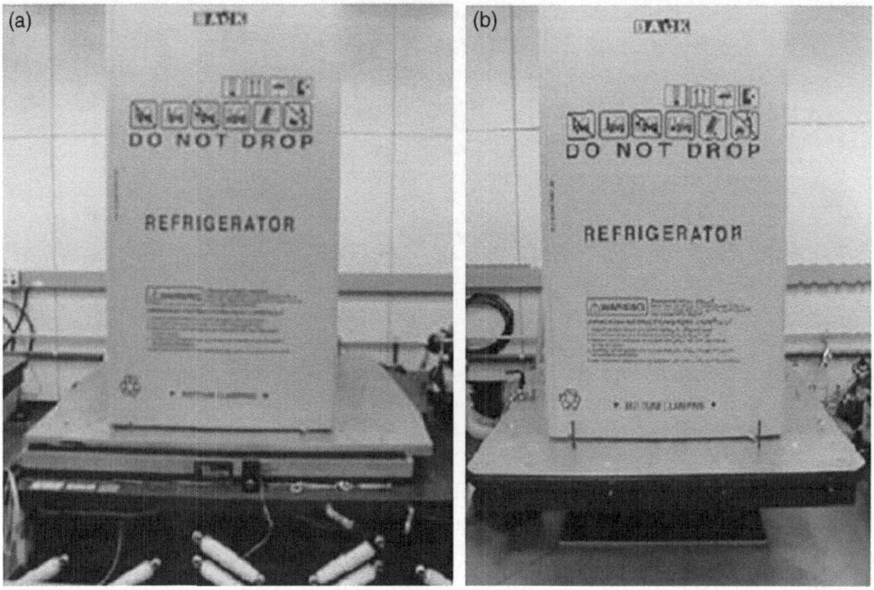

FIGURE 9.13 Vibration test equipment for the parametric ALT: (a) horizontal vibration (left ↔ right) and (b) vertical vibration (up ↔ down).

TABLE 9.2
ALT Conditions for Accelerated Testing of the Refrigerator

System Conditions	Worst Case	ALT	*AF*
Transmissibility, Q ($r = 1.0$, $\zeta = 0.1$)	–	5.1 (From Equation 9.20)	5.1①
Amplitude ratio of acceleration, R (a_1/a_0)	0.24 g	1 g	4.17②
Total AF (= (① × ②)^2)			452

The required testing time for three samples can be obtained from Equation (8.26) if the lifetime target and accelerated factor are given. Suppose that the shape parameter in the Weibull chart was 2.0, the test time obtained from Equation (8.26) was approximately 130 min for three pieces, respectively. If the refrigerator fails less than once during 130 min, the refrigerator for a total travel distance of 7,200 km (7 days) was designed to ensure B1 life with about a 60% level of confidence.

Three refrigerators in the first parametric ALT were fractured from horizontal vibration (left ↔ right) within 40 min. Figure 9.14 shows the failed products from the field and the fractured samples from the ALT, respectively. The photos show that the shapes and locations of the failures in the ALT were similar to those seen in the field.

A graphical analysis of the ALT results and field data on the Weibull plot is shown in Figure 9.15. From the Weibull plot, the shape parameter was finally obtained to be 6.13. As seen in Figure 9.16a, the tearing in the first ALT occurred because of

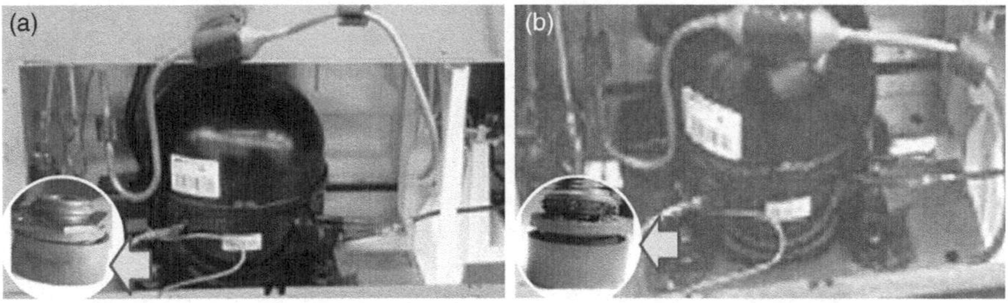

FIGURE 9.14 Failure of refrigerator tubes in the field and first ALT results: (a) field and (b) first ALT results.

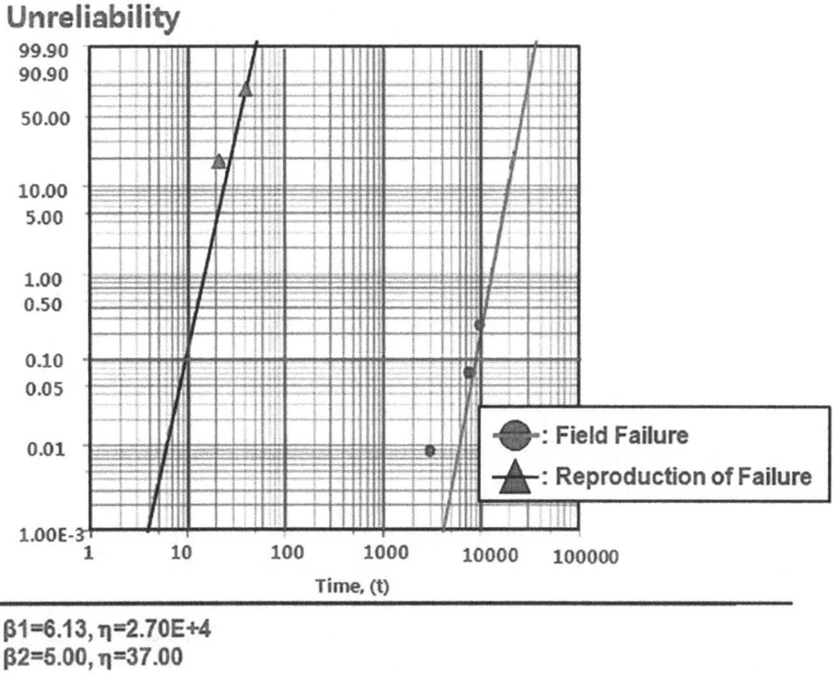

FIGURE 9.15 A graphical analysis of the ALT results and field data on a Weibull plot.

no support shape of the rubber mount for *X*-axis vibration (left ↔ right). We knew that parametric ALT was valid because it pinpoints the design weaknesses that were responsible for the failures in the field.

The design flaws of the refrigerator that might have an improper stiffness came from the shape of the compressor rubber and connecting tube design (Figure 9.16b). Fractured tubes between the compressor and condenser areas could be corrected by modifying the shape of the compressor rubber mount and the connecting tube design.

The rubber in the refrigerator was reshaped as follows: (1) a redesigned compressor rubber mount shape, C1 (Figure 9.16a) and (2) a redesigned connecting tube shape,

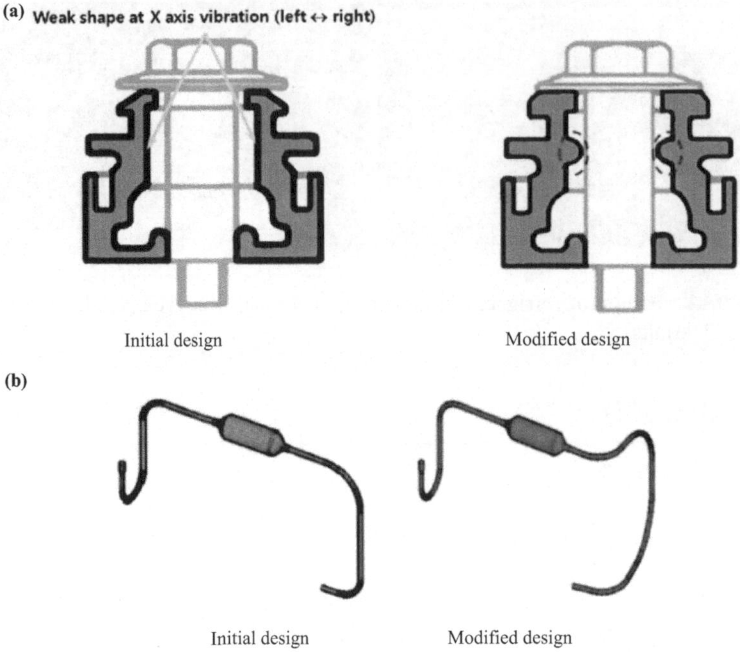

FIGURE 9.16 Problematic rubber and connecting tube shape in a refrigerator: (a) problematic rubber in the refrigerator and (b) problematic connecting tube shape in the refrigerator.

C2 (Figure 9.16b). With these modified parameters, second ALTs for the refrigerator were performed. The design targets of the newly designed samples were less than the target life of a B1 life for the total travel distance (7 days). The confirmed values of AF and β in Table 9.3 and Figure 9.13 were 452.0 and 6.41, respectively.

The recalculated test time in Equation (8.29) for three sample pieces was 40 min that would be the specification of parametric ALT. In the second ALT, because the refrigerators were not fractured until 60 min, we knew that the modified designs were effective to protect the refrigerator from the random vibrations. Moreover, when the refrigerator was given an acceleration of 1 g on a shaker table, we knew that the natural frequency of horizontal vibration (left ↔ right) increased from 5 to 8 Hz. It came from the increase of the damping force due to the design modifications (Figure 9.17).

Table 9.3 summarizes the results of the ALTs. Through two rounds of ALTs, the modified refrigerators with the targeted B1 life were expected to survive without failure for a total travel distance of 7,200 km (7 days).

9.3 HINGE KIT SYSTEM (HKS) IN A KIMCHI REFRIGERATOR

A refrigerator is designed to provide cold air from the evaporator to the freezer (or refrigerator) compartment to store food fresh. To close the refrigerator door conveniently, an HKS with a spring-damper mechanism was designed. Before launching the newly designed HKS, it was necessary to find possible design faults

TABLE 9.3
Results of ALTs in Residential-Sized Refrigerators During Transportation

	First ALT	Second ALT
Parametric ALT	**Initial Design**	**Final Design**
In 40 min, there is no fracture of the connecting tube in the refrigerator	20 min: 1/3 fracture 40 min: 2/3 fracture	40 min: 3/3 OK 60 min: 3/3 OK
Photo	<Machine room in refrigerator> 	
Action plans	C1: Shape of the compressor rubber C2: Connecting tube shape	

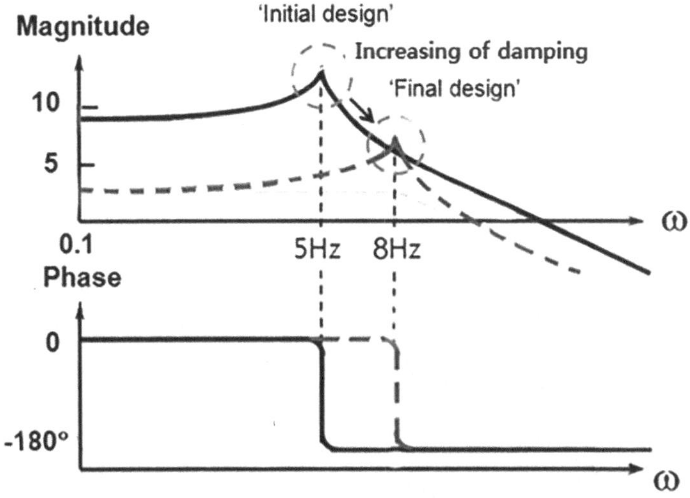

FIGURE 9.17 The change of the natural frequency of a product due to the design modifications.

and verify its reliability. The primary components in the HKS consisted of a kit cover, shaft, cam, spring, and an oil damper, as shown in Figure 9.18.

In the field, the HKS parts in a refrigerator were cracking and fracturing due to design failures. Engineers did not know which design tests were needed to simulate the actual customer usage and load conditions. To keep the product functioning for

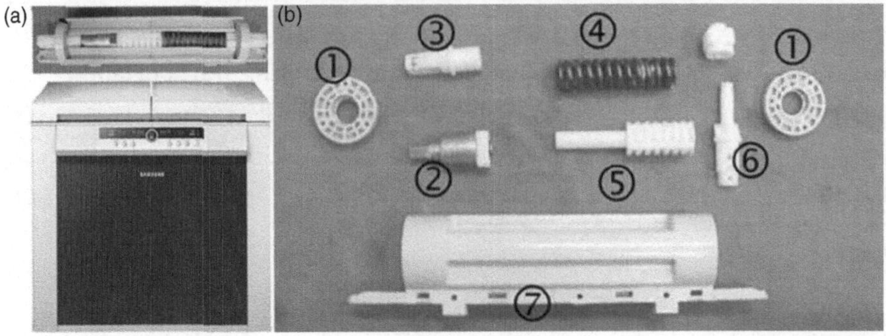

FIGURE 9.18 Refrigerator and mechanical parts of HKS: (a) Kimchi refrigerator and (b) mechanical parts of the HKS: (1) kit cover, (2) oil damper, (4) spring, (6) shaft, and (7) oil damper.

its expected lifetime, a manufacturer should design a mechanical system optimally and robustly. If there are design faults when the product structure is subjected to repetitive loads, the structure will fail before its desired lifetime. Therefore, an HKS product's actual lifetime depends on the faulty parts. To reproduce and correct the part(s), a design engineer requires a reliable methodology for identifying failing parts or systems in the design. The process consists of (1) a failure (or load) analysis for the returned product, (2) utilizing parametric ALTs with action plans, and (3) verifying if the reliability target of final designs is achieved.

Based on the identified consumer usage conditions in the field, the HKS was subjected to different loads during the opening and closing of the refrigerator door (Figure 9.19).

The moment balance around HKS can be expressed as

$$M_0 = W_{\text{door}} \times b = T_0 = F_0 \times R \tag{9.21}$$

If the accelerated weight on the refrigerator door was added, the moment balance around HKS can be modified as

$$\sum M = M_1 = M_0 + M_A = W_{\text{door}} \times b + W_A \times a = T_1 = F_1 \times R \tag{9.22}$$

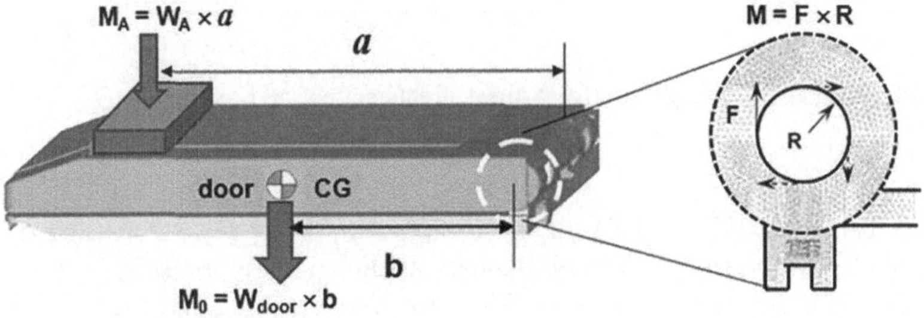

FIGURE 9.19 Design concept of HKS.

Under the same environmental conditions, the LS model in Equation (8.14) can be modified as follows:

$$TF = A(S)^{-n} = AT^{-\lambda} = A(F \times R)^{-\lambda} = B(F)^{-\lambda} \quad (9.23)$$

where A and B are constants.

We know that the product lifetime depends on the applied impact force. Therefore, the AF from Equation (8.15) can be redefined as

$$AF = \left(\frac{S_1}{S_0}\right)^n = \left(\frac{T_1}{T_0}\right)^\lambda = \left(\frac{F_1 \times R}{F_0 \times R}\right)^\lambda = \left(\frac{F_1}{F_0}\right)^\lambda \quad (9.24)$$

The environmental operating conditions of an HKS in a refrigerator varied from approximately 0°C to 43°C with a relative humidity ranging from 0% to 95%. The HKS can be subjected to between 0.2 and 0.24 g of acceleration. The number of door closing cycles is influenced by specific consumer usage patterns. Consumer data showed that the door system of the refrigerator was typically opened and closed between three and ten times per day in the Korean domestic market. With a life cycle design point of 10 years, the life of the HKS L_B^* incurred about 36,500 usage cycles for the worst-case scenario.

For this worst case, the impact force around the HKS was 1.10 kN, which was the expected maximum force applied by the typical consumer. For the ALT with an accelerated weight, the impact force on the HKS was 2.76 kN. Using a stress dependence of 2.0, the AF was found to be approximately 6.3 in Equation (9.24).

For the 10-year life with an accumulated failure rate of 1%, the test cycles for sample six pieces calculated in Equation (8.29) were 24,000 cycles if the shape parameter was supposed to be 2.0. This parametric ALT was designed to ensure a 10-year life with an accumulated failure rate of 1% in about a 60% level of confidence that it would fail less than once during 23,000 cycles. Figure 9.20 shows the experimental setup of the ALT with labeled equipment for the reliability design of HKS. Repetitive stress can be expressed as the duty effect due to the on/off cycles and HKS shortens part life.

In the first ALT, the housing of the HKS fractured at 3,000 and 15,000 cycles. Figure 9.21 shows a photograph comparing the failed product from the field and that from the first ALT, respectively. Figure 9.22 presents the graphical analysis of the ALT results and field data on a Weibull plot. The shape parameter in the first ALT was estimated to be 2.0. For the final design, the shape parameter from the Weibull plot was confirmed to be 2.0. As seen in the field and first ALT, they were very similar. By parametric ALT, we reproduced the problematic housing structure of HKS.

If there are faulty designs in the structure where the loads are applied, the HKS failure might fracture in its lifetime. Therefore, to have enough strength against repetitive loads, the fragile structure of the hinge kit housing system was redesigned. The notch was eliminated and the design was rounded outside and inside. Reinforced ribs were also added to the housing and decks (Figure 9.23).

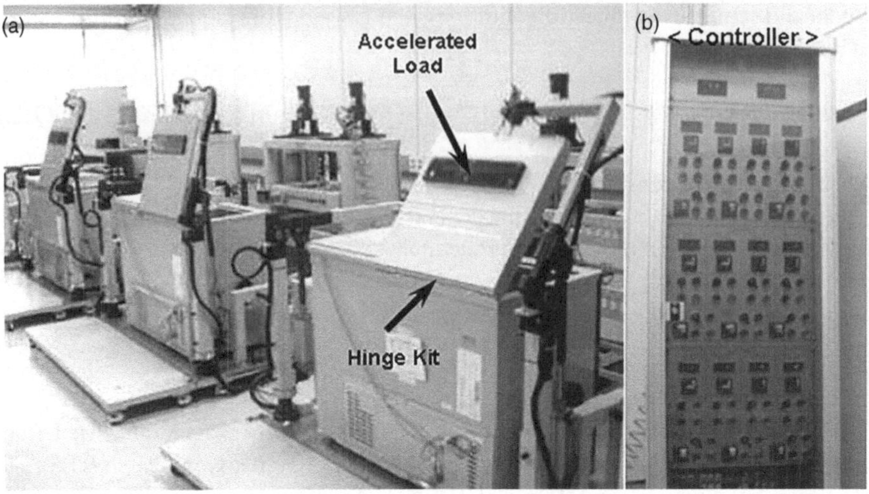

FIGURE 9.20 Equipment used in ALT and controller: (a) ALT equipment and (b) Controller.

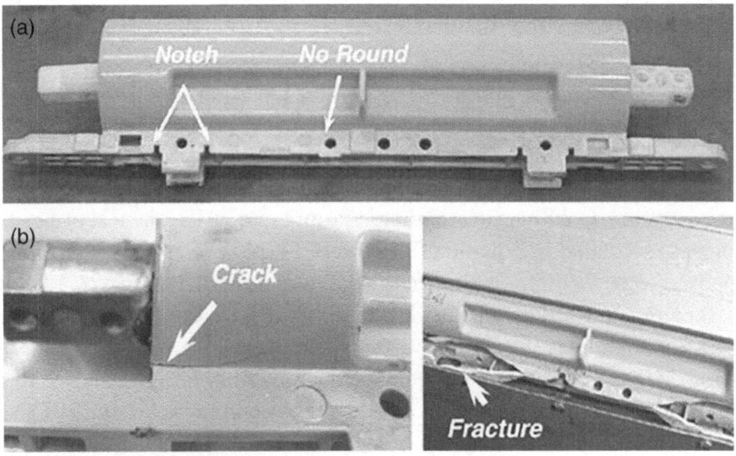

FIGURE 9.21 Failed products in the field and crack after the first ALT: (a) failed products in the field and (b) crack after the first ALT.

The maximum concentrated stresses of the housing hinge kit were approximately 21.2 MPa when finite element analysis was carried out. The high-stress risers came from the design flaws of sharp corners/angles, housing notches, and poorly enforced ribs. The corrective action plans were to implement fillets, add the enforced ribs, and remove the notching on the housing of the hinge kit. Applying the modified designs and analyzing them by the finite element analysis, we found that the stress concentrations in the housing of the hinge kit decreased from 21.2 to 18.9 MPa.

When breaking down the failed HKS samples, the damper oil in the hinge kit assembly leaked at 15,000 cycles (Figure 9.24). The root cause for this failure came

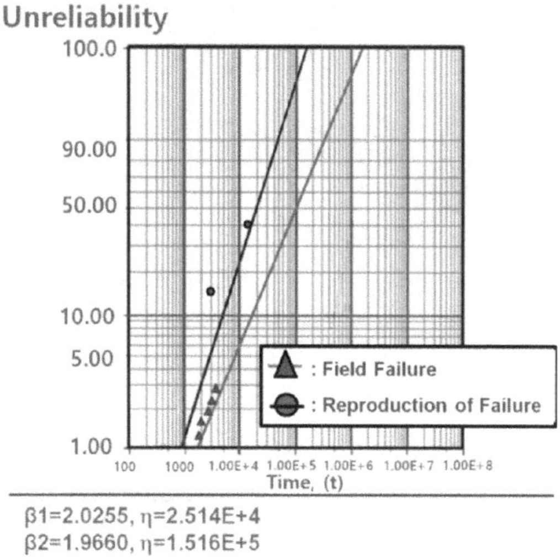

FIGURE 9.22 Field data and results of ALT on a Weibull chart.

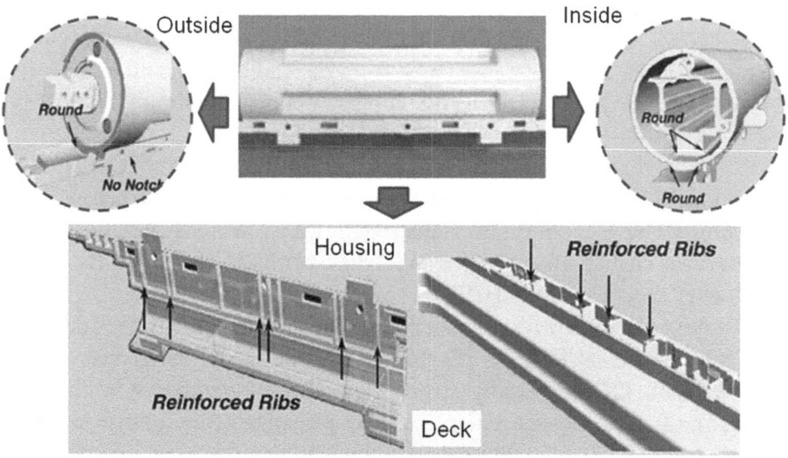

FIGURE 9.23 Redesigned HKS housing structure.

FIGURE 9.24 Spilled oil damper in the first ALT.

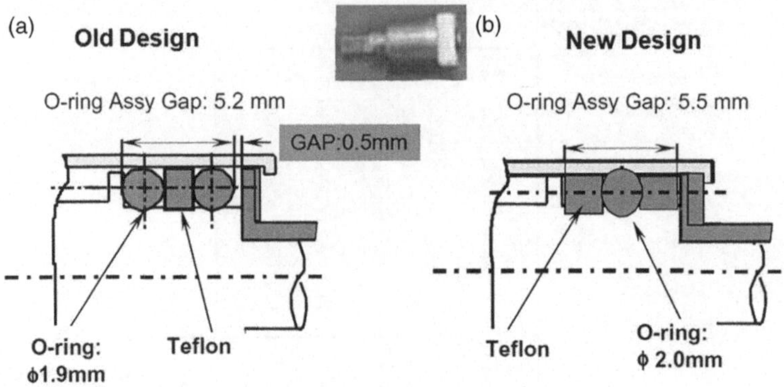

FIGURE 9.25 Redesigned oil damper: (a) old design and (b) new design.

from the oil damper sealing structure that had included an O-ring, Teflon, and an O-ring with a gap of 0.5 mm. It was determined that there might be interference between the O-ring and Teflon. To have the O-ring tightly held by the Teflon and have enough strength against impact, the sealing structure was modified as shown in Figure 9.25.

In the second ALTs, the hinge kit cover fractured at 8,000, 9,000, and 14,000 cycles (Figure 9.26). The root cause came from the choice of the material for the kit cover. When operating the HKS, the oil damper support (aluminum die casting) was striking the kit cover (plastic).

Consequently, the support started to crack and fracture at its end. The HKS failure for the second ALTs came from the type of materials used in the structure. As a corrective action, to have enough material strength for its own loading, the material of the kit cover was changed from plastic to Al die casting (Figure 9.27). By parametric ALT, the faulty kit cover of the HKS was modified.

To withstand the repetitive impact loads, the problematic HKS system in the field was redesigned as follows: (1) the housing design of HKS was reinforced, C1 (Figure 9.23); (2) the sealing structure in the oil damper was changed, C2 (Figure 9.25); (3) the kit cover material, C3, was changed from plastic to aluminum die casting (Figure 9.27). With these design changes, the refrigerator could be opened and closed as designed to meet the reliability requirement in the product lifetime because there were no problems until 23,000 mission cycles. Table 9.4 shows the summary of the results of the ALTs. Throughout three ALTs, the samples were guaranteed to be 10.0 years with an accumulated failure rate of 1%.

FIGURE 9.26 Structure of problematic products at the second ALT.

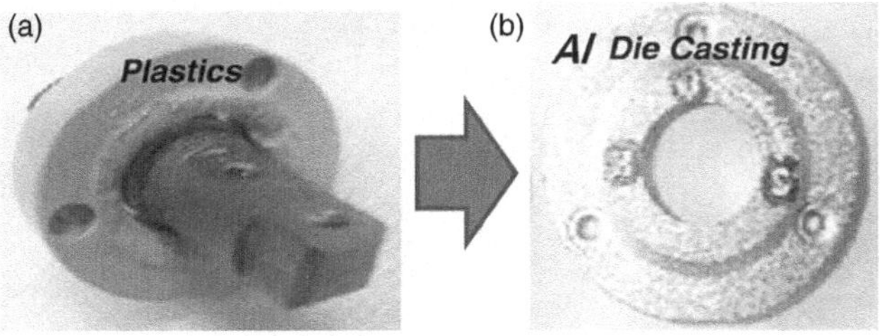

FIGURE 9.27 Redesigned kit cover: (a) old design and (b) new design.

TABLE 9.4
Results of ALTs in HKS of a Kimchi Refrigerator

	First ALT	Second ALT	Third ALT
	Initial Design	**Second Design**	**Final Design**
In 23,000 mission cycles, there is no problem in HKS	3,000 cycles: 2/6 fracture 15,000 cycles: 4/6 fracture	7,800 cycles: 1/6 fracture 9,200 cycles: 3/6 facture 14,000: 1/6 fracture 26,200: 1/6 fracture	23,000 cycles: 6/6 OK 41,000 cycles: 6/6 OK
Photo	HKS structure 		
Action plans	C1: Reinforced housing of HKS C2: Modified sealing structure of the oil damper	C3: Kit cover material (Plastic → Al die casting)	

9.4 REFRIGERATOR FREEZER DRAWER SYSTEM

To store food freshly, a refrigerator provides cold air from the evaporator to the freezer and refrigerator compartments through the traditional vapor-compression refrigeration cycle. Because customers want to have convenient access to the stored food, a freezer drawer system in a refrigerator is designed to handle the required food storage loads under expected consumer usage conditions over a refrigerator's

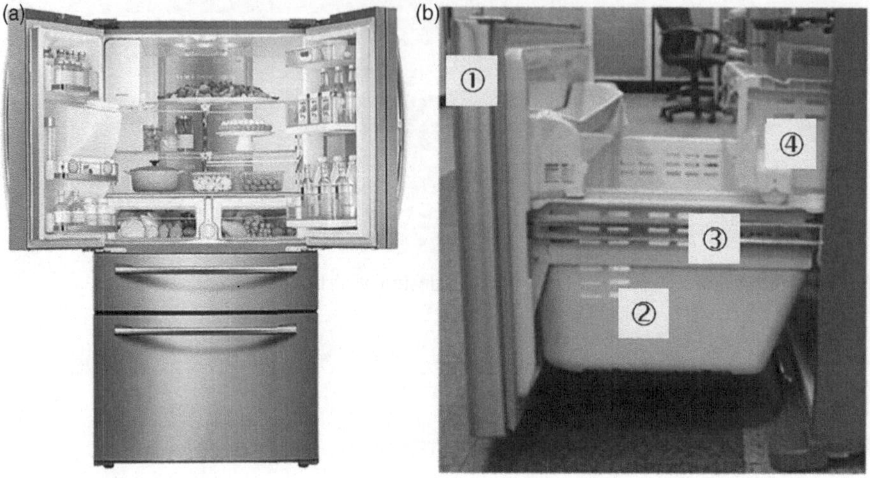

FIGURE 9.28 French door refrigerator and freezer drawer assembly: (a) French door refrigerator and (b) mechanical parts of the drawer: (1) handle, (2) drawer, (3) slide rail, and (4) pocket box.

lifetime. Storing food in a freezer drawer has the following repetitive handling steps: (1) opening the drawer to store the food in the drawer, (2) taking food out of the drawer, and (3) closing the draw. Figure 9.28 shows a French door refrigerator with a newly designed freezer drawer system.

French door refrigerators were being returned from the field because the handle of the freezer drawer fractured. Consequently, consumers were required to replace the refrigerators because they no longer functioned. Examination of the problematic refrigerators showed that the freezer drawer had serious design flaws in its structure. Thus, the freezer drawer had to be redesigned so that it could endure repetitive loading under customer operation conditions and improve its reliability (Figure 9.29).

When examining the consumer usage pattern of the freezer drawer, it was being subjected to repetitive food loads due to the opening and closing of the drawer. Because the problematic drawer had critical design flaws, engineers had to reproduce experimentally the conditions that produced the failures in the freezer drawer and correct them.

As seen in Figure 9.30, the force balance at the free-body diagram of the freezer drawer system can be expressed as follows:

$$F_{\text{draw}} = \mu W_{\text{load}} \tag{9.25}$$

As the applied stress of the freezer drawer system relied on the force proportionate to the food weights, the time to failure from Equation (8.16) can be expressed as follows:

$$TF = A(S)^{-n} = A\left(F_{\text{draw}}\right)^{-\lambda} = A\left(\mu W_{\text{load}}\right)^{-\lambda} \tag{9.26}$$

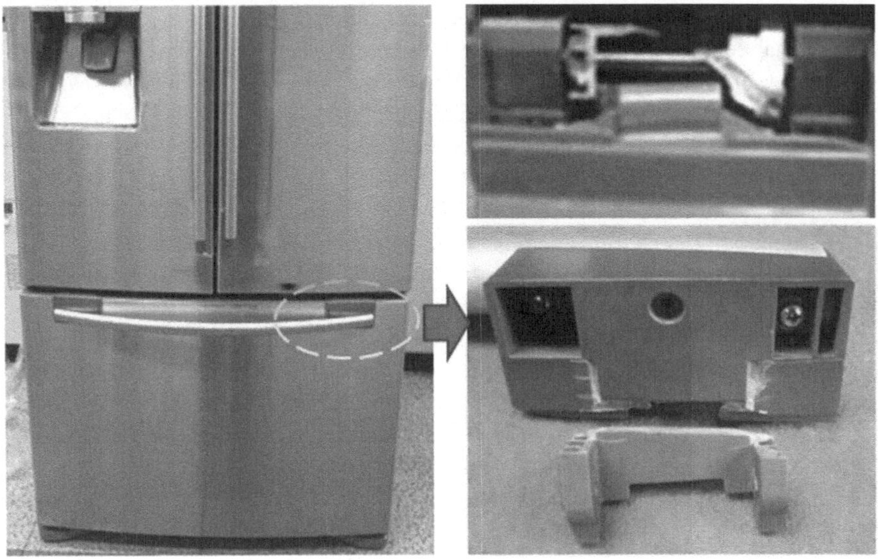

FIGURE 9.29 A damaged product after use.

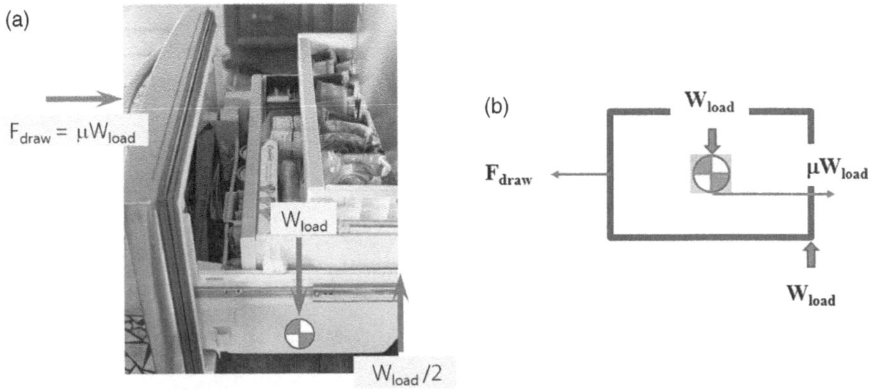

FIGURE 9.30 Functional design concept of the freezer drawer system: (a) Freezer drawer system and (b) free body diagram.

The AF in Equation (8.17) can be represented as follows:

$$AF = \left(\frac{S_1}{S_0}\right)^n = \left(\frac{F_1}{F_0}\right)^\lambda = \left(\frac{\mu W_1}{\mu W_0}\right)^\lambda = \left(\frac{W_1}{W_0}\right)^\lambda \tag{9.27}$$

For the freezer drawer system in the French door refrigerator, the typical working conditions for a customer range from 0°C to 43°C with a relative humidity ranging from 0% to 95%. Design conditions for transportation or operation assumed that the freezer drawer was subjected to 0.2 to 0.24 g of acceleration. In the United States, the operating cycles of the freezer drawer depended on the consumer usage profile. Data showed that consumers open and close the drawer system of a French door refrigerator between five and nine times per day. With a design life cycle of 10 years, the freezer drawer system incurred about 36,500 usage cycles.

Assuming the worst-case condition for the food weight in the drawer, the required force on the handle of the freezer drawer was 0.34 kN (35 kg_f). The applied food weight force for the ALT was 0.68 kN (70 kg_f). The load ratio was 2 (= 0.68 kN/0.34 kN). Using Equation (9.27) with a quotient, λ, of 2, the total AF was approximately 4.0.

For the lifetime target of a B1 life of 10 years, the mission cycles for three sample units calculated from Equation (8.29) were 67,000 cycles if the shape parameter β was supposed to be 2.0. This parametric ALT was designed to ensure a B1 life of 10 years with about a 60% level of confidence that it would fail less than once during 67,000 cycles. Figure 9.31 shows the experimental setup of the ALT with labeled equipment for the robust design of the freezer drawer. We carried out the ALT as a fatigue test under room temperature and humidity conditions.

French door refrigerators returned from the field had a primary failure mode with fractured handle due to repetitive food loading in the process of opening and closing of the freezer drawer. Field data indicated that the damaged products might have had a design flaw. Due to this faulty design, the repetitive pressure loads could create undue stresses on the drawer handle, causing it to fracture, and creating a failure of the drawer. To reproduce the failure mode of the drawer, parametric ALT was carried out.

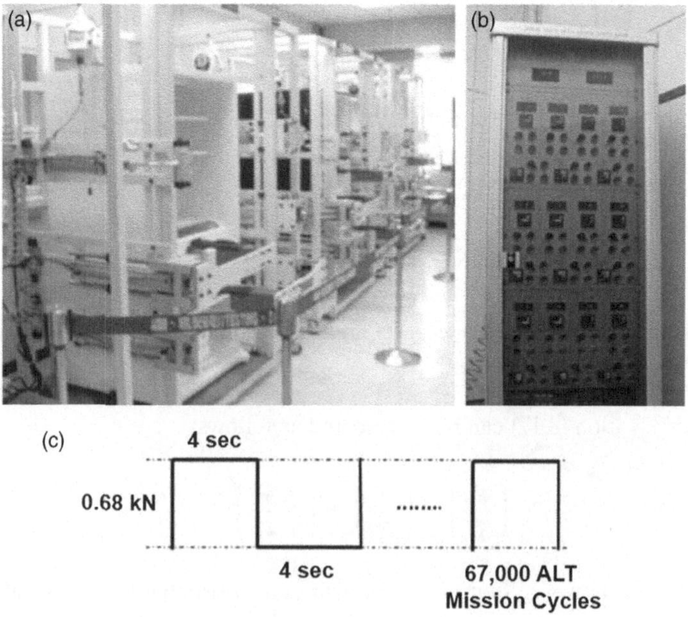

FIGURE 9.31 ALT equipment and duty cycles: (a) ALT equipment, (b) controller, and (c) duty cycles of repetitive food weight force on the drawer.

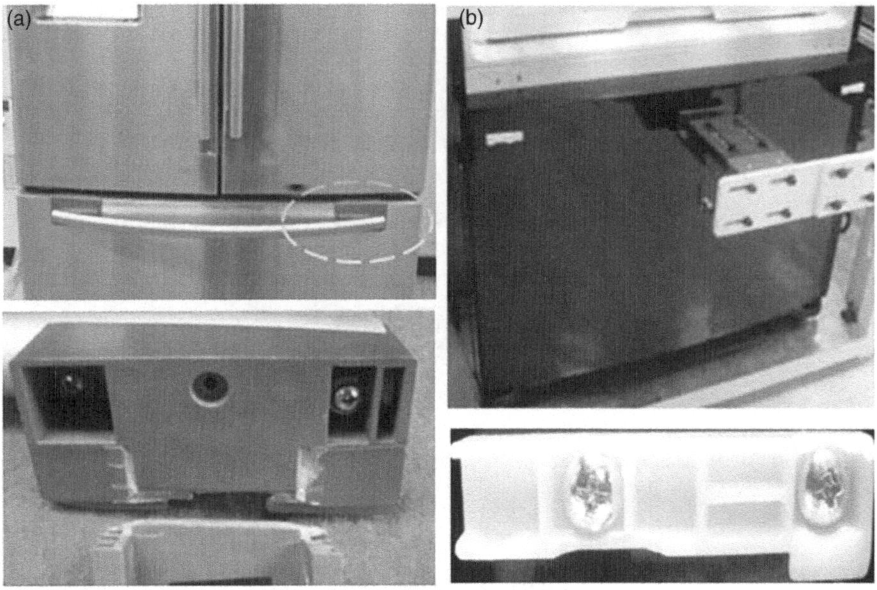

FIGURE 9.32 Failure of freezer drawer handles in the field and first ALT results: (a) field and (b) first ALT results.

In the first ALT, the handle of the drawer fractured at 7,000 and 8,000 cycles. Figure 9.32 shows a photograph comparing the failed product from the field and the first ALT, respectively. The failed shape of the first ALT was very similar to the ones from the field. The test results confirmed that the freezer drawer was not well designed for the opening and closing of its door. By parametric ALT, the faulty handle structure of the freezer drawer was reproduced.

The root causes of the fractured drawer handle came from the insufficient attached area of the handle to the drawer. This design flaw could cause the drawer handle to snap off suddenly when subjected to repetitive food loads. To prevent the drawer handle from fracturing due to the repetitive uses, the handle was redesigned as follows: (1) increasing the width of the reinforced handle, C1, Width 1 from 90 to 122 mm and (2) increasing the handle hooker size, C2, Width 2 from 8 to 19 mm (Figure 9.33).

In the second ALT, the slide rails of the drawer cracked at 15,000 and 16,000 cycles (Figure 9.34). When the food weight for the ALT was applied repetitively, the tests exposed the slide rails as a fragile component. The root cause of this failure originated from the shape of the corner of the slide rails. Consequently, the rail started to crack and fractured in its end. Corrective actions on the slide rail included: (1) increasing the rail fastening screw number, C3, from 1 to 2; (2) adding an inner chamber and plastic material, C4, from high impact polystyrene (HIPS) to acrylonitrile butadiene styrene (ABS); (3) thickening the boss, C5, from 2.0 to 3.0 mm; and (4) adding a new support rib, C6. By applying a parametric ALT, the faulty slide rail of the freezer drawer was modified (Figure 9.35).

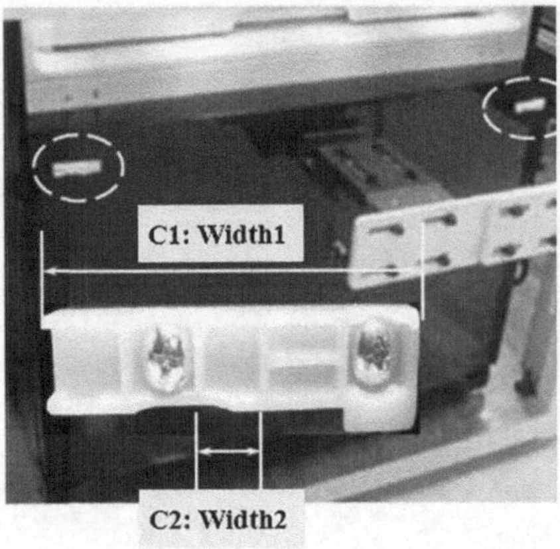

FIGURE 9.33 Redesigned freezer drawer handle.

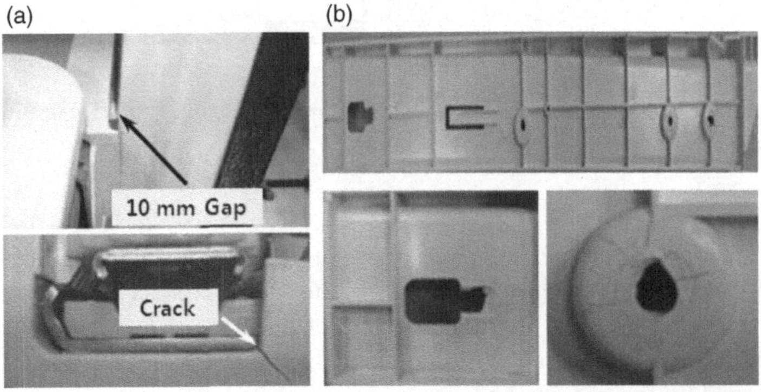

FIGURE 9.34 Failed slide rails in second ALT: (a) fractured slide rail in the freezer drawer and (b) detailed problem on the back of the rail.

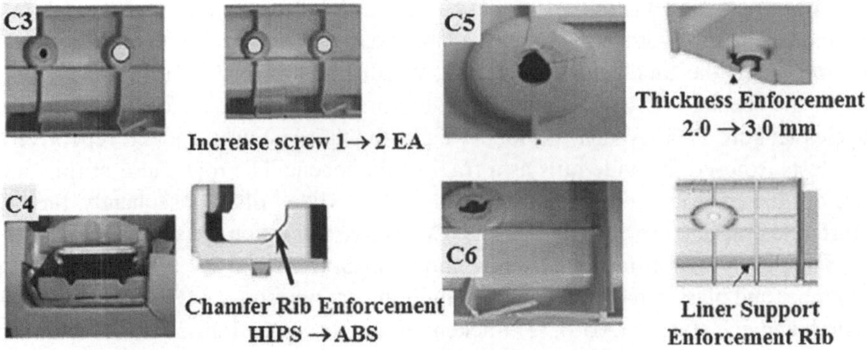

FIGURE 9.35 Redesigned slide rail.

TABLE 9.5
Results of ALT in Refrigerator Freezer Drawer System

	First ALT	**Second ALT**	**Third ALT**
Parametric ALT	**Initial Design**	**Second Design**	**Final Design**
In 32,000 cycles, there is no problem in the drawer	7,500 cycles: 2/6 fracture 12,000 cycles: 1/3 OK	16,000 cycles: 2/3 fracture (Rail system)	32,000 cycles: 3/3 OK 67,000 cycles: 3/3 OK
Photo	<Handle> 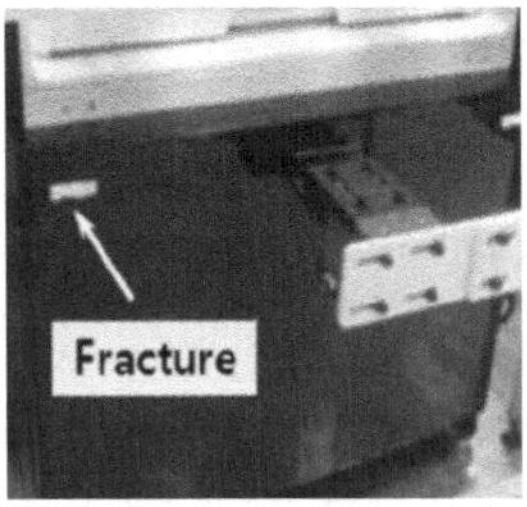	<Rail> 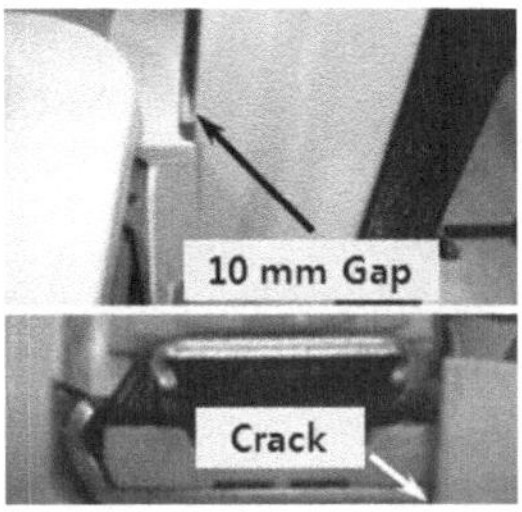	
Action plans	C1: Width 1: 90 mm → 122 mm C2: Width 2: L8 → L19.0	C3: Screw number: 1.0 → 2.0 C4: Corner chamfer and material (HIPS → ABS) C5: Thickening the boss: 2.0 → 3.0 mm C6: Adding a new support rib	

In the third ALT, there were no design problems with the freezer drawer system until the test was carried out to 67,000 cycles. It was concluded that the modifications to the design identified from the first and second ALT were effective in extending the lifetime of the drawer system. Table 9.3 summarizes the parametric ALT results. With the modified design parameters, the final freezer drawer system was guaranteed to achieve its lifetime target of a B1 life of 10 years (Table 9.5).

9.5 COMPRESSOR SUCTION REED VALVE

A refrigerator provides cold air from the evaporator to the freezer and refrigerator compartments through the traditional vapor-compression refrigeration cycle. The cycle consists of a compressor, a condenser, an expansion device (usually a capillary tube), and an evaporator. The refrigerant enters the compressor in the refrigeration cycle from the evaporator where it is compressed. It then leaves the compressor and enters the condenser. The compressor increases the refrigerant pressure from the evaporator before discharging it to the condenser. A suction reed valve allows the refrigerant to stream into the compressor during the reciprocating operation of the piston (Figure 9.36).

French door refrigerators were being returned from the field with no cooling. As a consequence, consumers had to replace their refrigerators because they no

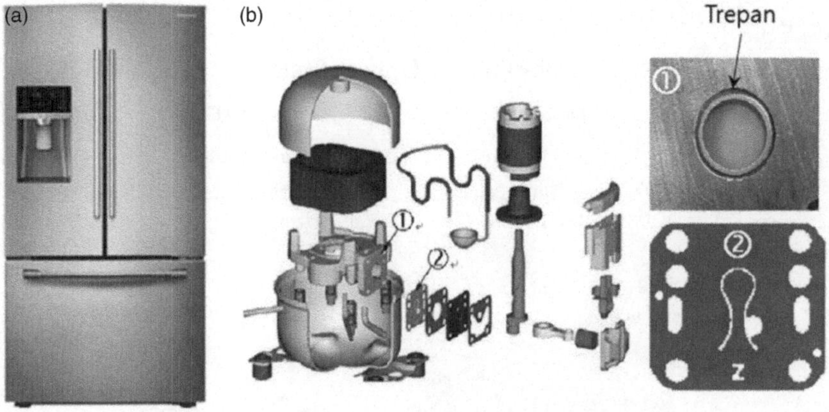

FIGURE 9.36 French door refrigerator with the newly designed compressor: (a) French door refrigerator and (b) Mechanical parts of a reciprocating compressor: (1) valve plate and (2) suction reed valve.

longer functioned. When the refrigerators returned from the field were disassembled, the suction reed valve in the compressor was found to have fractured. Scanning electron microscopy (SEM) was used to inspect the fractured suction reed valves. The fracture started in a void in the valve and propagated to the end (Figure 9.37).

Based on the expected consumer usage conditions in the field, we knew that the compressors were subjected to repetitive pressure loads due to normal refrigerator on/off operations. The refrigerator compressors had critical design flaws. Thus, engineers had to reproduce the compressor failures experimentally and then correct them.

Before carrying out a parametric ALT, the compressor was analyzed from the standpoint of the vapor-compression cycle to determine the proper pressure loads to subject it to in the tests. A reciprocating compressor is a complicated mechanical system that is an integral part of the refrigeration cycle. The compressor takes refrigerant vapor from the evaporator, compresses it, and transfers the high-temperature and high-pressure refrigerant to the condenser. In the condenser, heat energy is transferred from the hot refrigerant to the relatively warm air in the environment outside the refrigerator. The high-temperature liquid refrigerant leaves the exit of the condenser. The refrigerant then flows through the capillary tube(s) where the pressure is decreased. Some liquid refrigerant flashes into cold vapor. A mixture of liquid and vapor leaves the capillary tube(s) and flows into the evaporator. As heat is absorbed from the cold refrigerated space, the refrigerant liquid/vapor mixture is converted to superheated vapor at a low temperature. The refrigerant vapor then flows back into the compressor (Figure 9.38).

In evaluating the design of a refrigeration cycle, it was necessary to determine both the condensing temperature, T_c, and evaporating temperature, T_e. The mass flow rate of a refrigerant in a compressor can be modeled as

$$\dot{m} = PD \times \frac{\eta_v}{v_{suc}} \tag{9.28}$$

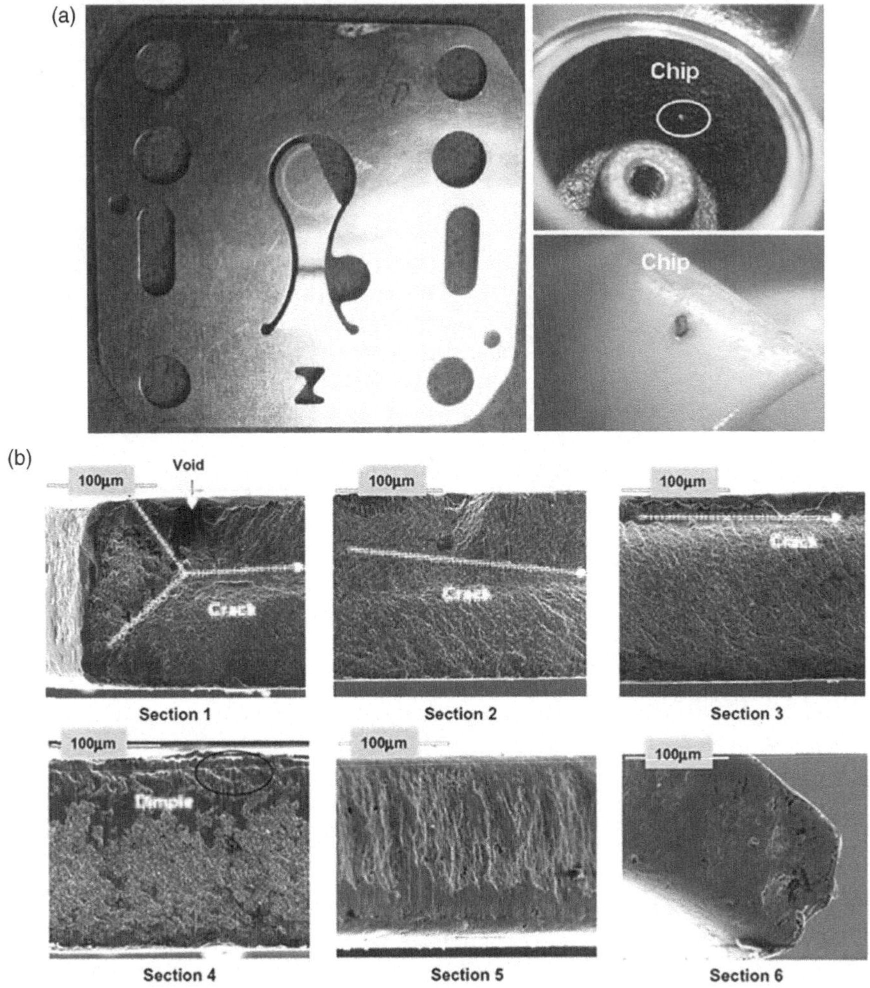

FIGURE 9.37 A damaged suction reed valve after use: (a) fractured compressor suction reed valve from the field and (b) crack propagation direction from bottom to top.

The mass flow rate of a refrigerant in a capillary tube can be modeled as follows:

$$\dot{m}_{\text{cap}} = A\left[\frac{-\int_{P_3}^{P_4} \rho\, dP}{\frac{2}{D} f_m \Delta L + \ln\left(\frac{\rho_3}{\rho_4}\right)}\right]^{0.5} \tag{9.29}$$

By conservation of mass, the mass flow rate can be determined as follows:

$$\dot{m} = \dot{m}_{\text{cap}} \tag{9.30}$$

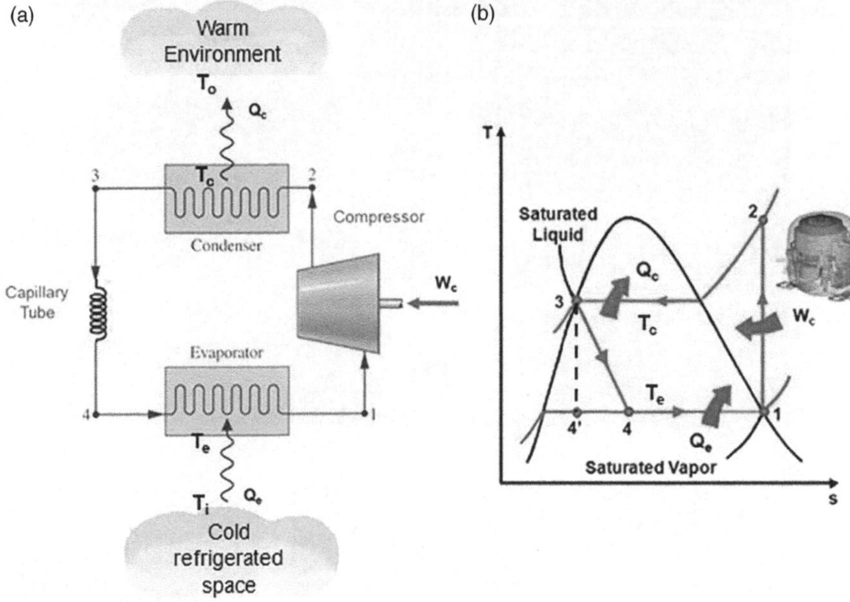

FIGURE 9.38 Functional design concept of the compressor system in the refrigeration cycle: (a) refrigeration cycle and (b) *T–S* diagram.

The energy balance in the condenser can be described as

$$Q_c = \dot{m}(h_2 - h_3) = (T_c - T_o)/R_c \tag{9.31}$$

The energy balance in the evaporator can be described as

$$Q_e = \dot{m}(h_1 - h_4) = (T_i - T_e)/R_e \tag{9.32}$$

When nonlinear Equations (9.30) through (9.32) are solved, the mass flow rate, $\dot{m}$, evaporator temperature, T_e, and condenser temperature, T_c, can be obtained. Because the saturation pressure, P_{sat}, is a function of temperature, the evaporator pressure, P_e (or condenser pressure P_c), can be obtained as follows:

$$P_e = f(T_e) \text{ or } P_c = f(T_c) \tag{9.33}$$

One source for accelerated testing of a compressor system may come from the pressure difference between suction pressure, P_{suc}, and discharge pressure, P_{dis}.

$$\Delta P = P_{\text{dis}} - P_{\text{suc}} \cong P_c - P_e \tag{9.34}$$

Another source is the compressor dome temperature. For a compressor system, the time-to-failure, *TF*, from Equation (8.16) can be modified as

$$TF = A(S)^{-n} \exp\left(\frac{E_a}{kT}\right) = A(\Delta P)^{-\lambda} \exp\left(\frac{E_a}{kT}\right) \tag{9.35}$$

The AF from Equation (8.17) can be redefined as

$$AF = \left(\frac{S_1}{S_0}\right)^n \left[\frac{E_a}{k}\left(\frac{1}{T_0} - \frac{1}{T_1}\right)\right] = \left(\frac{\Delta P_1}{\Delta P_0}\right)^\lambda \left[\frac{E_a}{k}\left(\frac{1}{T_0} - \frac{1}{T_1}\right)\right] \tag{9.36}$$

For a reciprocating compressor in a French door refrigerator, the typical working conditions for a customer range from 0°C to 43°C with a relative humidity ranging from 0% to 95%. Design conditions for transportation or operation assumed that the compressor was subjected to 0.2–0.24 g of acceleration. As the compressor operates, the suction reed valve will open and close to allow the refrigerant to flow into the compressor. A reciprocating compressor is expected to cycle on and off 22–98 times per day. With a product life cycle design point of 10 years, the compressor may experience as many as 357,700 usage cycles.

As seen in Table 9.6, from the test data of the worst case, the differential pressure expected by the consumer in the operating compressor was 1.27 MPa, and the compressor dome temperature was 90°C. To carry out accelerated testing, the compressor dome temperature was increased to 120°C and the differential pressure was increased to 2.94 MPa. With a cumulative damage exponent, λ, of 2, the total AF was approximately 7.32 using Equation (9.36).

For the lifetime target of a B1 life of 10 years, the test cycles for 20 sample units calculated from Equation (8.29) were 109,300 cycles if the shape parameter β was supposed to be 2.0. This parametric ALT was designed to ensure a B1 life of 10 years with about a 60% level of confidence that it would fail less than once during 109,300 cycles.

Figure 9.39 shows the experimental setup of the ALT with labeled equipment for evaluating the design of the compressor. The setup consisted of an evaporator, a compressor, a condenser, and a capillary tube in a simplified vapor-compression refrigeration system. A fan and two 60-W lamps maintained the temperature within the insulated (fiberglass) box. A thermal switch attached to the compressor top controlled a 51 m^3/h axial fan. The test conditions and test limits were set up on the control board. As the test began, the high- and low-side pressures could be observed on the pressure gauge (duty cycle).

TABLE 9.6
Refrigerator ALT Conditions in the Refrigeration Cycle

System Conditions		Worst Case	ALT	*AF*
Pressure (MPa)	High-side	1.27	2.94	5.33①
	Low-side	0.0	0.0	
	ΔP	1.27	2.94	
Temperature (°C)	Dome	90	120	1.36②
Total AF (= (① × ②))				7.32

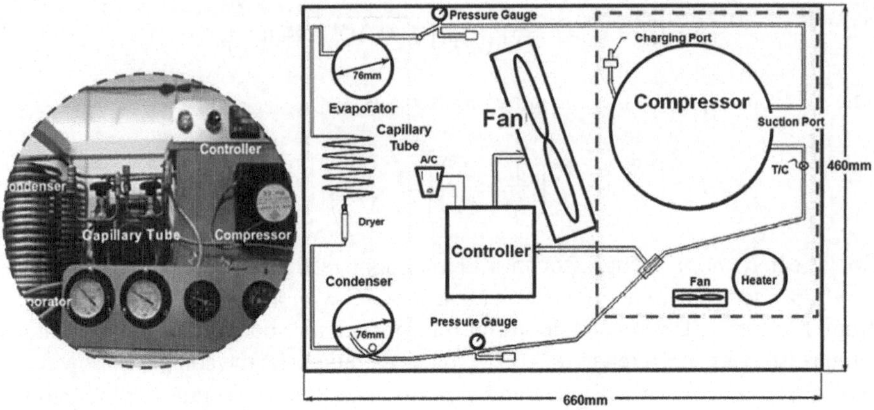

FIGURE 9.39 Equipment used in ALT.

French door refrigerators returned from the field had a primary failure mode with no cooling because the reciprocating compressor did not operate. Field data indicated that the damaged products might have had a design flaw. Due to this faulty design, the repetitive pressure loads could create undue stresses on the suction reed valve, causing it to fracture, and causing the compressor to fail. To reproduce the failure mode of the compressor, parametric ALTs were carried out.

In the first ALT, a compressor locked at 8,700 cycles. Disassembly of the problematic compressors from the field and the first ALT showed that the suction reed valves fractured in some regions where they overlapped with the valve plate. Figure 9.40 shows a photograph comparing a failed product from the field and that from the first ALT, respectively. The failed shape of the first ALT was very similar to the ones from the field. The test results confirmed that the compressor was not well designed to open and shut the suction reed valve.

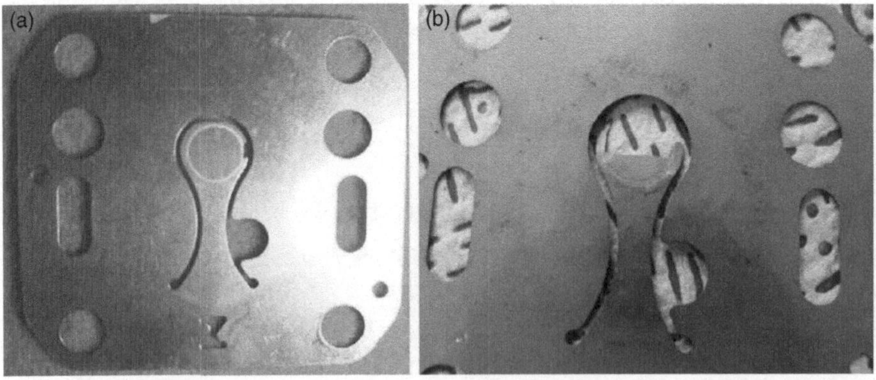

FIGURE 9.40 Failed suction reed valve products from the field and after the first ALT: (a) failed products in the field and (b) crack after the first ALT.

The root causes of the fractured suction reed valve came from the following improper designs: (1) an overlap with the valve plate (Figure 9.41a); (2) a sharp edge on the valve plate (Figure 9.41b); and (3) weak material (0.178 *t*) used in the design of suction reed valve. These design flaws could cause the compressor to lock up suddenly when subjected to repetitive pressure loads.

Figure 9.42 presents the graphical analysis of the ALT results and field data on a Weibull plot. The shape parameter in the first ALT was estimated to be 2.0. For the final design, the shape parameter from the Weibull plot was confirmed to be 1.95.

To prevent the fracture of the suction reed valve from the repetitive pressure stresses over the product's lifetime, the valve plate was redesigned as follows: (1) the trepan (see Figure 9.5b), C1, was increased from 0.73 to 1.25 mm and (2) a ball peening process, C2, was added to eliminate the sharp edge of the valve plate. Additionally, the suction reed was redesigned as follows: (3) the thickness of the suction reed valve (SANDVIK 20C), C3, was changed from 0.178 to 0.203 mm and (4) the tumbling process time, C4, was extended from 4 to 14 hours to decrease the residual stress (Table 9.7).

In the second ALT, there were no design problems with the compressor until the test was carried out to 49,000 cycles. We, therefore, concluded that the modifications to the design found from the first ALT were effective. Table 9.8 summarizes the parametric ALT results. With the modified design parameters, the final compressor system samples were guaranteed to achieve their lifetime target of a B1 life of 10 years.

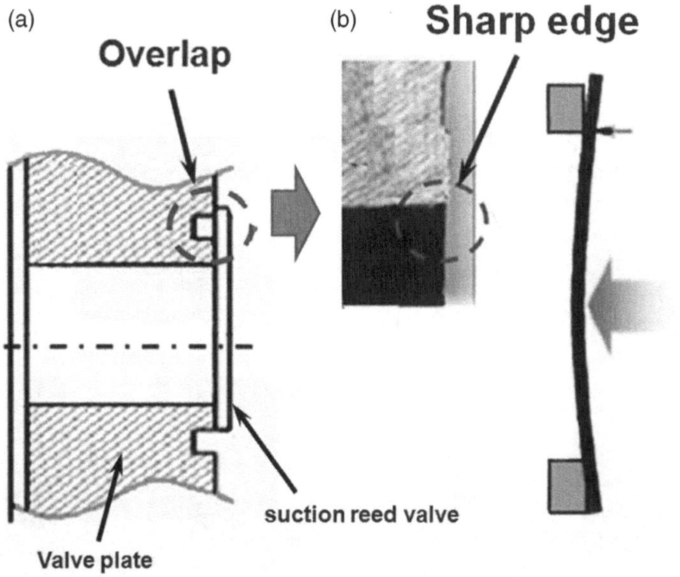

FIGURE 9.41 Structure of failing suction reed valve and valve plate from the field: (a) overlapped suction reed valve and (b) valve plate with a sharp edge.

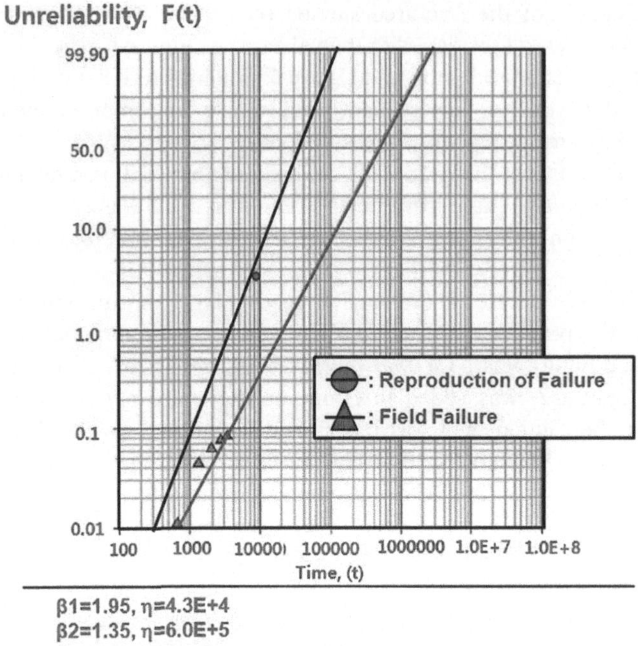

FIGURE 9.42 Field data and results of ALT on a Weibull chart.

TABLE 9.7
Redesigned Valve Plate and Suction Reed Valve

Valve Plate

C1: Trepan:0.73 mm → 1.25 mm

C2: Ball peening and brush process

Suction Reed Valve

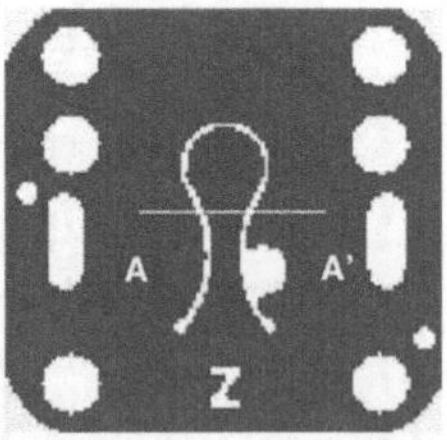

Section A–A′

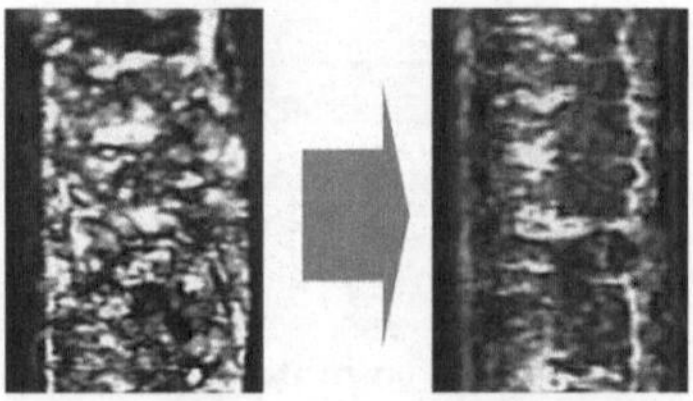

C3: 0.178 *t* → 0.203 *t* (SANDVIK 20C thickness)

C4: Tumbling process time: 4 hours → 14 hours

TABLE 9.8
Results of the ALTs in Compressor Suction Reed Valve

	First ALT		Second ALT
Parametric ALT	**Initial Design**		**Final Design**
In 49,000 mission cycles, there is no problem in the compressor	8,687 cycles: 1/20 locking		49,000 cycles: 100/100 OK
Photo	<Valve plate> 	<Suction reed valve> 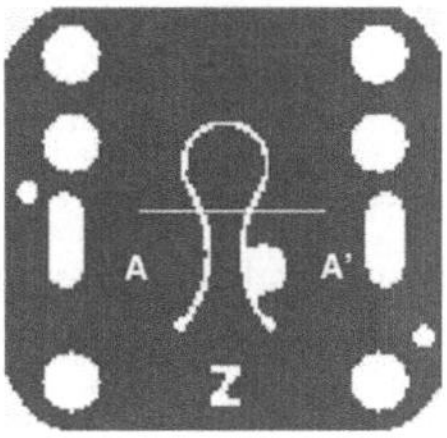	–
Action plans	C1: Trepan size: 0.73 mm → 1.25 mm C2: Adding ball peening and brush process C3: SANDVIK 20C: 0.178 *t* → 0.203 *t* C4: Extending tumbling: 4 hours → 14 hours		–

9.6 FAILURE ANALYSIS AND REDESIGN OF THE EVAPORATOR TUBING

Figure 9.43 shows the Kimchi refrigerator with the aluminum cooling evaporator tubing suggested for cost saving. When a consumer stores the food in the refrigerator, the refrigerant flows through the evaporator tubing in the cooling enclosure to maintain a constant temperature and preserve the freshness of the food. To perform this function, the tube in the evaporator needs to be designed to reliably work under the operating conditions it is subjected to by the consumers who purchase and use the Kimchi refrigerator. The evaporator tube assembly in the cooling enclosure consists of an inner case (1), an evaporator tubing (2), a lockring (3), and an adhesive tape (4), as shown in Figure 9.43b.

In the field, the evaporator tubing in the refrigerators had been pitting, causing loss of the refrigerant in the system and resulting in the loss of cooling in the refrigerator. The data on the failed products in the field were important for understanding the usage environment of consumers and pinpointing design changes that needed to be made to the product (Figure 9.44).

Field data indicated that the damaged products might have had design flaws. The design flaws combined with the repetitive loads could cause failure. The pitted surfaces of a failed specimen from the field were characterized by SEM and EDX spectrum (Figure 9.44). We found a concentration of chlorine in the pitted surface (Table 9.9). When ion liquid chromatography was used to measure the chlorine concentration, the result for the tubing having had the cotton adhesive tape was 14 ppm. In contrast, the chlorine concentration for tubing having had the generic transparent

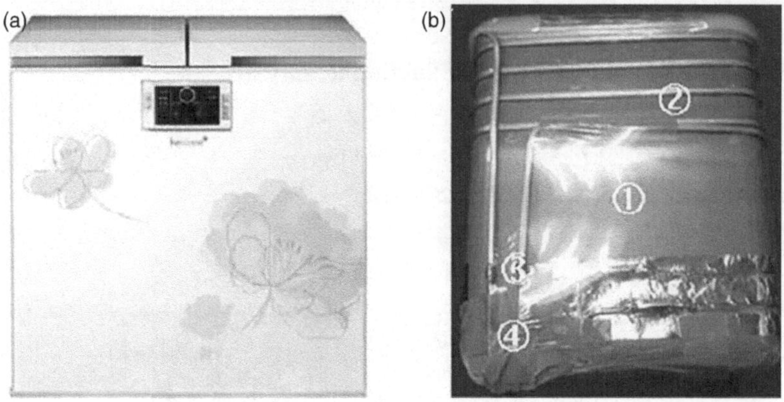

FIGURE 9.43 Kimchi refrigerator and the cooling evaporator assembly: (a) Kimchi refrigerator and (b) mechanical parts of the HKS: inner case (1), evaporator tubing (2), lockring (3), and cotton adhesive tape (4).

FIGURE 9.44 A damaged product after use: (a) pitted evaporator tube in the field and (b) X-ray photography showing pitting corrosion on the evaporator tube.

TABLE 9.9
Chemical Composition of the No Pitting and Pitting Surfaces

	No Pitting		Pitting	
	Weight	Atomic	Weight	Atomic
O	11.95	19.65	25.92	37.39
Al	97.29	90.74	69.29	59.61
Cl	0.33	0.23	3.69	2.41
Si	0.42	0.39	0.66	0.55
Ca	100.00		0.70	0.40
K			0.50	0.30
Na			0.34	0.34
Totals			100.00	

tape was 1.33 ppm. It was theorized that the high chlorine concentration found on the surface must have come from the cotton adhesive tape.

As mentioned in Figure 9.45, the evaporator tubing assembly in the cooling enclosure of the Kimchi refrigerator consists of many mechanical parts. Depending on the consumer usage conditions, the evaporator tubing experienced repetitive thermal duty loads due to the normal on/off cycling of the compressor to satisfy the thermal load in the refrigerator. Because the refrigerant temperatures are often below the dew point temperature of the air, condensation can occur on the external surface of the tubing.

Figure 9.46 shows a robust design schematic overview of the cooling evaporator system. Figure 9.47 shows the failure mechanism of the crevice (or pitting) corrosion that occurs because of the reaction between the cotton adhesive tape and the aluminum evaporator tubing. As a Kimchi refrigerator operates, water acts as an electrolyte and will condense between the cotton adhesive tape and the aluminum tubing. The crevice (or pitting) corrosion will begin.

The crevice (or pitting) corrosion mechanism on the aluminum evaporator tubing can be summarized as (1) passive film breakdown by Cl^- attack; (2) rapid metal dissolution: $Al \rightarrow Al^{+3} + 3e^-$; (3) electro-migration of Cl into pit; (4) acidification by the hydrolysis reaction: $Al^{+3} + 3H_2O \rightarrow Al(OH)_3 \downarrow + 3H^+$; (5) large cathode: external surface, small anode area: pit; (6) and the large voltage drop (i.e., "IR" drop, according to Ohm's law $V = I \times R$, where R is the equivalent path resistance and I is the average current) between the pit and the external surface is the driving force for the propagation of pitting.

The number of Kimchi refrigerator operation cycles is influenced by specific consumer usage conditions. In the Korean domestic market, the compressor can be expected to cycle on and off 22–99 times per day to maintain the proper temperature inside the refrigerator.

Because the corrosion stress of the evaporator tubing depends on the corrosive load (F) that can be expressed as the concentration of the chlorine, the LS model from Equation (8.16) can be modified as

$$TF = A(S)^{-n} = A(F)^{-\lambda} = A(\text{Cl}\%)^{-\lambda} \tag{9.37}$$

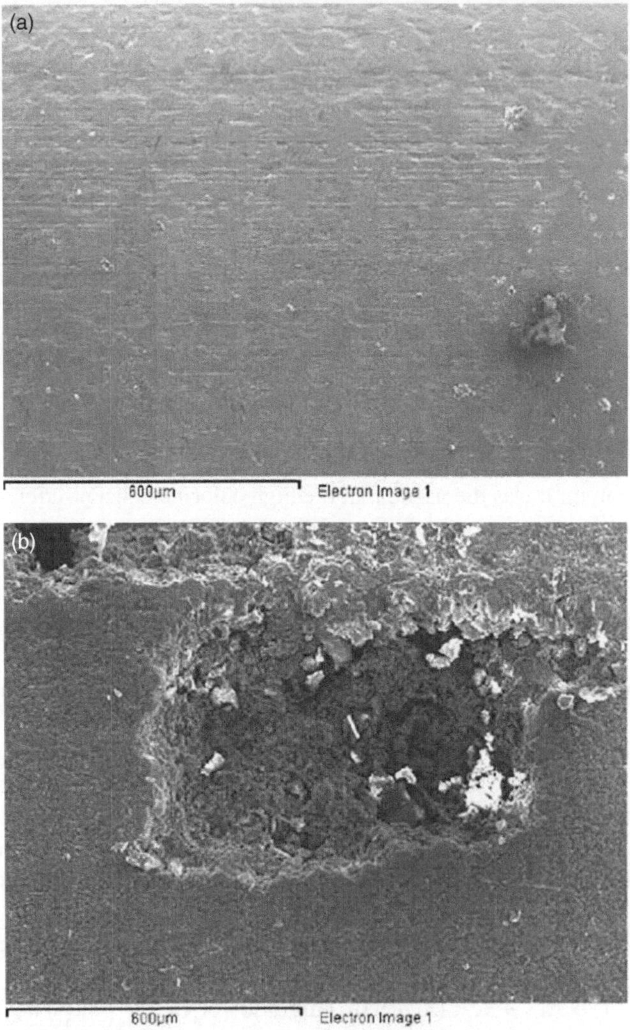

FIGURE 9.45 SEM fractography showing pitting corrosion on the evaporator tube: (a) no pitting and (b) pitting.

The AF can be derived as

$$AF = \left(\frac{S_1}{S_0}\right)^n = \left(\frac{F_1}{F_0}\right)^\lambda = \left(\frac{\text{Cl}_1\%}{\text{Cl}_0\%}\right)^\lambda \tag{9.38}$$

The compressor in a Kimchi refrigerator is expected to cycle on average 22–99 times per day. With a life cycle design point of 10 years, the Kimchi refrigerator incurs 359,000 cycles. The chlorine concentration of the cotton adhesive tape was 14 ppm. To accelerate the pitting of the evaporator tubing, the chlorine concentration of the

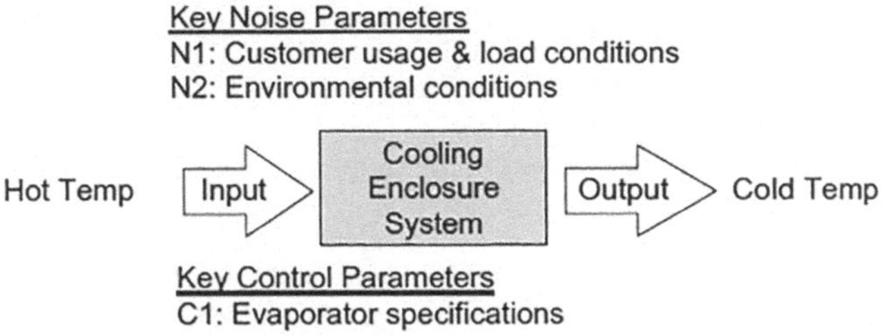

FIGURE 9.46 Robust design schematic of a cooling enclosure system.

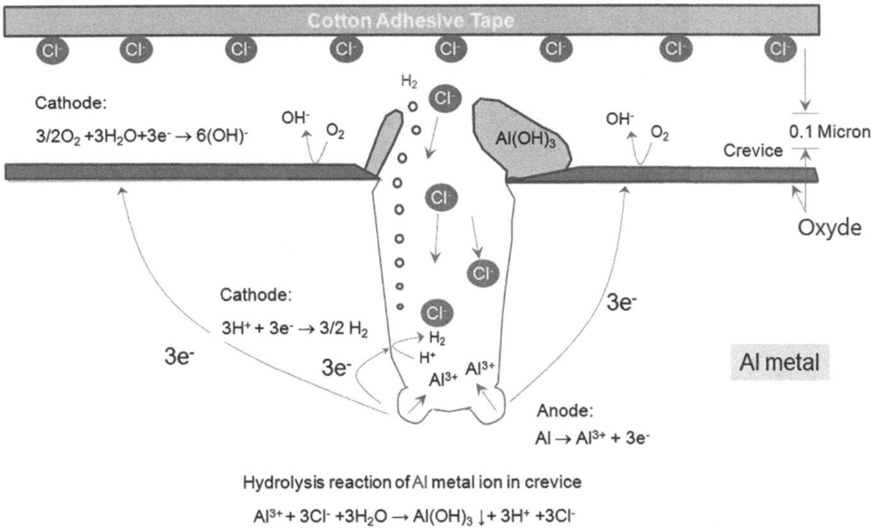

FIGURE 9.47 An accelerating corrosion in the crevice due to low PH, high Cl^- concentration, de-passivation, and IR drop.

cotton tape was adjusted to approximately 140 ppm by adding some salt. Using a stress dependence of 2.0, the AF was found to be approximately 100 in Equation (9.38).

For a B1 life of 10 years, the test cycles and test sample numbers with the shape parameter $\beta = 6.41$ calculated in Equation (8.35) were 4,700 cycles and 19 pieces, respectively. The ALT was designed to ensure a B1 of 10 years life with about a 60% level of confidence that it would fail less than once during 4,700 cycles. Figure 9.48a shows the Kimchi refrigerators in ALT and an evaporator tubing in the enclosure contained a 0.2 M NaCl water solution. Figure 9.48b shows the duty cycles for the corrosive force (F) due to the chlorine concentration.

Figure 9.49 shows the failed product from the field and from the ALT respectively. In the photos, the shape and location of the failure in the ALT were similar to those seen in the field. Figure 9.50 shows a graphical analysis of the ALT results and field data on a Weibull plot. These methodologies were valid in pinpointing the weak designs responsible for failures in the field and were supported by two

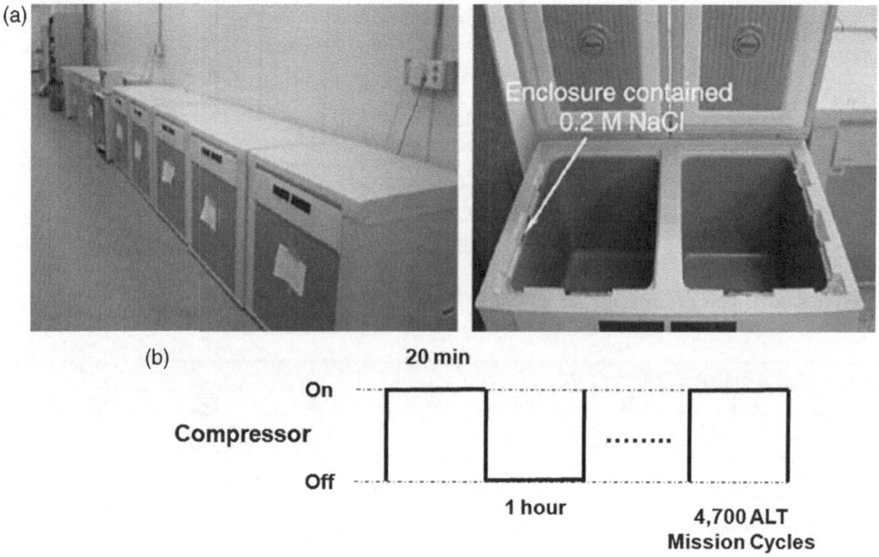

FIGURE 9.48 Kimchi refrigerators in ALT and duty cycles of repetitive corrosive load *F*: (a) Kimchi refrigerators in testing with 0.2 M NaCl water solution on the evaporator and (b) Duty cycles of repetitive corrosive load *F*.

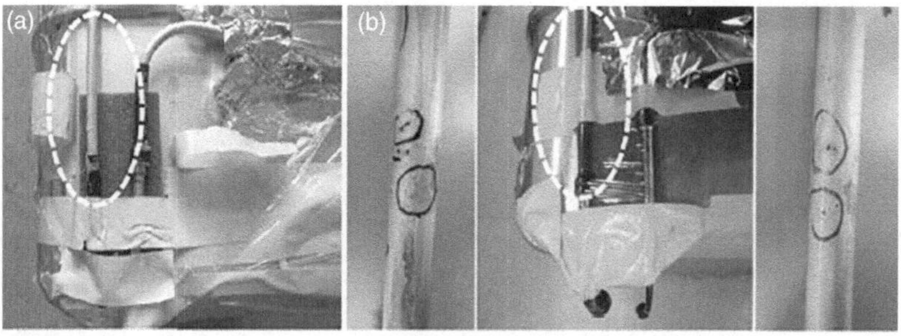

FIGURE 9.49 Failed products in the field and ALT: (a) failed product from the field and (b) failed product from the ALT.

findings in the data. The location and shape, from the Weibull plot, the shape parameters of the ALT, ($\beta 1$), and market data, ($\beta 2$) were found to be similar.

The pitting of the evaporator tubing in both the field products and the ALT test specimens occurred in the inlet/outlet of the evaporator tubing (Figure 9.51). Based on the modified design parameters, corrective measures taken to increase the life cycle of the evaporator tubing system included: (1) extending the length of the contraction tube (C1) from 50.0 to 200.0 mm and (2) replacing the cotton adhesive tape (C2) with the generic transparent tape.

Figure 9.52 shows a redesigned evaporator tubing with high corrosive fatigue strength. The confirmed values of AF and β in Figure 9.49 are 100.0

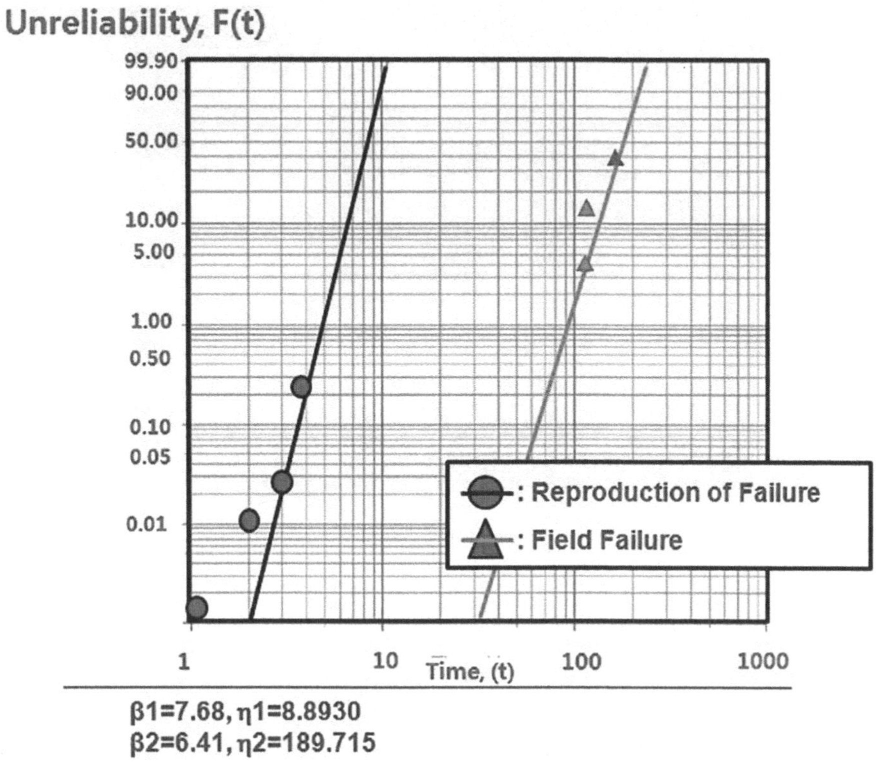

FIGURE 9.50 Field data and results of ALT on a Weibull chart.

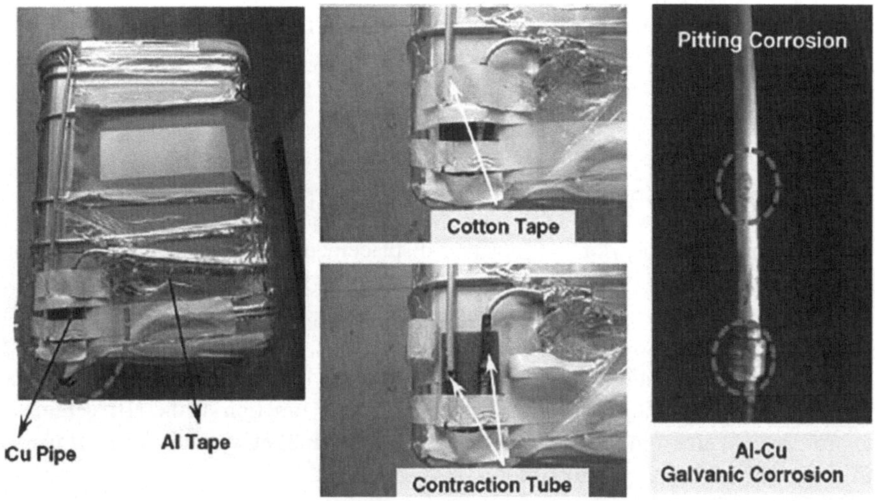

FIGURE 9.51 Structure of pitting the corrosion tubing in the field and the ALT test specimens.

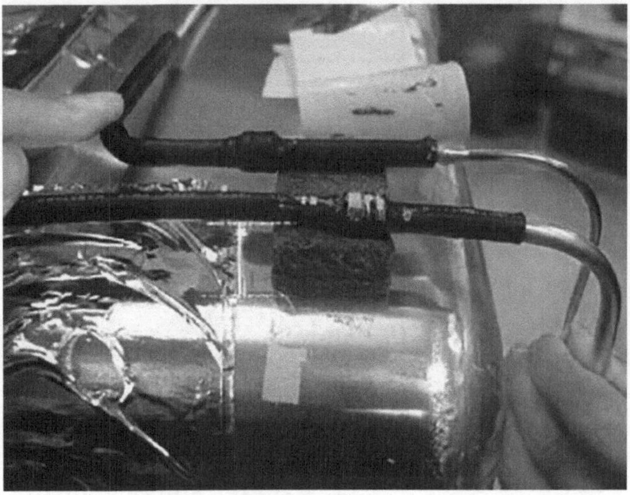

FIGURE 9.52 A redesigned evaporator tubing.

and 6.41, respectively. The test cycles and sample size recalculated in Equation (8.29) were 5,300 and 9 EA, respectively. Based on the target BX life, two ALTs were performed to obtain the design parameters and their proper levels. In the two ALTs, the outlet of the evaporator tubing was pitted in the first test and was not pitted in the second test. The repetitive corrosive force in combination with the high chlorine concentration of the cotton tape and the crevice between the cotton adhesive tape and the evaporator tubing contained the condensed water as an electrolyte may have been pitting.

With these modified parameters, the Kimchi refrigerator can preserve the food for a longer period without failure. Figure 9.53 and Table 9.10 show the graphical results of ALT plotted in a Weibull chart and the summary of the results of the ALTs, respectively. Throughout the two ALTs, the B1 life of the samples increased by over 10.0 years.

9.7 IMPROVING THE NOISE OF A MECHANICAL COMPRESSOR

A reciprocating compressor is a positive-displacement machine that uses a piston to compress a gas and deliver it at high pressure through a slider-crank mechanism. A refrigerator system, which operates using the basic principles of thermodynamics, consists of a compressor, a condenser, a capillary tube, and an evaporator. The vapor compression refrigeration cycle receives work from the compressor and transfers heat from the evaporator to the condenser. The main function of the refrigerator is to provide cold air from the evaporator to the freezer and refrigerator compartments. Consequently, it keeps the stored food fresh.

To improve its energy efficiency, the designer would choose the good performance of the compressor. Figure 9.54 shows a reciprocating compressor with a redesigned rotor and stator. The redesign was developed to improve the energy efficiency and

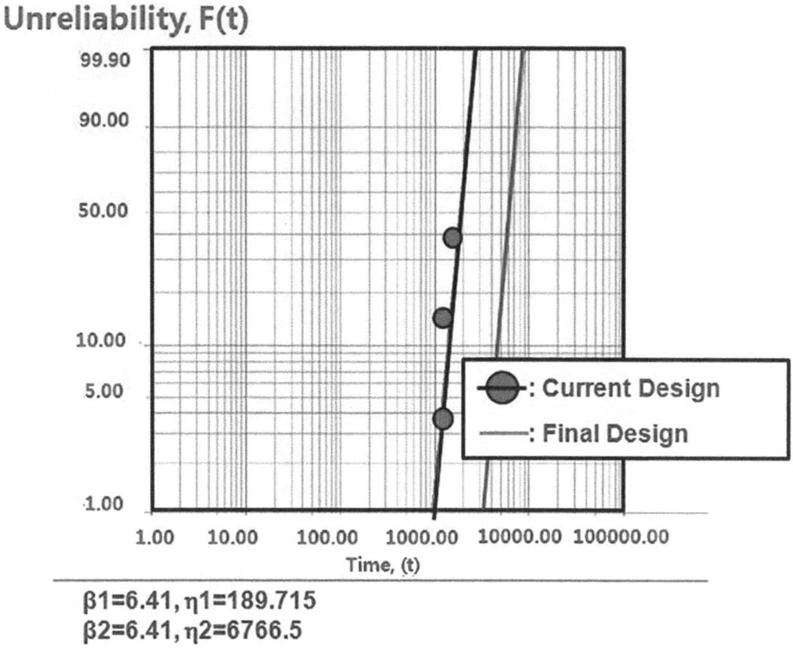

FIGURE 9.53 Results of ALT plotted in a Weibull chart.

TABLE 9.10
Results of ALTs in Failure Analysis and Redesign of the Evaporator Tubing

	First ALT	Second ALT
	Initial Design	**Second Design**
In 5,300 cycles, corrosion of the evaporator pipe is less than 1	1,130 cycles: 1/19 pitting 1,160 cycles: 2/19 pitting 1,690 cycles: 4/19 pitting 1,690 cycles: 11/19 OK	5,300 cycles: 9/9 OK
Evaporator pipe structure		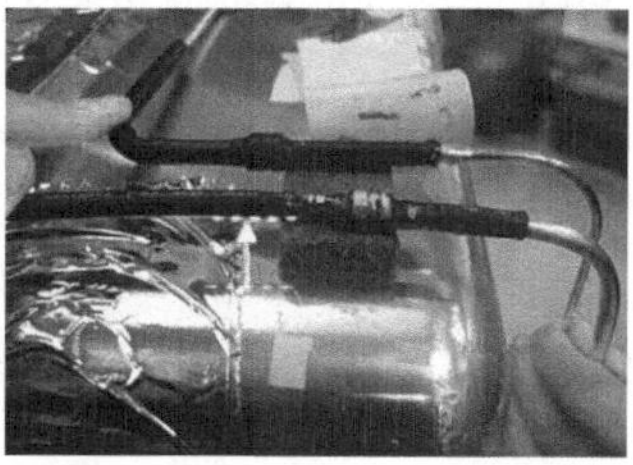
Material and specification	Length of the contraction tube C1: 50.0 mm → 200.0 mm Adhesive tape type C2: cotton type → generic transparent tape	

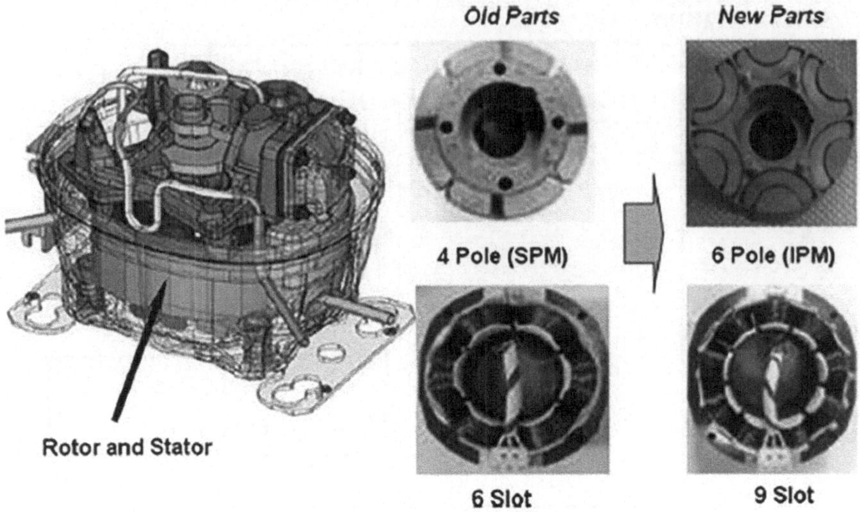

FIGURE 9.54 Reciprocating compressors with redesigned rotor and stator.

reduce the noise from the compressors in a side-by-side refrigerator. For these applications, the compressor needed to be designed robustly to operate under a wide range of consumer usage conditions (Figure 9.55).

As seen in Figure 9.56, the reciprocating compressor in the refrigerators had been making noise in the field, causing the consumer to request a replacement of their refrigerator. One of the specific causes of compressor failure during the operation was the compressor suspension spring. When the sound level during compressor shutdown of problematic refrigerators in the field was recorded, the result was approximately 46 dB (6.2 sones). The design flaws of the suspension spring in the problematic compressor were the number of turns and the mounting spring diameter. When the compressor would stop suddenly, the spring sometimes would not grab the stator frame tightly and would cause the noise.

After identifying the missing control parameters related to the newly designed compressor system, it was important to modify the defective compressor either

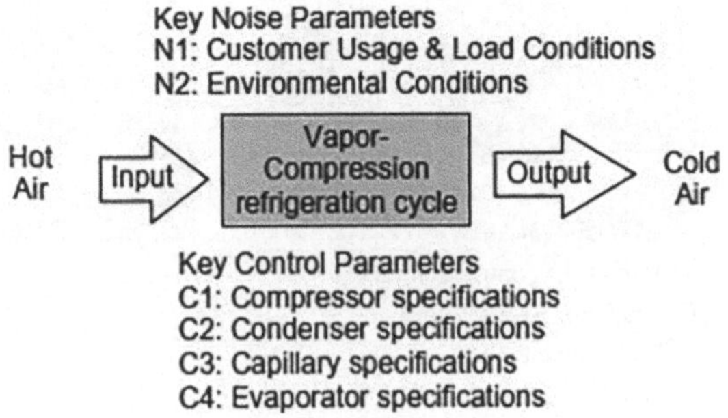

FIGURE 9.55 Parameter diagram of the refrigeration cycle.

(a)

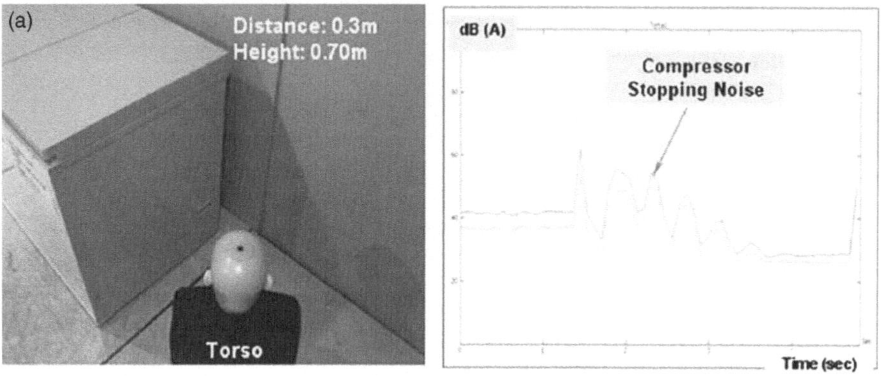

(b)

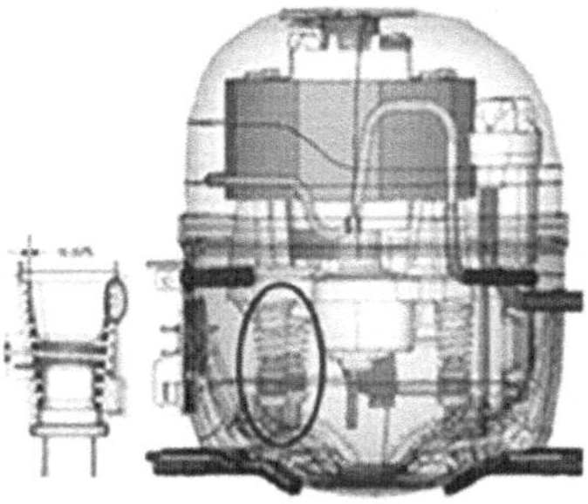

	Current	Modified
Suspension Spring		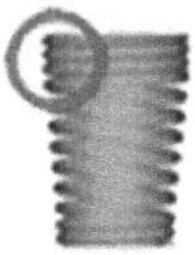
Height, mm	30.2	31.2
Turn	7/ Upper 2 / Lower 2	7/ Upper 3 / Lower 2
Diameter, ϕ	15.55	15.35
Weight	6.9 kg	

FIGURE 9.56 Stopping noise of the reciprocating compressor: (a) compressor stopping noise recorded with torso and (b) reciprocating compressor and the design flaws of a suspension spring.

through redesign of components or change of the material used in the components. Failure analysis of marketplace data and ALT can help confirm the missing key control parameters and their levels in a newly designed compressor system.

To evaluate the ride quality of the newly designed stator frame and piston–cylinder assembly mounted on the compressor shell, the compressor can be modeled to two degrees of freedom (Figure 9.57). Though it is simply four state variables in its model, it serves the purpose of figuring out the compressor motion in operation. The assumed model of the vehicle consists of the sprung mass and the unsprung mass, respectively. The sprung mass m_s represents the stator frame and piston–cylinder assembly, and the unsprung mass m_{us} represents the rotor–stator assembly. The main suspension is modeled as a spring k_s and a damper c_s in parallel, which connects the unsprung to the sprung mass. The compressor suspension spring on its shell is modeled as a spring k_{us} and represents the transfer of the road force to the unsprung mass.

$$m_s\ddot{x}_s + c_s(\dot{x}_s - \dot{x}_{us}) + k_s(x_s - x_{us}) = F\sin\omega t \tag{9.39}$$

$$m_{us}\ddot{x}_{us} + c_s(\dot{x}_{us} - \dot{x}_s) + (k_{us} + k_s)x_{us} - k_s x_s = 0 \tag{9.40}$$

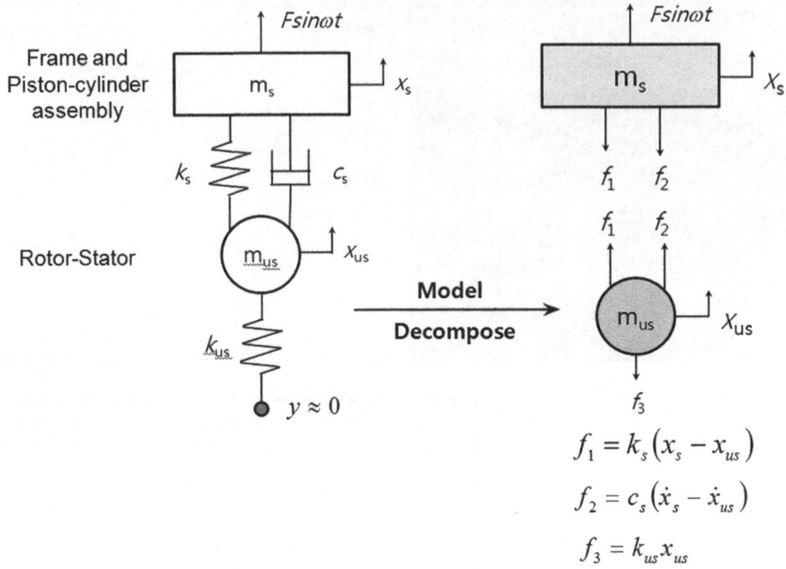

FIGURE 9.57 A compressor assembly model and its decomposition.

So the above equations of motion can be concisely represented as follows:

$$\begin{bmatrix} m_s & 0 \\ 0 & m_{us} \end{bmatrix}\begin{bmatrix} \ddot{x}_s \\ \ddot{x}_{us} \end{bmatrix}+\begin{bmatrix} c_s & -c_s \\ -c_s & c_s \end{bmatrix}\begin{bmatrix} \dot{x}_s \\ \dot{x}_{us} \end{bmatrix}$$

$$+\begin{bmatrix} k_s & -k_s \\ -k_s & k_{us}+k_s \end{bmatrix}\begin{bmatrix} x_s \\ x_{us} \end{bmatrix}=\begin{bmatrix} F\sin\omega t \\ 0 \end{bmatrix} \tag{9.41}$$

As a result, Equation (9.41) can be expressed in a matrix form:

$$[M]\ddot{X}+[C]\dot{X}+[K]X=F \tag{9.42}$$

When Equation (9.42) is numerically integrated, we can obtain the time response of the state variables. At this time, we have to understand the excited force due to the compressor operation.

In a refrigeration cycle design, it is necessary to determine both the condensing pressure, P_c, and the evaporating pressure, P_e. One indicator of the internal stresses on components in a compressor depends on the pressure difference between suction pressure, P_{suc}, and discharge pressure, P_{dis}, previously mentioned in Section 9.5.

Because an excited force due to the reciprocating motion of piston comes from the pressure difference between the condenser and evaporator, the general LS model from Equation (8.16) can be modified as follows:

$$TF = A(S)^{-n} = A(F)^{-\lambda} = A(\Delta P)^{-\lambda} \tag{9.43}$$

The AF can be derived as follows:

$$AF = \left(\frac{S_1}{S_0}\right)^n = \left(\frac{F_1}{F_0}\right)^\lambda = \left(\frac{\Delta P_1}{\Delta P_0}\right)^\lambda \tag{9.44}$$

The normal number of operating cycles for 1 day was approximately 24; the worst case was 74. Under the worst case, the objective compressor cycles for 10 years would be 270,100 cycles. From the ASHRAE Handbook test data for R600a, the normal pressure was 0.40 MPa at 42°C and the compressor dome temperature was 64°C. For the accelerated testing, the AF for pressure at 1.39 MPa was 12.6 with a quotient, λ, of 2. The total AF was approximately 12.6 (Table 9.11).

The parameter design criterion of the newly designed compressor can be more than the lifetime target – B1 10 years. Assuming the shape parameter β was 2.0, the test cycles for 100 test samples calculated in Equation (8.29) were 21,400 cycles, respectively. The ALT was designed to ensure a B1 life of 10 years with about a 60% level of confidence that it would fail less than once during 21,400 cycles. For the ALT experiments, a simplified vapor compression refrigeration system was fabricated (Figure 9.58a). Figure 9.58b shows the duty cycles for the repetitive pressure difference ΔP.

Figure 9.59 shows the stopping noise and vibration of a compressor from the ALT. In the chart, the peak noise level and vibration of a normal sample in the compressor were 52 dB and 0.09 *g* when it stopped. On the other hand, for the failed sample #1, the peak noise levels and vibration were 65 dB and 0.52 *g*. For the failed sample #2, the peak noise levels and vibration were 70 dB and 0.60 *g*. Considering that the vibration specifications called for less than 0.2 *g*, the failed sample vibrations violated the specification. When the problematic samples in ALT equipment were mounted on the test refrigerator, the vibration also was reproduced with 0.25 *g* and violated the specification. In the field, the consumer would request the failed samples to be replaced. Figure 9.60 presents the graphical analysis of the ALT results and field data on a Weibull plot. For the shape parameter, the estimated value on the chart was 1.9.

When the failed samples were cut apart, a scratch was found inside the upper shell of the compressor where the stator frame had hit the shell. The gap between the frame and the shell was measured to be 2.9 mm. The design gap specification should have been more than 6 mm to avoid the compressor hitting the shell for the worst case. It was concluded that the stopping noise came from the hitting (or interference)

TABLE 9.11
ALT Conditions in Vapor Compression Cycles for R600a

System Conditions		Worst Case	ALT	*AF*
Pressure, MPa	High side	0.40	1.39	12.6
	Low side	0.02	0.04	
	ΔP	0.38	1.35	

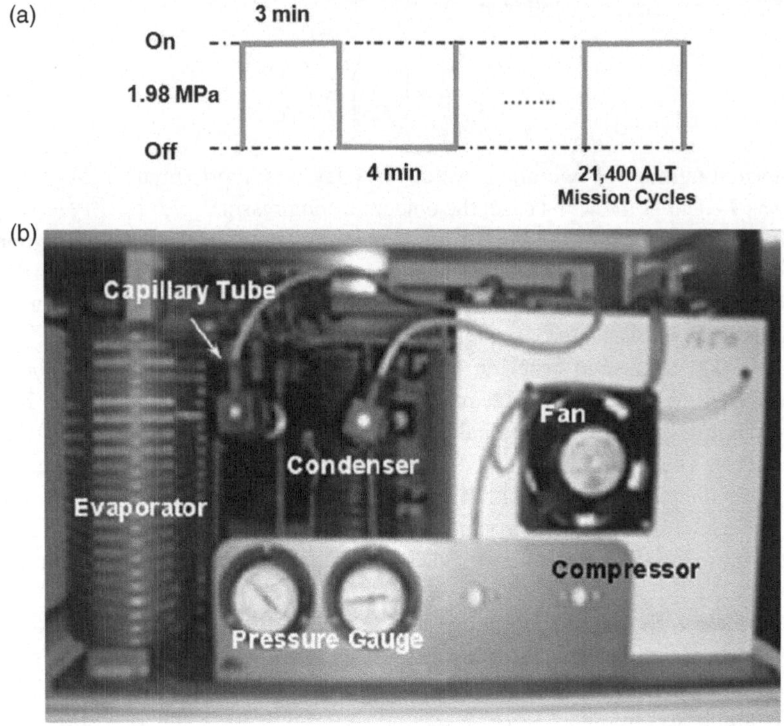

FIGURE 9.58 Duty cycles and equipment used in ALT: (a) duty cycles of repetitive pressure difference on the compressor and (b) equipment used in ALT.

between the stator frame and the upper shell. Thus, the tests pinpointed the design flaws in the compressor (see Figure 9.61a). For the shape parameter, the estimated value on the chart was 1.9 from the graphical analysis of the ALT results and field data on a Weibull plot. The vital missing parameter in the design phase of the ALT was a gap between the stator frame and the upper shell. These design flaws may make noise when the compressor stops suddenly. To reduce the noise problems in the frame, the shape of the stator frame was redesigned. As the test setup of the compressor assembly was modified to have more than a 6-mm gap, the gap size increased from 2.9 to 7.5 mm (Figures 9.61b and 9.62).

The parameter design criterion of the newly designed samples was more than the lifetime target – a B1 life of 10 years. The confirmed value, β, on the Weibull chart was 1.9. When the second ALT proceeded, the test cycles recalculated in Equation (8.29) for 100 sample pieces were 21,400, respectively. In the second ALT, no problems were found with the compressor in 21,400 cycles. We expected that the modified design parameters are effective (Figure 9.62). Table 9.11 provides a summary of the ALT results. With the improved design parameters, the B1 life of the samples in the second ALT lengthens more than 10.0 years (Table 9.12).

(a)

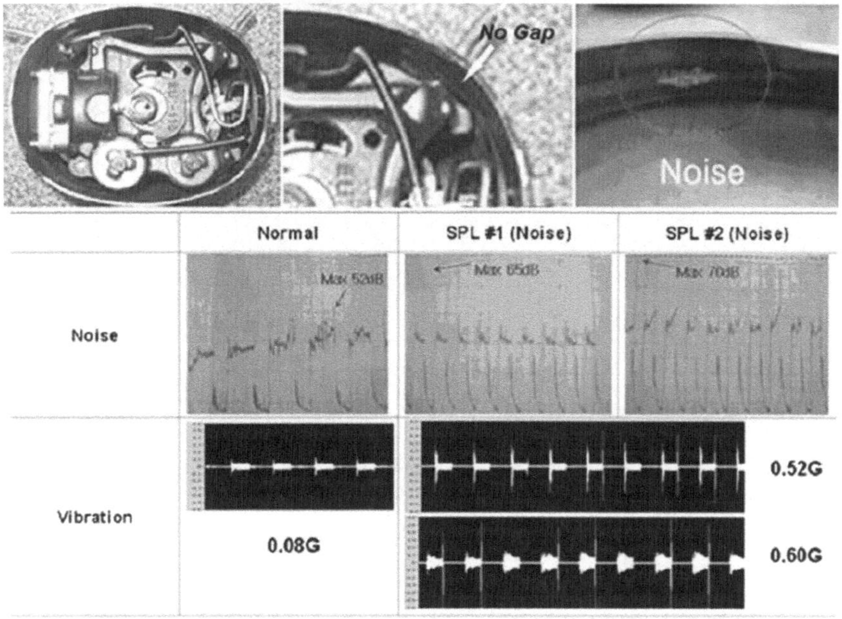

(b)

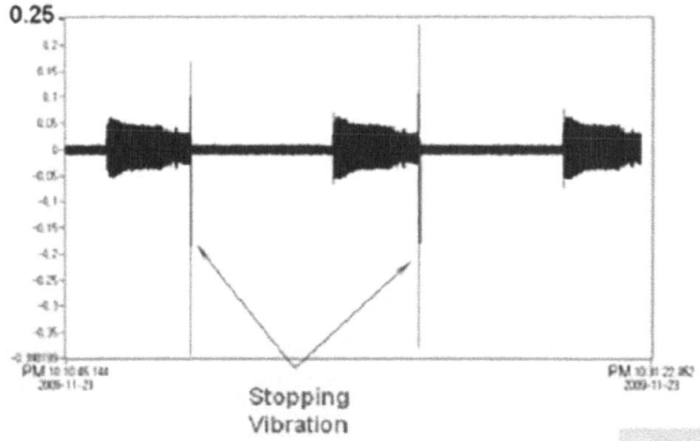

Fan Off Condition

	25°C	32°C	43°C (3800RPM)
3min On/3min Off	No	No	Reproduced
3min On/5min Off	No	No	Reproduced
10min On/10min Off	No	No	No

FIGURE 9.59 Failed products in the first ALT: (a) noise in ALT equipment and (b) noise in the refrigerator.

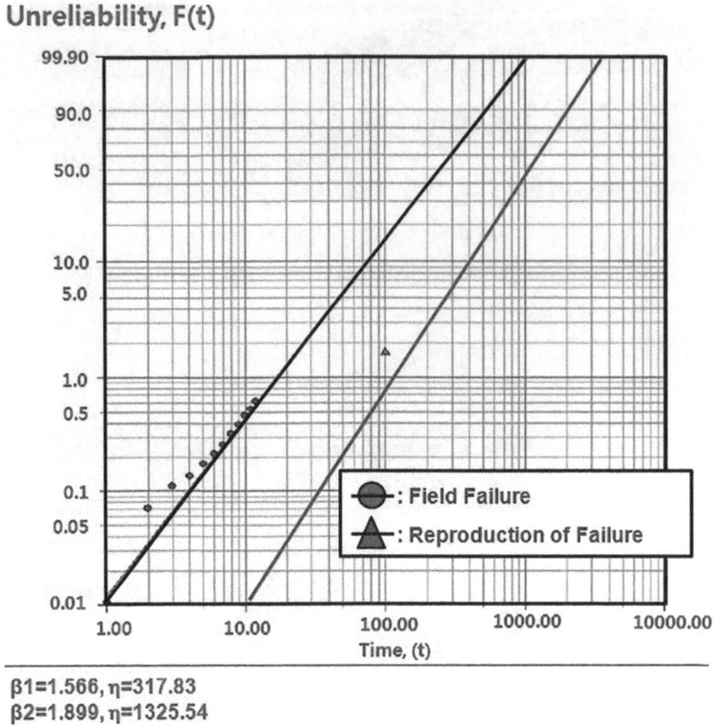

FIGURE 9.60 Field data and results of ALT on a Weibull chart.

9.8 REFRIGERATOR COMPRESSOR SUBJECTED TO REPETITIVE LOADS

To store food, the refrigerator utilizing a vapor-compression refrigeration cycle has to provide cold air from the evaporator to the freezer and refrigerator compartments. The refrigeration unit consists of a compressor, a condenser, a capillary tube, and an evaporator. The compressor takes the refrigerant from the evaporator and then compresses it, which transfers it to the condenser in the refrigeration cycle. In the process, a reciprocating compressor increases the refrigerant pressure from that in the evaporator to that in the condenser and is subjected to repeated stresses by the operation of the crankshaft. As customers demand refrigerators that use less energy, it is necessary to improve the overall energy efficiency of the refrigerator. One way to improve efficiency is to redesign the compressor in the refrigerator. The primary components in a domestic compressor consist of the crankshaft, piston assembly, stator and its frame, valve plate, and suction reed valve (Figure 9.63).

In the field, compressors in the refrigerators had been locked due to wear, which was the continuing loss of material from the surface of a crankshaft and causing loss of the cooling function. Based on the consumer usage conditions, we knew that compressors were subjected to repetitive pressure loads during the refrigerator operation (Figure 9.64).

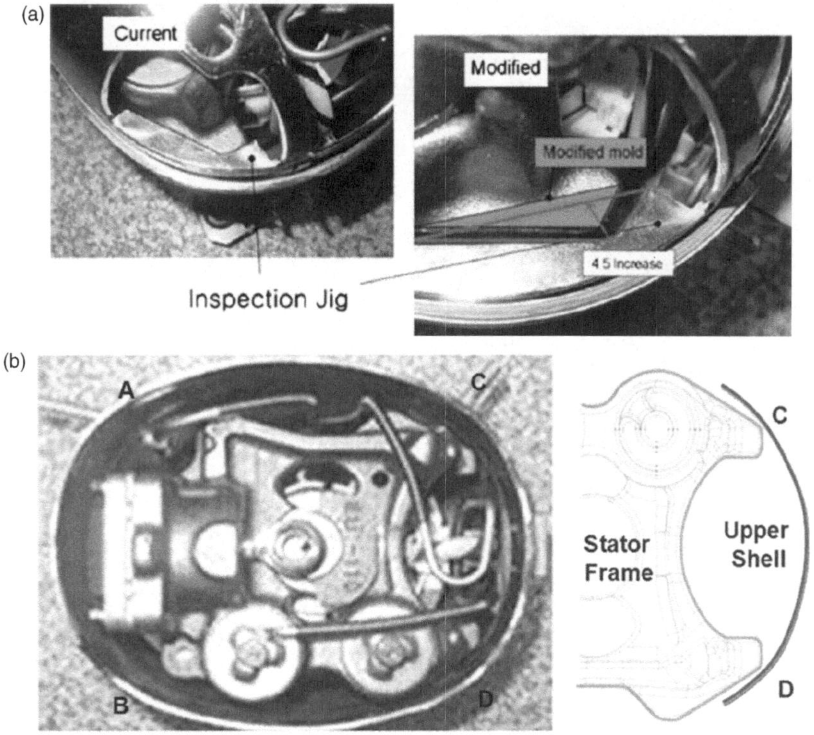

Gap between frame and upper shell (Spec.: 6mm↑)			
A	B	C	D
8.5	9.1	3.2	2.9

FIGURE 9.61 Modified inspection jig and gaps: (a) modified inspection jig and (b) the gap between the stator frame and the upper shell.

In evaluating the design of a refrigeration cycle, it was necessary to determine both the condensing temperature, T_c, and evaporating temperature, T_e. The mass flow rate of refrigerant in a compressor can be modeled as follows:

$$\dot{m} = PD \times \frac{\eta_v}{v_{suc}} \tag{9.45}$$

The mass flow rate of refrigerant in a capillary tube can be modeled as

$$\dot{m}_{cap} = A\left[\frac{-\int_{P_3}^{P_4} \rho\, dP}{\frac{2}{D} f_m \Delta L + \ln\left(\frac{\rho_3}{\rho_4}\right)}\right]^{0.5} \tag{9.46}$$

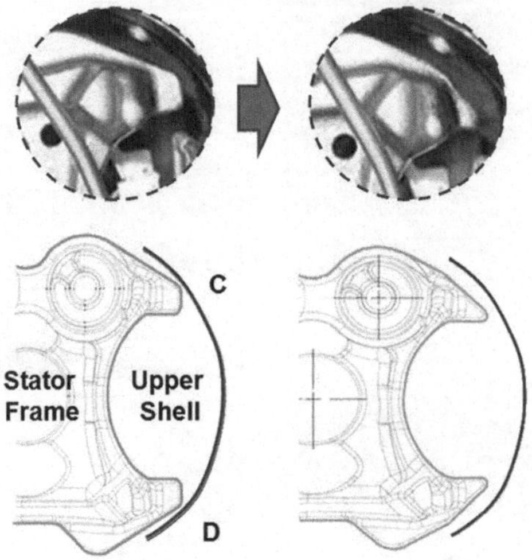

Gap between frame and upper shell (Spec.: 6mm↑)				
	A	B	C	D
Old Part	8.5	9.1	3.2	2.9
New Part	8.3	8.8	7.7	7.5

FIGURE 9.62 Redesigned stator frame in the second ALT.

TABLE 9.12
Results of ALTs in Improving the Noise of Mechanical Compressor

<table>
<tr><th></th><th>First ALT</th><th>Second ALT</th></tr>
<tr><th></th><th>Initial Design</th><th>Second Design</th></tr>
<tr><td>In 21,400 cycles, locking is less than 1</td><td>100 cycles: 2/100 noise
100 cycles: 99/100 OK</td><td>21,400 cycles: 100/100 OK
20,000 cycles: 100/100 OK</td></tr>
<tr><td>Compressor structure</td><td>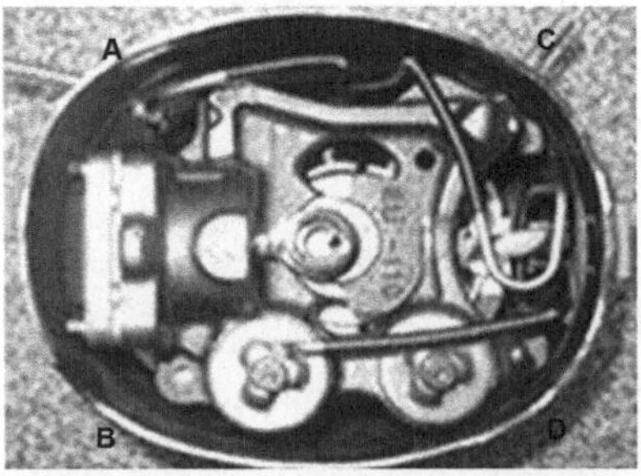

<table>
<tr><th colspan="4">Gap between frame and upper shell (Spec.: 6mm↑)</th></tr>
<tr><td>A</td><td>B</td><td>C</td><td>D</td></tr>
<tr><td>8.5</td><td>9.1</td><td>3.2</td><td>2.9</td></tr>
</table></td><td></td></tr>
<tr><td>Material and specification</td><td>C1: Modification of the frame shape</td><td></td></tr>
</table>

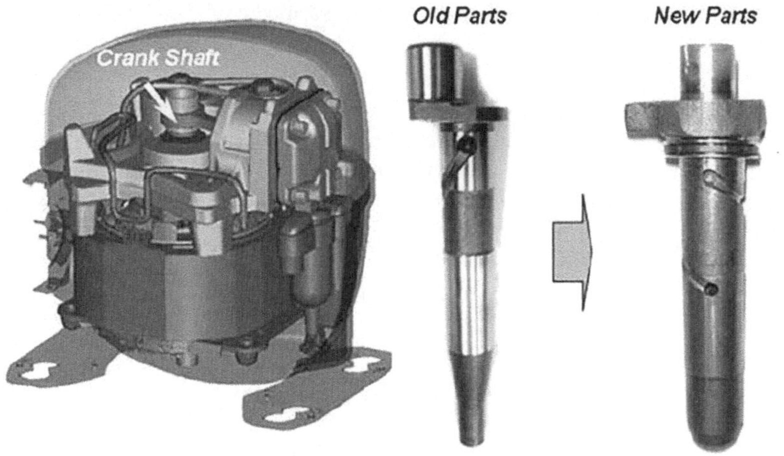

FIGURE 9.63 Redesigned compressor and crankshaft.

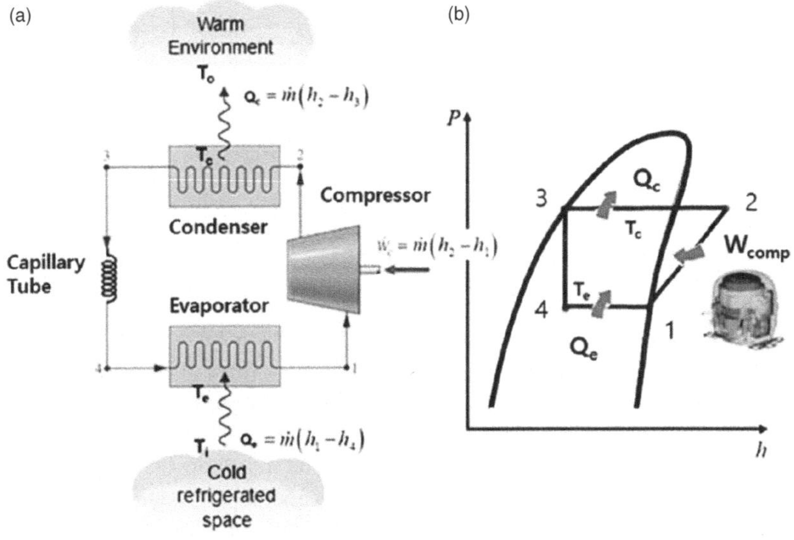

FIGURE 9.64 Functional design concept of the compressor system in the refrigeration cycle: (a) refrigeration cycle and (b) *P*–*h* diagram.

By conservation of mass, the mass flow rate can be determined as follows:

$$\dot{m} = \dot{m}_{\text{cap}} \tag{9.47}$$

The energy balance in the condenser can be described as

$$Q_c = \dot{m}(h_2 - h_3) = (T_c - T_o)/R_c \tag{9.48}$$

The energy balance in the evaporator can be described as

$$Q_c = \dot{m}(h_1 - h_4) = (T_i - T_e)/R_e \tag{9.49}$$

When nonlinear Equations (9.47) through (9.49) are solved, the mass flow rate, $\dot{m}$, evaporator temperature, T_e, and condenser temperature, T_c, can be obtained. Because the saturation pressure, P_{sat}, is a function of temperature, the evaporator pressure, P_e (or condenser pressure P_c), can be obtained as follows:

$$P_e = f(T_e) \text{ or } P_c = f(T_c) \tag{9.50}$$

One source of stress in a refrigeration system may come from the pressure difference between suction pressure, P_{suc}, and discharge pressure, P_{dis}.

$$\Delta P = P_{dis} - P_{suc} \cong P_c - P_e \tag{9.51}$$

For a compressor system, the time-to-failure in Equation (9.7), *TF*, can be modified as

$$TF = A(S)^{-n} \exp\left(\frac{E_a}{kT}\right) = A(\Delta P)^{-n} \exp\left(\frac{E_a}{kT}\right) \tag{9.52}$$

The *AF* from Equation (9.52) can be derived as

$$AF = \left(\frac{S_1}{S_0}\right)^n \left[\frac{E_a}{k}\left(\frac{1}{T_0} - \frac{1}{T_1}\right)\right] = \left(\frac{\Delta P_1}{\Delta P_0}\right)^\lambda \left[\frac{E_a}{k}\left(\frac{1}{T_0} - \frac{1}{T_1}\right)\right] \tag{9.53}$$

The operating conditions for the compressor in a refrigerator were approximately 0°C–43°C with a relative humidity ranging from 0% to 95%, and 0.2–0.24 *g* of acceleration. The number of compressor operation is influenced by specific consumer usage conditions. The reciprocating compressor occurs about 10–24 times per day in the Korean domestic market. With a product life cycle design point of 10 years, the compressor approximately undergoes 87,600 usage cycles.

For this worst case, the pressure was 1.07 MPa and the compressor dome temperature was 90°C. For the ALT, the pressure was 1.96 MPa and the dome temperature was 120°C. Using a cumulative damage exponent of 2.0 (λ), the *AF* from Equation (9.30) was found to be approximately 4.58 (Table 9.13).

For the lifetime target – a B1 life of 10 years, the test cycles calculated from Equation (8.29) for one hundred sample units were 19,000 cycles if the shape parameter was assumed to be 2.0. This parametric ALT was designed to ensure a B1 life of 10 years with about a 60% level of confidence that it would fail less than once during 19,000 cycles. Figure 9.65 shows the experimental setup of the ALT with labeled equipment for the robust design of the compressor. It consisted of an evaporator, a compressor, a condenser, and a capillary tube in a simplified vapor-compression refrigeration system.

The inlet to the condenser section was at the top and the condenser outlet was at the bottom. The condenser inlet was constructed with quick coupling and had a

TABLE 9.13
ALT Conditions in Vapor Compression Cycles

System Conditions		Worst Case	ALT	*AF*
Pressure, MPa	High side	1.07	1.96	3.36③
	Low side	0.0	0.0	(= (①/②)^2)
	ΔP	1.07①	1.96②	
Temp., °C	Dome Temp.	90	120	1.37④
Total *AF* (= ③ × ④)		–		4.58

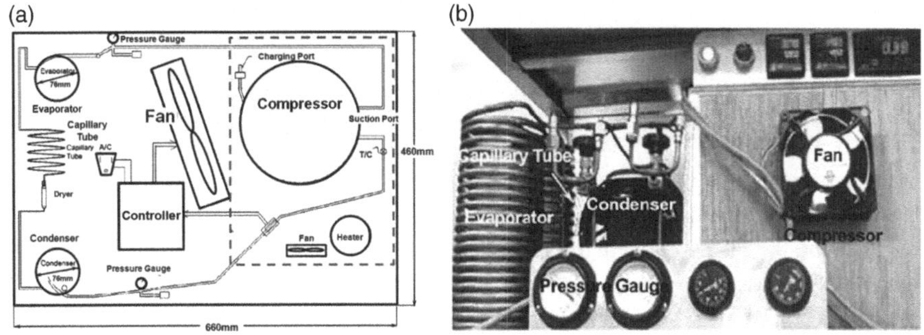

FIGURE 9.65 Equipment used in ALT: (a) a drawing of the test system and (b) photograph.

high-side pressure gauge. A ten-gram refrigerator dryer was installed vertically at the condenser inlet. A thermal switch was attached to the condenser tubing at the top of the condenser coil to control the condenser fan. The evaporator inlet was at the bottom. At a location near the evaporator outlet, pressure gauges were installed to enable access to the low side for evacuation and refrigerant charging.

The condenser outlet was connected to the evaporator outlet with a capillary tube. The compressor was mounted on rubber pads and was connected to the condenser inlet and evaporator outlet. A fan and two 60-Watt lamps maintained the room temperature within an insulated (fiberglass) box. A thermal switch attached on the compressor top controlled a 51 m^3/h axial fan. Repetitive stress can be expressed as the duty effect cycles and the compressor shortens part life due to the on/off.

In a French door refrigerator, it was found that two compressors with the newly designed crankshafts were locked due to wear at 10,504 cycles. Field data indicated that the damaged products may have had a design flaw. That is, there was the problem of oil lubrication when a crankshaft was repetitively subjected to the tribological stress on its surface and relative movement with a solid counter body such as a connecting rod. Therefore, we found the severely wore crankshaft on its top and body when the failed product was broken down (field and first ALT). At the initial phase, due to lubrication problem and direct contact, the repetitive pressure loads in the compressor could create undue adhesive wear on the surface of the crankshaft. As the wear particles were detached and blocked the refrigerant line in the compressor, it eventually causes the compressor to lock. An engineer should reproduce these design problems and modify them before the product launches. Figure 9.66 shows a photograph comparing the failed product from the field and that from the first ALT, respectively.

Figure 9.67 presents the graphical analysis of the ALT results and field data on a Weibull plot. For the shape parameter, the estimated value in the previous

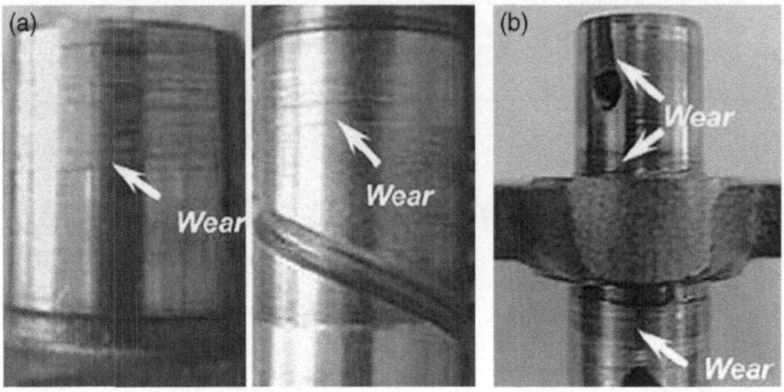

FIGURE 9.66 Failed product in the field and first ALT: (a) failed product in the field and (b) failed sample in the first ALT.

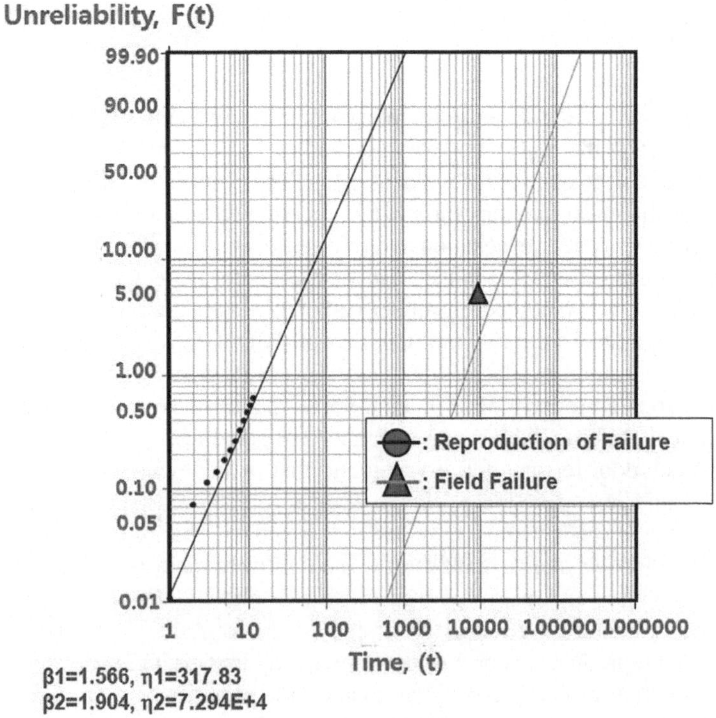

FIGURE 9.67 Field data and results of ALT on a Weibull chart.

ALT was 1.9. It was concluded that the methodologies used were valid in pinpointing the weaknesses in the original design of the units sold in the market because (1) the location and shape of the locking crankshaft from both the field and ALT were similar and (2) on the Weibull, the shape parameters of the ALT results, $\beta 1$, and market data, $\beta 2$, are very similar. Consequently, we know that this parametric ALT is effective to reproduce the field failure and modify it.

In the first ALT, the two compressors were locked at 7,500 and 10,000 cycles. When the locked compressors from the field and first ALT were cut apart, severe wear was found in some regions of the crankshaft where there was no lubrication – the moving area between the shaft and connecting rod as well as the rotating area between the crankshaft and block. As shown in Figure 9.67b, the tests confirmed that the compressor was not well designed to ensure proper lubrication. Whenever the compressor was starting, we recognized that the poor lubrication in combination with the repetitive pressure loads might cause problematic wear on the surface of the crankshaft. The defective shape of the first ALT was very similar to that of the ones from the field.

The root causes of the crankshaft wear came from the improper designs: (1) the lack of oil in areas that should have lubricated (Figure 9.67a), (2) low starting RPM (Figure 9.68b), and (3) weak crankshaft material (FCD450) (Figure 9.69). These compressor design flaws could cause the compressor to lock up suddenly when subjected to repetitive pressure loads. To improve the lubrication problems on the surface of the crankshaft, it was redesigned by increasing the starting RPM from 1,650 to 2,050 RPM, the relocated lubrication holes, new groove, and new shaft material FCD500 (Figure 9.70).

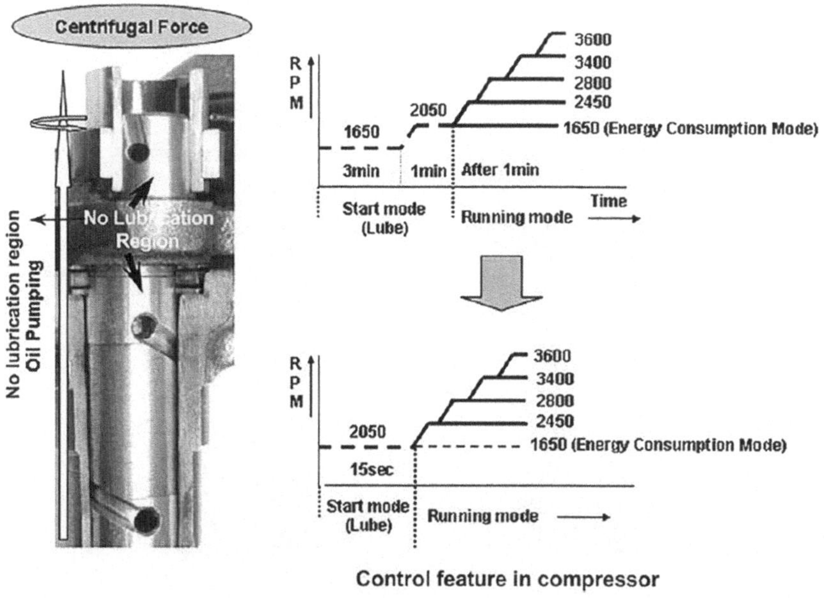

FIGURE 9.68 No lubrication region in the crankshaft and low starting RPM (1,650 RPM).

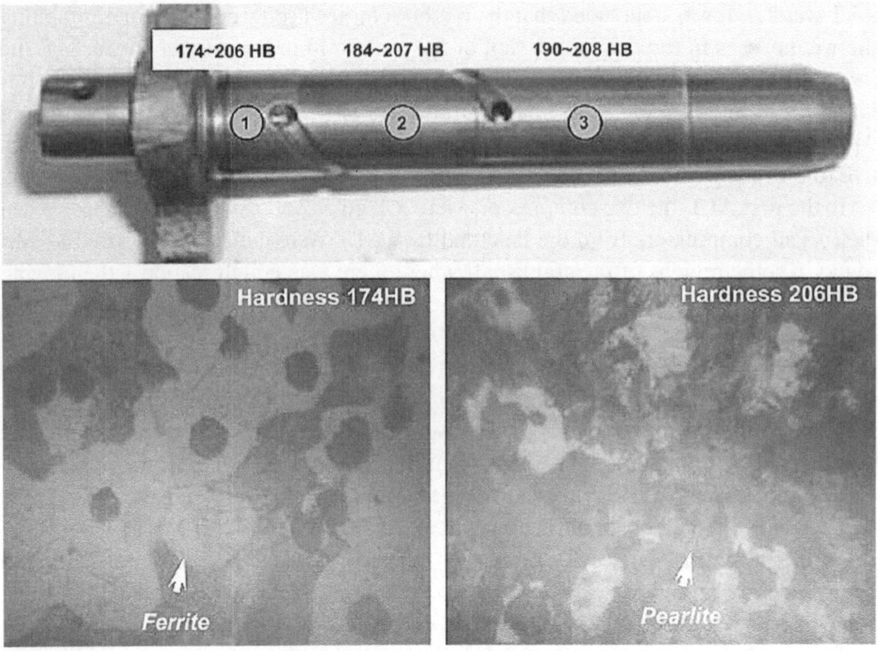

FIGURE 9.69 A large variation of hardness (FCD450) in the crankshaft.

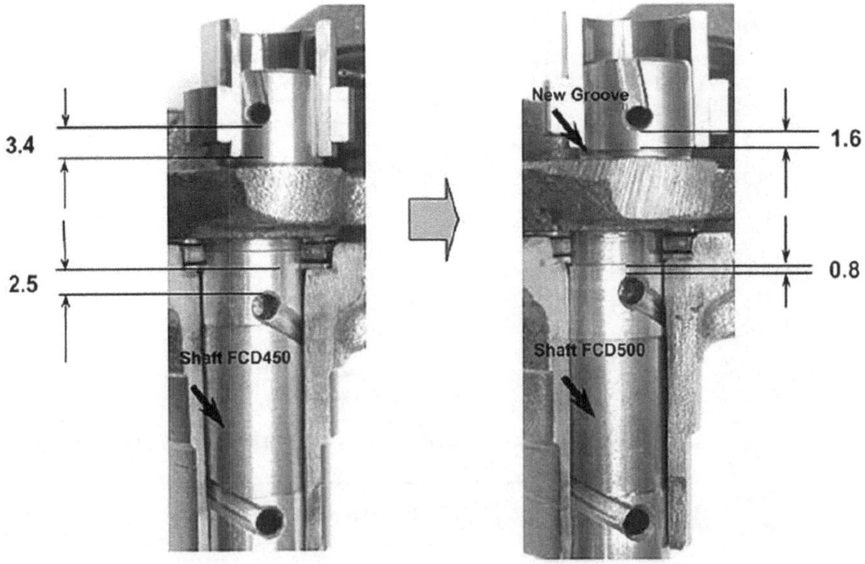

FIGURE 9.70 Redesigned crankshaft in the first ALT: relocating the lubrication holes, adding the new groove, and changing material (FCD450–FCD500).

For the lifetime target – a B1 life of 10 years, the test cycles recalculated from Equation (8.29) for one hundred sample units were 19,000 cycles if the shape parameter was supposed to be 2.0. This parametric ALT was designed to ensure a B1 life of 10 years with about a 60% level of confidence that it would fail less than once during 19,000 cycles.

In the second ALTs, when the compressors were broken down, the crankshaft wear due to interference between the crankshaft and a thrust washer was found at 19,000 cycles (Figure 9.71). To improve the design problem in the second ALT, we increased the minimum clearance between the crankshaft and washer from 0.141 to 0.480 mm (Figure 9.72).

To withstand the design problems of crankshaft due to the repetitive pressure loads, we can summarize the improved design as follows: (1) changing the starting RPM, C1, from 1,650 to 2,050 RPM; (2) relocating the lubrication holes and adding the new groove, C2; (3) changing the crankshaft material, C3, from FCD450 to FCD500; and (4) increasing the minimum clearance between crankshaft and washer, C4, from 0.141 to 0.480 mm (Table 9.14).

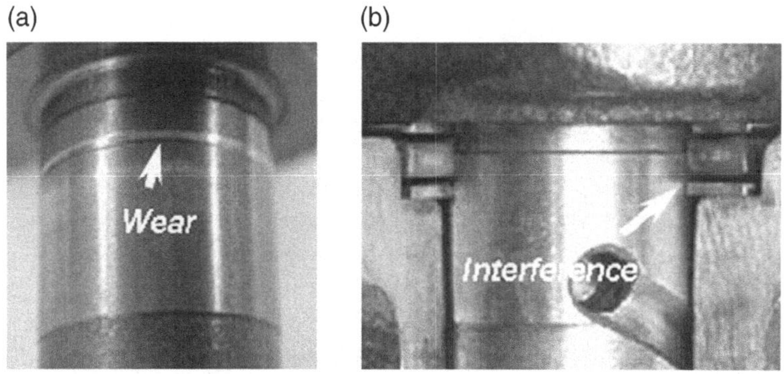

FIGURE 9.71 Failed products in the second ALT: (a) failed products in the second ALT and (b) the root cause.

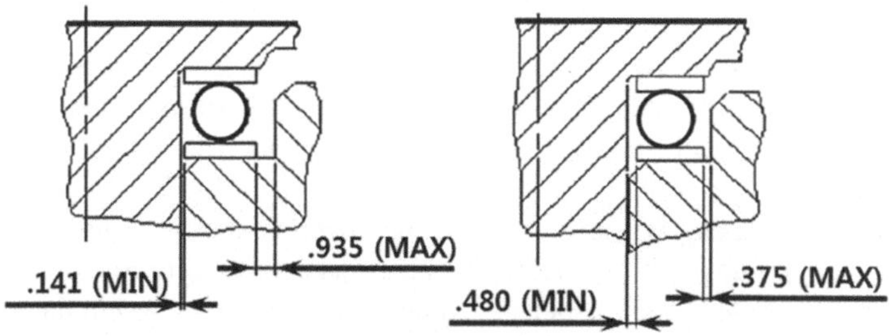

FIGURE 9.72 Redesigned crankshaft in the second ALT: modified minimum clearance between the crankshaft and washer (0.141 mm → 0.480 mm).

TABLE 9.14
Results of ALTs in Refrigerator Compressor Subjected to Repetitive Loads

	First ALT	Second ALT	Third ALT
	Initial Design	Second Design	Final Design
In 19,000 mission cycles, there is no problem in the compressor	10,504 cycles: 2/100 locking 19,000 cycles: 28/100 OK	19,000 cycles: 2/100 wear 19,000 cycles: 28/100 OK	19,000 cycles: 100/100 OK 20,000 cycles: 100/100 OK
Photo	<Crank shaft> 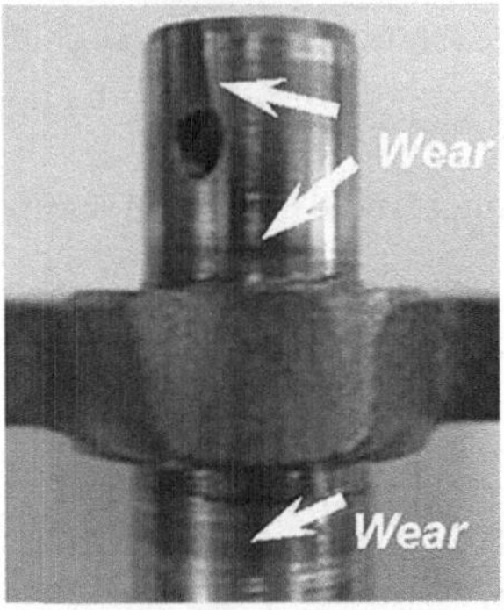	<Crank shaft> 	–
Action plans	C1: Starting RPM (1,650 → 2,050) C2: Relocating hole and one new groove C3: Material (FCD450 → FCD500)	C4: Clearance (0.141 → 0.480 mm) (Modification of washer dimension)	

9.9 WATER DISPENSER LEVER IN A REFRIGERATOR

As customers want to have a water-dispensing function, Figure 9.73 shows the BMF refrigerator with the newly designed water dispensing system. As shown in Figure 9.73b, it consists of the dispenser cover, spring, and dispenser lever. To dispense water for a product lifetime, the dispenser system needs to be designed to withstand the operating conditions subjected to it by the consumers who use the BMF refrigerator. Dispensing water in the BMF refrigerator has the following operating steps: (1) press the lever and (2) dispense water. The customer will dispense water if drinking the cooling water. Consequently, the water dispenser system will have a variety of repetitive mechanical loads when the consumer uses it, though depending on the consumer usage conditions.

The water dispensing system in the field had been fracturing, causing end-users to replace their product. When the refrigerator was subjected to repetitive stresses, we knew that the failed water dispensing system came from the design flaws. Market data also showed that the returned products had critical design flaws in the structure, including stress raisers – sharp corner angles and thin ribs – in water dispensing. The design flaws that could not withstand the repetitive impact loads on the water

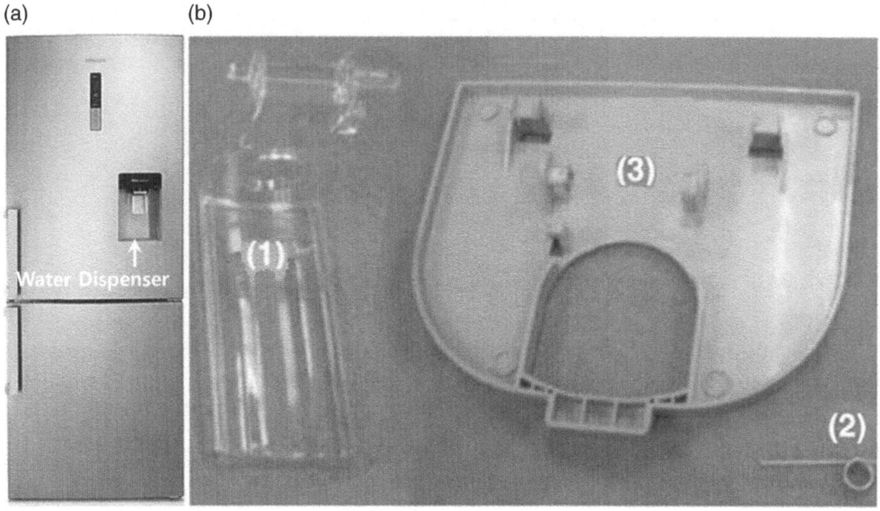

FIGURE 9.73 BMF refrigerator and dispenser assembly: (a) BMF refrigerator and (b) mechanical parts of the dispenser lever assembly: dispenser cover (1), spring (2), and dispenser lever (3).

dispensing could cause a crack to occur. Thus, the reliability design of the new water dispensing system is required to robustly withstand repetitive loads under customer usage conditions (Figures 9.74 and 9.75).

The mechanical dispensing system works like the lever mechanism that has been used for thousands of years. That is, small effort from one end of the beam will lift a large load with minimal effort. In the United States, the typical consumer claims the refrigerator to release water 4–20 times per day. The mechanical lever assembly consists of many mechanical structural parts. To properly operate the water dispensing system under such customer conditions, it should be robustly designed.

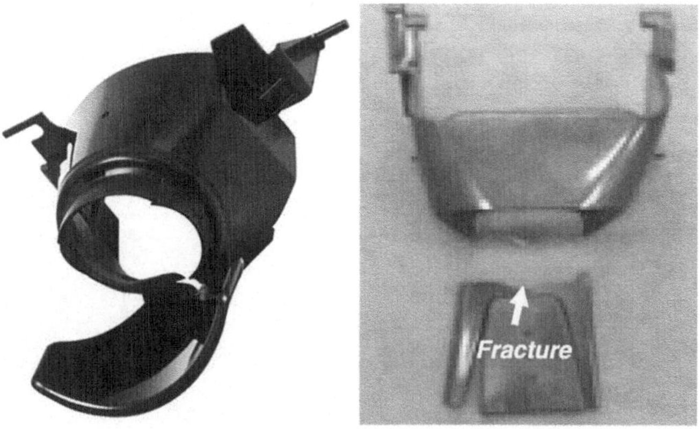

FIGURE 9.74 A damaged product after use.

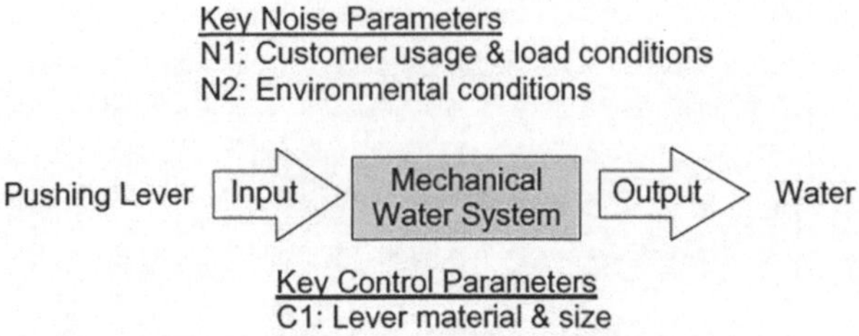

FIGURE 9.75 Robust design schematic of water dispensing.

When a cup touches the lever to release the water, water will dispense. Depending on the customer usage conditions, the lever assembly goes through repetitive impact loads in the water dispensing process. The concentrated stress of the pressing cup reveals stress raisers such as sharp corner angles. If there is a void (design fault) in the structure that causes the inadequacy of strength (or stiffness) when under dynamic loads are applied, the compressor starts making problems before it reaches its expected lifetime (Figure 9.76).

Because the stress of the lever hinge depends on the applied pressure force of the consumer, the LS model from Equation (8.16) can be modified as

$$TF = A(S)^{-n} = A(F)^{-\lambda} \tag{9.54}$$

The AF can be derived as

$$AF = \left(\frac{S_1}{S_0}\right)^n = \left(\frac{F_1}{F_0}\right)^\lambda \tag{9.55}$$

The lifetime target of the newly designed water dispensing system was over a B1 life of 10 years. The operating conditions and cycles of the dispenser system were examined, based on the customer usage conditions. Under lifetime target – a B1 life

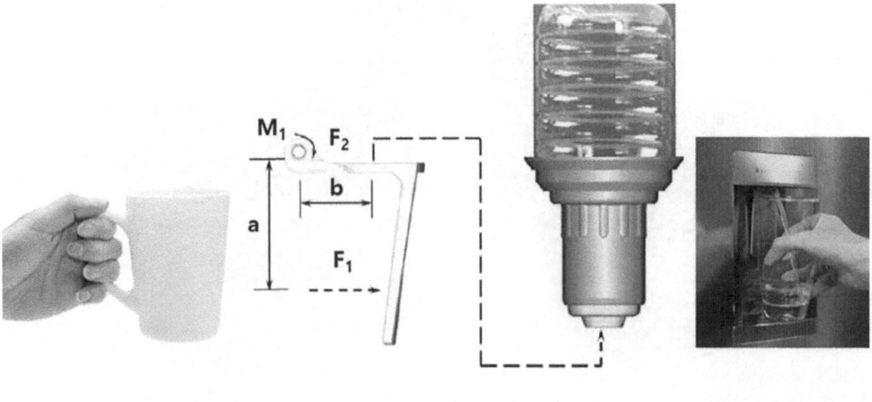

FIGURE 9.76 Design concept of a mechanical dispensing system.

of 10 years, if the objective number of life cycles L_{BX} and AF are given, the actual required test cycles h_a could be obtained from Equation (8.29). ALT equipment was made and conducted in accordance with the operation procedure of the dispenser system. From parametric ALT, we could find the design missing parameters.

For the water dispensing system in the BMF refrigerator, the working conditions of customer were about 0°C–43°C with a relative humidity ranging from 0% to 95%, and 0.2–0.24 g of acceleration. The water dispensing happens approximately 4–20 times per day. With a product life cycle design point of 10 years, the water dispensing system approximately undergoes 73,000 usage cycles.

The maximum force expected by the consumer in dispensing water was 19.6 N. For accelerated testing, the applied force makes double to 39.2 N. With a quotient, λ, of 2, the total AF was approximately 4.0 using Equation (9.55). To find out the missing design parameter of the newly designed water dispensing system, the lifetime target can be set to more than the B1 life of 10 years. Presumed the shape parameter β was 2.0, the actual test cycles calculated in Equation (8.29) were 65,000 cycles for eight sample units. If parametric ALT for water dispensing system fails less than once during 65,000 cycles, it will be guaranteed to have a B1 life of 10 years with about a 60% level of confidence (Figure 9.77).

Figure 9.78 shows the failed product from the field and the first ALT, respectively. The failure sites in the field and the first ALT occurred at the hinge and front corner of the dispenser lever as a result of high impact stress. Figure 9.79 presents a graphical analysis of the ALT results and field data display in the Weibull distribution.

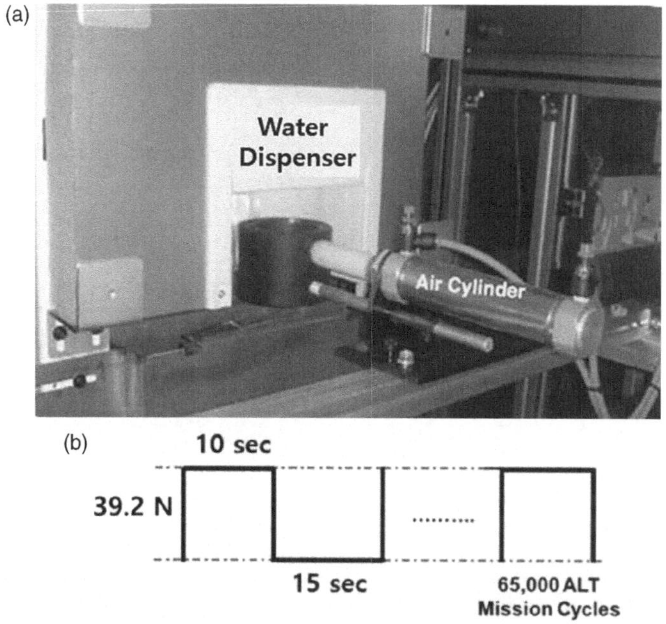

FIGURE 9.77 Accelerated life testing: (a) test equipment of a water dispenser used in ALT and (b) duty cycles of repetitive load F.

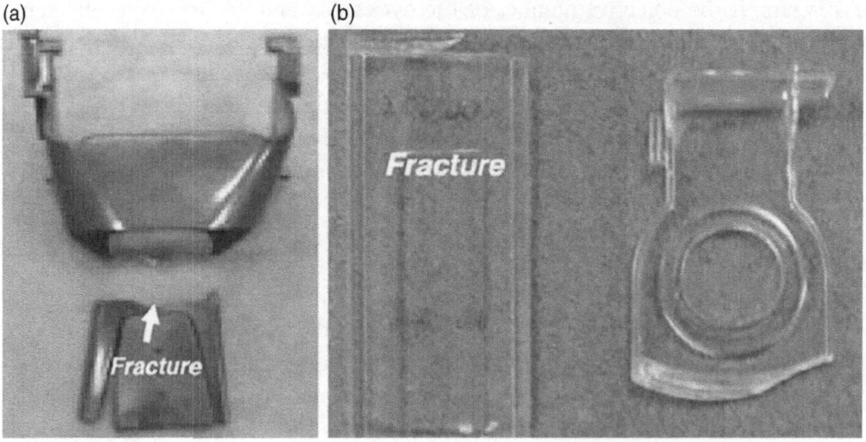

FIGURE 9.78 Failed products in the field and ALT: (a) failed product in the field and (b) failed sample in ALT.

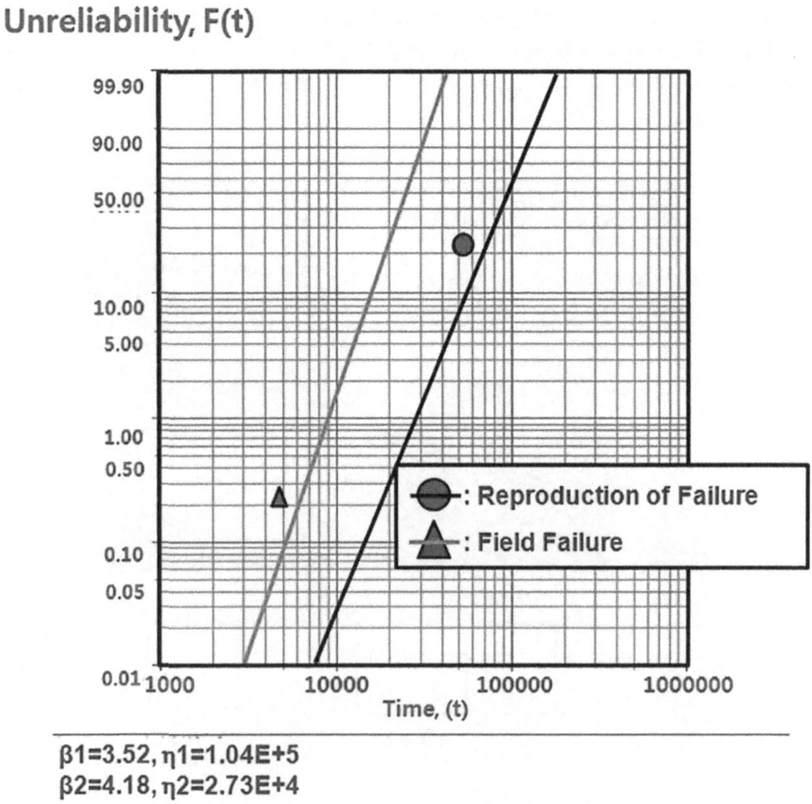

FIGURE 9.79 Photograph of the ALT results and field data and Weibull plot.

For the shape parameter, the estimated value on the chart was 2.0. For the final design, the shape parameter was determined to be 3.5. It was concluded that the methodologies used were valid in pinpointing the weaknesses in the original design of the units sold in the market because (1) the location and shape of the failed dispenser lever from both the field and ALT were similar and (2) on the Weibull, the shape parameters of the ALT results, $\beta 1$, and market data, $\beta 2$, are very similar. Consequently, we know that this parametric ALT is effective to reproduce the field failure and modify it.

As seen in Figures 9.78b and 9.80, the fracture of the dispenser lever in the field and ALTs occurred in its hinge and front corner that have a high-stress raiser. We knew that the dispenser structure has no rounding at the corner and not enough thickness to withstand the repetitive impact loading that is caused by end usage. When a problematic part such as the dispenser lever is subjected to repeated cyclic loadings, it produces permanent deformation and cracks. After repetitive loading, we also recognized that it propagates to the end. Fatigue fracture can be characterized by two factors: (1) the stress due to loads on the structure and (2) the type of materials (or shape) used in the product.

The repetitive applied force in combination with the structural flaws may have caused cracking and fracture of the dispenser lever. The design flaw of sharp corners/angles resulting in stress risers in high-stress areas can be corrected by implementing fillets on the hinge rib and front corner as well as increasing the hinge rib thickness. Through a finite element analysis, it was determined that the concentrated stresses resulting in fracture at the shaft hinge and the front corner were 9.37 and 5.66 MPa, respectively.

To withstand the fatigue fracture of the dispenser lever due to the repetitive impact stresses, the dispenser lever was redesigned as follows: (1) increasing the rib rounding of the hinge, C1, from R0 to R2.0 mm; (2) increasing the front corner rounding, C2, from R0 to R1.5 mm; (3) increasing the rib thickness of hinge, C3, from T1 to T1.8; (4) increasing the front side rounding, C4, from R0 to R11 mm; and (5) thickening the front lever, C5, from T3 to T4 mm; (Table 9.15).

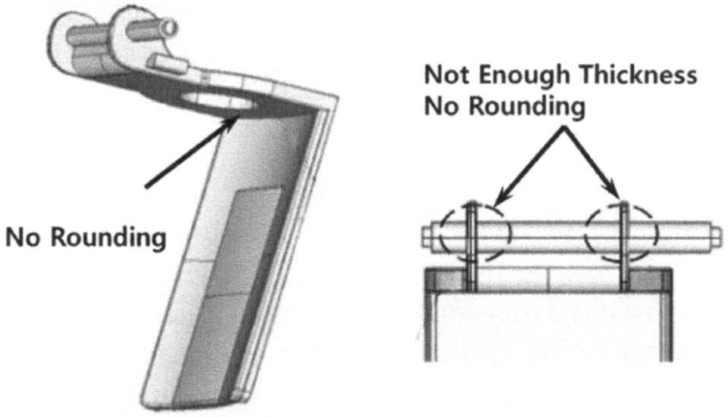

FIGURE 9.80 Structure of a failing dispenser lever in the field and ALTs.

TABLE 9.15
Redesigned Dispenser Lever

<Back>

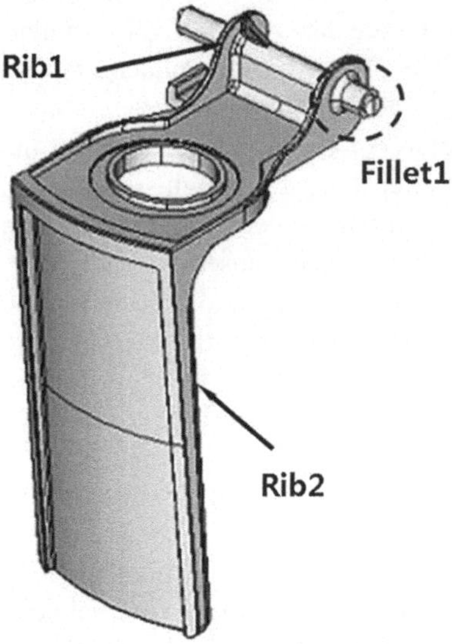

<Front>

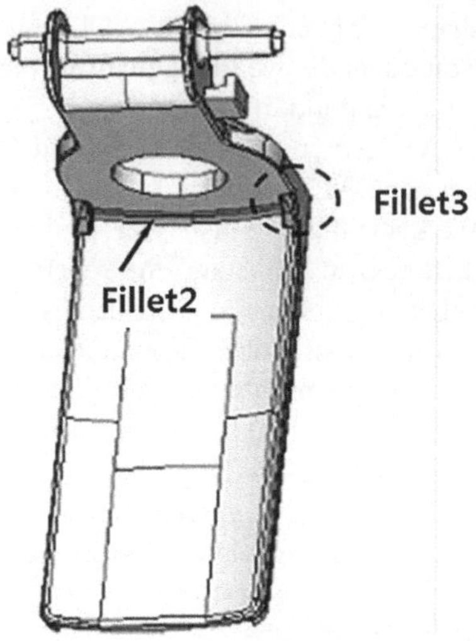

Rib1C1: T1 mm → T1.8 mm (first ALT)
Rib2 C2: T3 mm → T4 mm (third ALT)
Fillet1 C3: R0 mm → R1.5 mm (first ALT) → R2.0 mm (second ALT)

Fillet2 C4: R0 mm → R1.5 mm (first ALT)
Fillet3 C5: R0 mm → R8 mm (first ALT)

As the design flaws become better, the parameter design criterion of the newly designed samples was secured to have more than the reliability target – a B1 life of 10 years. The confirmed value, β, on the Weibull chart was found to be 3.5. For the lifetime target of the second ALT – a B1 life of 10 years, the actual test cycles in Equation (8.29) for eight sample units were 38,000. In the third ALT, there were no design problems with the water dispensing system until the test was carried out to 68,000 cycles. We, therefore, concluded that the modified design parameters found from the first and second ALT were effective.

Based on the modified design parameters, corrective measures taken to increase the life cycle of the dispenser system included: (1) increasing the hinge rib rounding, C1, from 0.0 to 2.0 mm; (2) increasing the front corner rounding, C2, from 0.0 to 1.5 mm; (3) increasing the front side rounding, C3, from 0.0 to 11.0 mm; (4) increasing the hinge rib thickness, C4, from 1.0 to 1.8 mm; and (5) increasing the front lever thickness, C5, from 3.0 to 4.0 mm (Tables 9.15). With these modified parameters, the BMF refrigerator can repetitively dispense water for a longer period without failure.

Table 9.16 shows the summary of a tailored set of parametric ALTs. With the modified design parameters, final samples of the water dispensing system were guaranteed to the lifetime target – a B1 life of 10 years.

TABLE 9.16
Results of ALTs in Water Dispenser Lever of a Refrigerator

	First ALT	Second ALT	Third ALT
Parametric ALT	**Initial Design**	**Second Design**	**Final Design**
In 38,000 mission cycles, there is no problem in the dispenser lever	25,000 cycles: 2/8 fracture	38,000 cycles: 8/8 OK 56,000 cycles: 8/8 OK 67,500 cycles: 1/8 fracture	56,000 cycles: 8/8 OK 68,000 cycles: 1/8 fracture 92,000 cycles: 7/8 OK
Photo	<Dispenser lever> 	<Dispenser lever> 	<Dispenser lever> 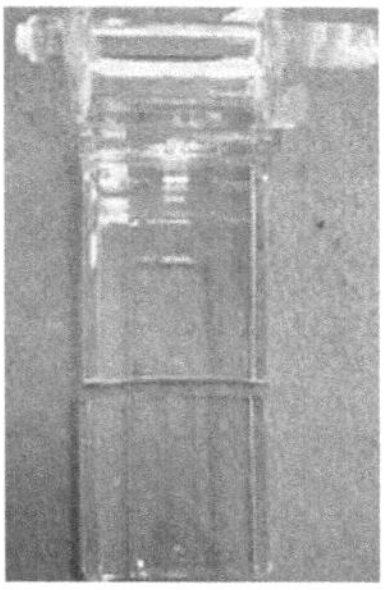
Action plans	Rib1: T1.0 mm → T1.8 mm Fillet 1: R0.0 mm → R1.5 mm Fillet 2: R0.0 mm → R1.5 mm	Fillet 1: R1.5 mm → R2.0 mm Fillet 2: R8.0 mm → R11.0 mm	Rib2: T3.0 mm → T4.0 mm

9.10 FRENCH REFRIGERATOR DRAWER SYSTEM

Figure 9.81 shows the French refrigerator with the newly designed drawer system that consists of a box-rail mechanism. When a consumer put food inside the refrigerator, they want to have convenient access to it and have the food stay fresh. For this to occur, the draw system needs to be designed to withstand the operating conditions it is subjected to by users. The drawer assembly consists of a box, left/right of the guide rail, and a support center, as shown in Figure 9.81b.

In the field, parts of the drawer system of a French refrigerator were failing due to cracking and fracturing under unknown consumer usage conditions. Thus, the data on the failed products in the field were important for understanding the usage environment of consumers and helping to pinpoint design changes that needed to be made in the product (Figure 9.82).

Field data indicated that the damaged products might have had structural design flaws, including sharp corner angles and weak ribs that resulted in stress risers in high-stress areas. These design flaws that were combined with the repetitive loads on the drawer system could cause a crack to occur and thus cause failure.

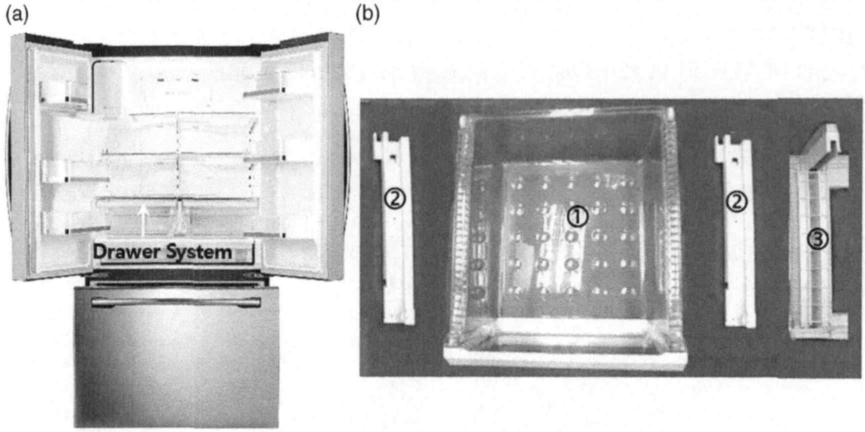

FIGURE 9.81 French refrigerator with the newly designed drawer system: (a) French refrigerator and (b) mechanical parts of a drawer: (1) a box, (2) left/right of the guide rail, and (3) a support center.

FIGURE 9.82 A damaged product after use.

The drawer assembly consists of many mechanical structural parts. Depending on the consumer usage conditions, the drawer assembly receives repetitive mechanical loads when the drawer is open and closed. Putting and storing food in the drawer involves two mechanical processes: (1) the consumer opens the drawer to store or take out the stored food and (2) then closes the drawer by force.

Figure 9.83 shows the functional design concept of the drawer system and its robust design schematic overview. As the consumer stores the food, the drawer system helps to keep the food fresh. The stress due to the weight load of the food is concentrated on the drawer box and its support rails. Thus, it is important to overcome these repetitive stresses when designing the drawer.

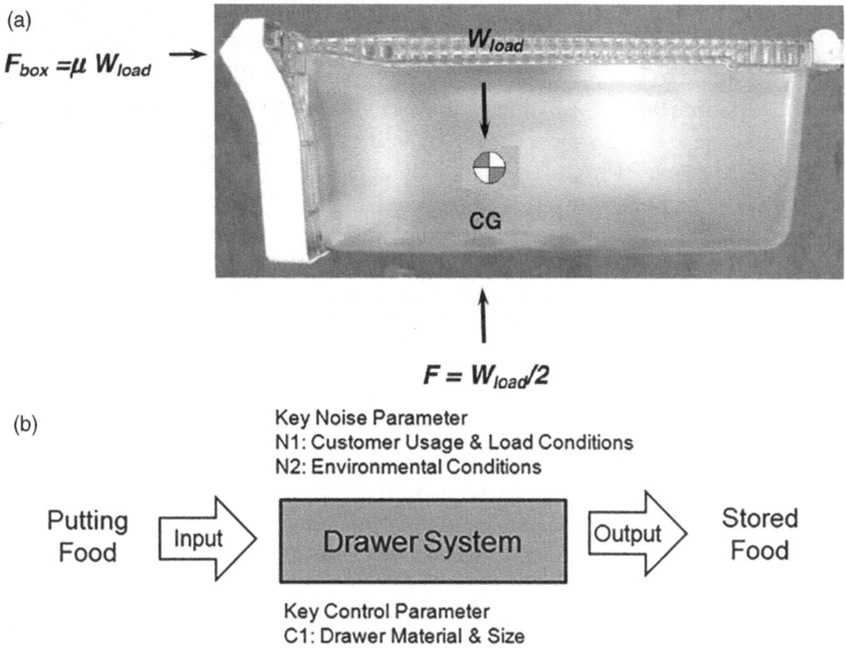

FIGURE 9.83 Design concept and robust design schematic of a mechanical drawer system: (a) design concept of a mechanical drawer system and (b) robust design schematic of the drawer system.

The number of drawer open and close cycles will be influenced by consumer usage conditions. In the United States, the typical consumer requires the drawer system of a French refrigerator to open and close between five and nine times per day.

The force balance around the drawer system can be represented as follows:

$$F_{\text{box}} = \mu W_{\text{load}} \tag{9.56}$$

Because the stress of the drawer system depends on the applied force from the weight of the food, the LS model from Equation (8.15) can be expressed as

$$TF = A(S)^{-n} = A(F_{\text{box}})^{-\lambda} = A(\mu W_{\text{load}})^{-\lambda} \tag{9.57}$$

The AF can be derived as

$$AF = \left(\frac{S_1}{S_0}\right)^n = \left(\frac{F_1}{F_0}\right)^\lambda = \left(\frac{\mu W_1}{\mu W_0}\right)^\lambda = \left(\frac{W_1}{W_0}\right)^\lambda \tag{9.58}$$

The opening and closing of the drawer system occur on an estimated average of five to nine times per day. With a life cycle design point of 10 years, the drawer would have about 32,900 usage cycles. For the worst case, the weight force on the drawer is 0.59 kN, which is the maximum force applied by the typical consumer. The applied weight force for the ALT was 1.17 kN. Using a stress dependence of 2.0, the *AF* was found to be approximately 4.0 using Equation (9.58).

For the lifetime target – a B1 life of 10 years, the test cycles calculated in Equation (8.29) for six test sample pieces were 39,000 cycles, respectively. The ALT was designed to ensure a B1 life of 10 years with about a 60% level of confidence that it would fail less than once during 39,000 cycles.

Figure 9.84 shows the experimental setup of the ALT with the test equipment and the duty cycles for the opening and closing force *F*. The control panel on the top of the testing equipment started and stopped the equipment and indicated the completed test cycles and the test periods, such as sample on/off time. The drawer opening and closing force, *F*, was controlled by the accelerated weight load in the drawer system. When the start button in the control panel gave the start signal, the simple hand-shaped arms held the drawer system. The arms then pushed and pulled the drawer with the accelerated weight force (1.17 kN) (Figure 9.85).

Figure 9.86 shows the failed product from the field and from the ALT, respectively. In the photos, the shape and location of the failure in the ALT were similar to those seen in the field. Figure 9.87 presents the graphical analysis of the ALT results and field data on a Weibull plot.

The shape parameter in the first ALT was estimated at 2.0. For the final design, the shape parameter was obtained from the Weibull plot and was determined to be 3.6. These methodologies were valid in pinpointing the weak designs

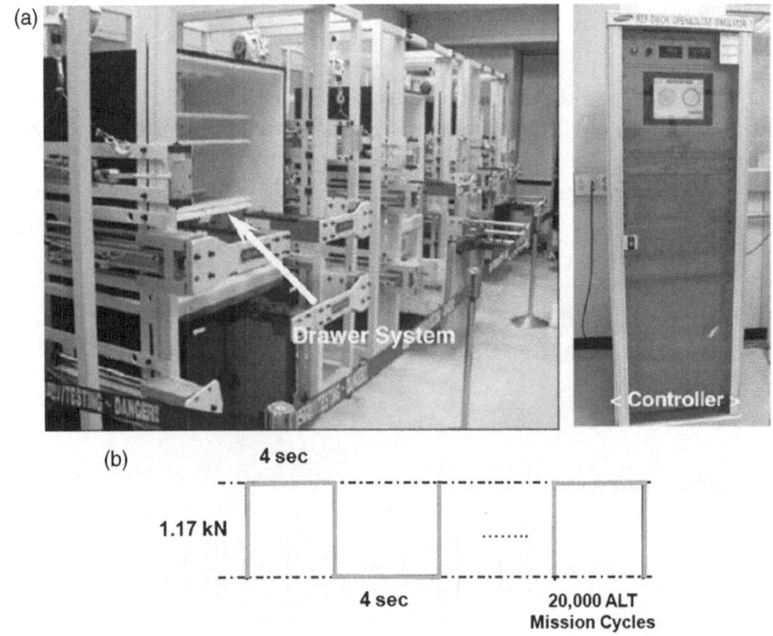

FIGURE 9.84 Equipment used in ALT and duty cycles: (a) equipment used in ALT and (b) duty cycles of repetitive load *F* on the drawer system.

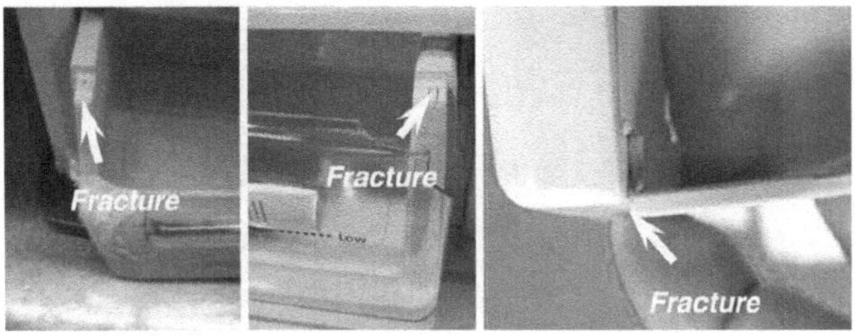

FIGURE 9.85 Failed products in the field (left) and second ALT (right).

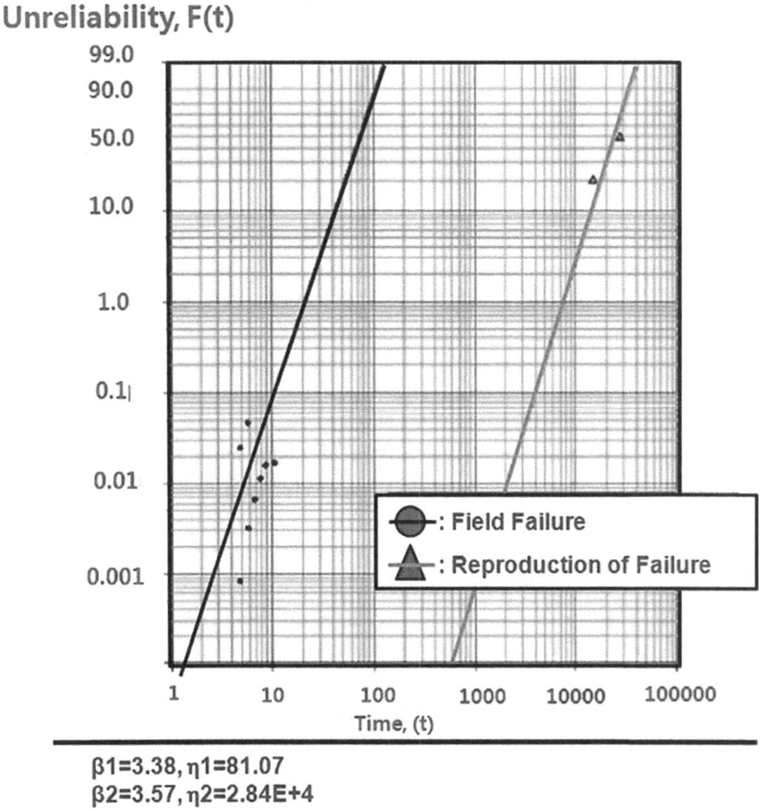

FIGURE 9.86 Field data and results of ALT on a Weibull chart.

responsible for failures in the field and were supported by two findings in the data. In the photo, the shape and location of the broken pieces in the failed market product are identical to those in the ALT results. Moreover, the shape parameters of the ALT ($\beta 1$) and market data ($\beta 2$) were found to be similar to the Weibull plot. Consequently, we know that this parametric ALT is effective to save the testing time and sample size.

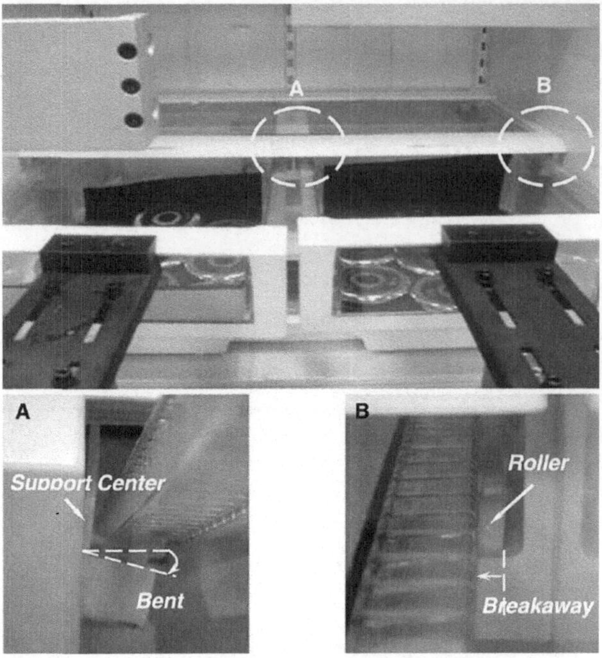

FIGURE 9.87 Structural problems of the left, right, and center support rails in loading.

Initially, when the accelerated load of 12 kg was put into the drawer, the center support rail was bent and the rollers on the left and right rail were broken away (Figure 9.88). The design flaws of the bent center rail and the breakaway roller resulted in not sliding drawers. The rail systems could be corrected by adding reinforced ribs on the center support rail as well as extruding the roller support to 7 mm (see center support rail in Table 9.17).

The fracture of the drawer in both the field products and the ALT test specimens occurred in the intersection areas of the box and its cover (Figure 9.89). The structural design flaws in combination with the repetitive food loading forces may have caused the fracturing of the drawer. The design flaws of no corner rounding and poorly enforced ribs resulted in the high-stress areas. They can be corrected by implementing the fillets and thickening the enforced ribs (see box in Table 9.17).

FIGURE 9.88 Structure of a failing drawer system in the field.

TABLE 9.17
Redesigned Box and Center Support Rail

Box

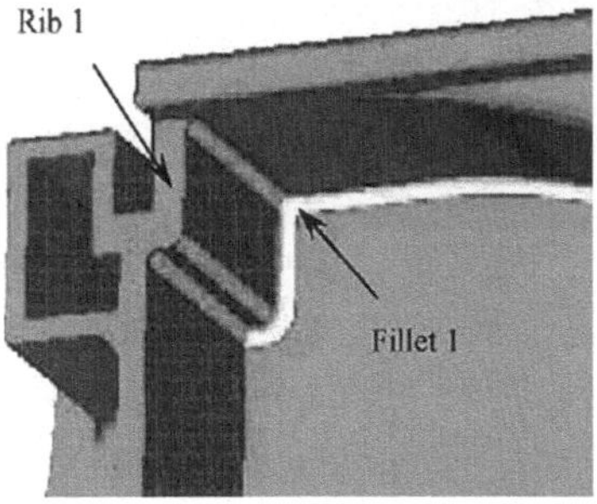

C1: Rib1 T2.0 mm → T3.0 mm
C2: Fillet R0.0 mm → R1.0 mm

Center Support Rail

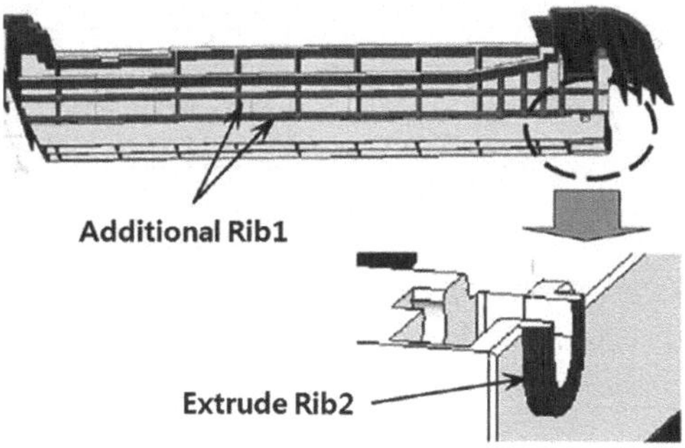

C3: Rib2 added new rib
C4: Extending Rib1 L0.0 mm → L2.0 mm

Guide rail (left/right)

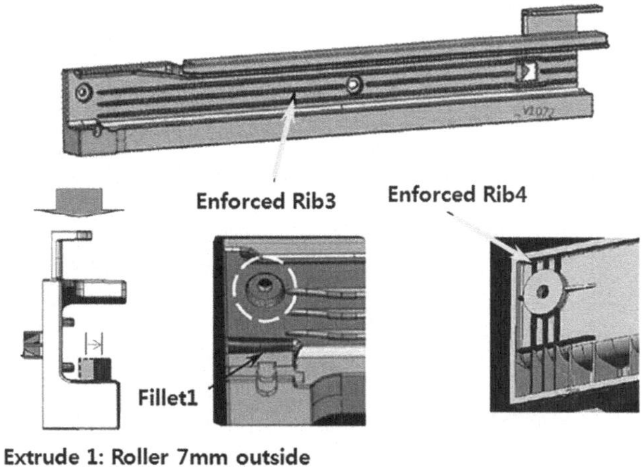

C5: Rib3 (newly added rib, loading test)
C6: Extruder roller L0.0 mm → L7.0 mm
C7: Fillet R3 mm → R4 mm
C9: Rib4 newly added back rib

In the first ALT, due to repetitive stresses, the left and right rails of the drawer system cracked (Figure 9.89a), and the roller of the support center was sunken (Figure 9.89b) in the first ALTs. Thus, the guide rail systems (left and light) were corrected by design changes such as corner rounding and inserting ribs. To strengthen a guide rib, it extruded 2 mm from the center support rail and added new ribs (see center support rail and guide rail in Table 9.17).

The confirmed values of AF and β in Figure 9.92 were 4.0 and 3.6, respectively. The test cycles recalculated in Equation (8.29) for six sample pieces were 20,000,

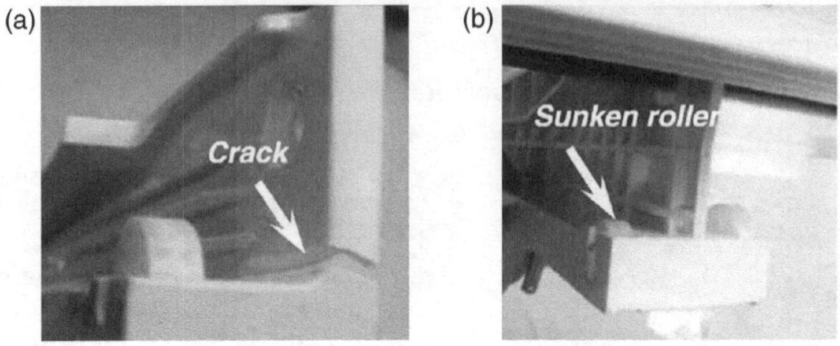

FIGURE 9.89 Structural problems of the left/right guide rail (a) and center support rails (b) in the first ALT.

TABLE 9.18
Results of ALTs in French Refrigerator Drawer System

	First ALT	Second ALT	Third ALT
	Initial Design	**Second Design**	**Final Design**
In 20,000 cycles, there is no problems in the drawer system	3,900 cycles: 3/6 fail 3,900 cycles: 3/6 OK	15,000 cycles: 2/6 fail 29,000 cycles: 1/6 fail 29,000 cycles: 3/6 OK	20,000 cycles: 6/6 OK 45,000 cycles: 6/6 OK
Drawer structure	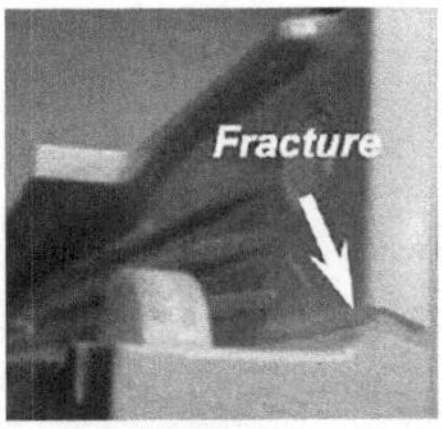	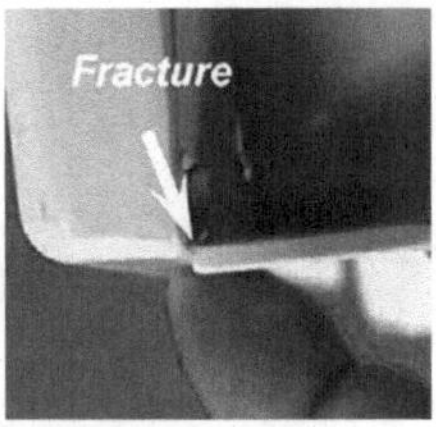	
Material and specification	Redesigned rail C1: Rib3 newly added rib C2: Extrude1: L0.0 mm → L7.0 mm C3: Fillet2: R3 mm → R4 mm C4: Rib4: newly added back	Redesigned box C5: Rib1 T2.0 mm → T3.0 mm C6: Fillet1 R0.0 mm → R1.0 mm	

respectively. Based on the targeted B1 life of 10 years, third ALTs were performed to obtain the design parameters and their proper levels.

Table 9.18 gives a summary of the results of the ALTs, respectively. With these modified parameters, the drawers in the French refrigerator can be smoothly opened and closed for a longer period without failure.

9.11 IMPROVING THE LIFETIME OF A HINGE KIT SYSTEM IN A REFRIGERATOR

When consumers open and close a Kimchi refrigerator door, they should be able to accomplish this with minimal effort. Originally, HKS was the spring mechanism that could decrease the speed of door closing due to door weight. Because its damping effects at 0°–10° (door angle) were not up to much, the spring-damper mechanism in HKS was changed to improve this. Before launching the newly designed HKS, we had to find the design faults and verify its reliability. The hinges of a door are a component of the door that is subjected to repetitive use over the life of the refrigerator. A new HKS was designed for the Kimchi refrigerator (Figure 9.90a) to improve the ease of opening and closing the door for the consumer. The HKS shown in Figure 9.88b consists of a kit cover, shaft, spring, and an oil damper.

The functional loss of the original HKS had been reported often by owners of the refrigerator. Thus, exact data analysis was required to find out the root cause of the defective HKS and what parameter in the HKS needed to be redesigned.

Figure 9.91 shows a damaged HKS, which has two cracks that appeared after a period of use. It was not known under what usage conditions the failure occurred. When comprehensive data from the field were reviewed, it was concluded that the root cause of the HKS failure was a structural design flaw – no round of torsional shaft. Moreover, due to the repetitive loading of the opening and closing of the door, this design defect eventually led to creating the cracks of HKS. Figure 9.92 shows the robust design schematic overview of the HKS. Depending on the consumer usage conditions, an HKS was subjected to different loads during the opening and closing of the refrigerator door.

Because the HKS is a relatively simple structure, it can be modeled with a simple force-moment equation (see Figure 9.93). As the consumer opens or closes the

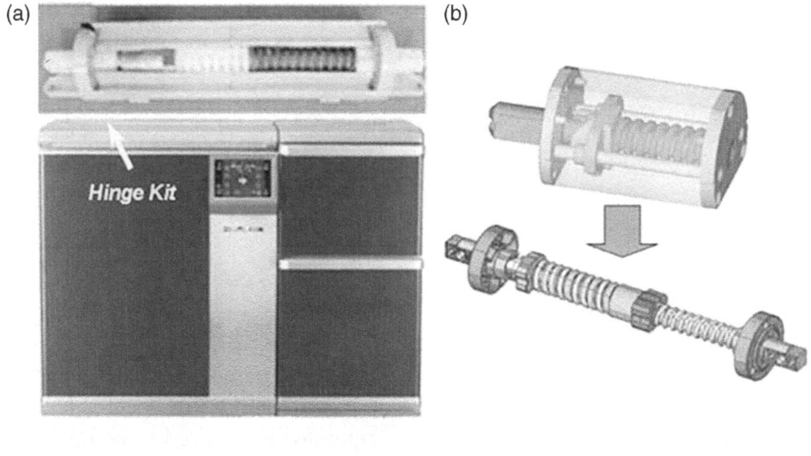

FIGURE 9.90 Commercial Kimchi refrigerator and its HKS: (a) commercial Kimchi refrigerator and (b) HKS.

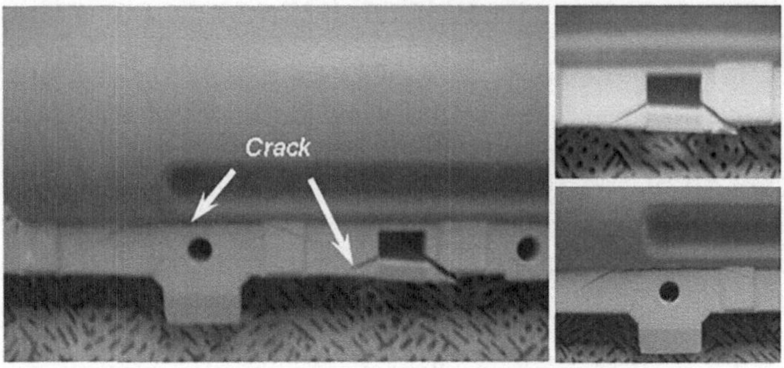

FIGURE 9.91 View of a damaged hinge kit system after a period of use.

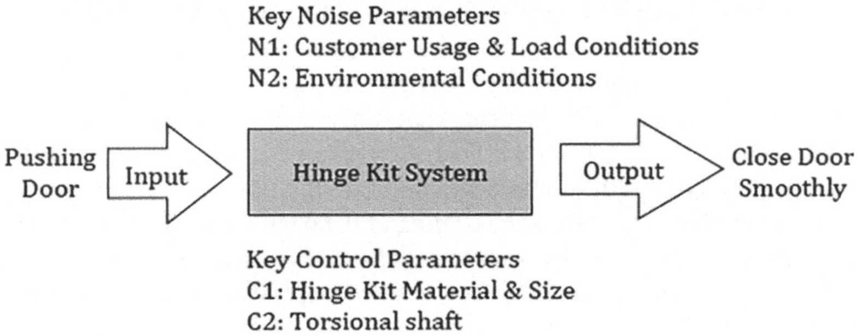

FIGURE 9.92 Robust design schematic of HKS.

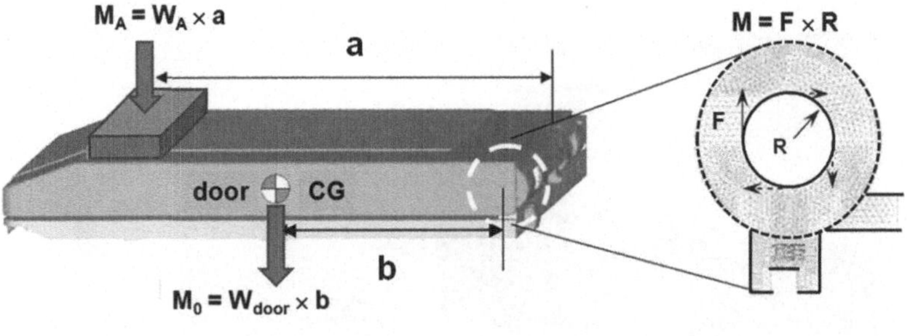

FIGURE 9.93 Design concept of HKS.

refrigerator door, the stress due to the weight momentum of the door is concentrated on HKS. The number of door closing cycles will be influenced by specific consumer usage conditions. The door system of the refrigerator is required to be opened and closed between three and ten times per day in the Korean domestic market.

The moment balance around the HKS can be represented as

$$M_0 = W_{\text{door}} \times b = T_0 = F_0 \times R \tag{9.59}$$

The moment balance around the HKS with an accelerated weight can be represented as

$$M_1 = M_0 + M_A = W_{\text{door}} \times b + W_A \times a = T_1 = F_1 \times R \tag{9.60}$$

Because F_0 is the impact force under normal conditions and F_1 is the impact force in accelerated weight, the stress on the HKS depends on the applied impact. Under the same temperature and load, the LS model from Equation (8.16) can be modified as

$$TF = A(S)^{-n} = AT^{-n} = A(F \times R)^{-n} \tag{9.61}$$

The AF can be derived as

$$AF = \left(\frac{S_1}{S_0}\right)^n = \left(\frac{T_1}{T_0}\right)^n = \left(\frac{F_1 \times R}{F_0 \times R}\right)^n = \left(\frac{F_1}{F_0}\right)^n \tag{9.62}$$

Generally, the operating conditions for the HKS in a Kimchi refrigerator were approximately 0°C–43°C with a relative humidity ranging from 0% to 95%, and 0.2–0.24 g of acceleration. The closing of the door occurred on an estimated average of three to ten times per day. With a life cycle design point of 10 years, HKS incurs about 36,500 usage cycles.

For the worst case, the impact force around the HKS was 1.10 kN, which was the maximum force applied by the typical consumer. The impact force for the ALT with accelerated weight was 2.76 kN. Using a stress dependence of 2.0, the AF was found to be approximately 6.3 in Equation (9.62).

For the B1 life of 10 years, the mission cycles in the ALT calculated from Equation (8.29) for six sample units were 34,000 cycles, respectively. ALT would be designed to ensure a B1 life of 10 years with about a 60% level of confidence if it fails less than once during 34,000 cycles. Figure 9.94 shows the experimental setup of the ALT

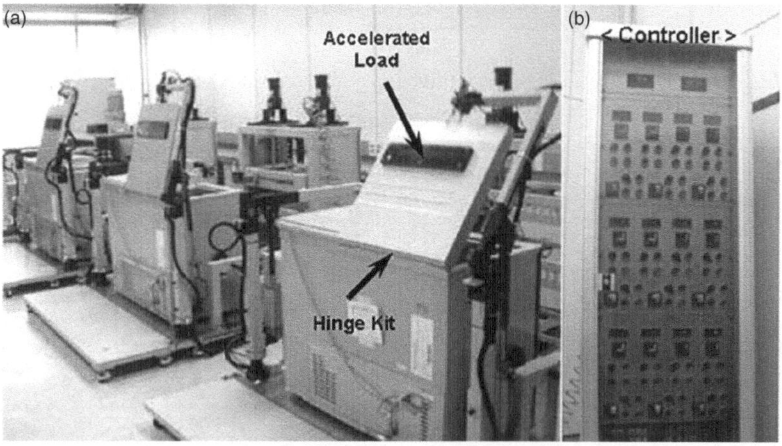

FIGURE 9.94 Equipment used in ALT and controller: (a) ALT equipment and (b) controller.

with labeled equipment for the robust design of HKS. Figure 9.95 shows the duty cycles for the impact force *F*.

The control panel was used to operate the testing equipment – the number of test time, starting or stopping the equipment, and the other. When the start button in the control panel gave the start signal, the simple hand-shaped arms held and lifted the Kimchi refrigerator door. As the door was closing, it was applied to the HKS with the maximum mechanical impact force due to the accelerated load (2.76 kN).

Figure 9.96 shows a photograph comparing the failed product from the field and first ALT, respectively. As shown in the picture, the shape and location of the failure in the ALT were similar to those seen in the field. Figure 9.97 presents the graphical analysis of the ALT results and field data on a Weibull plot. The shape parameter in the first ALT was estimated at 2.0. From the Weibull plot, the shape parameter was confirmed to be 2.1. The defective shape of the ALT was very similar to that of the field. From the Weibull plot, the shape parameters of the ALT and market data were found to be similar. As supported by two findings in the data, these methodologies were valid in pinpointing the weak designs responsible for failures in the field, which determined the lifetime.

The fracture of the HKS in both the field products and the ALT test specimens occurred in the housing and support of the HKS (Figure 9.98). The missing design variables of the HKS in the design phase came from no support structure. The repetitive applied force in combination with the structural flaws may have caused the fracturing of the HKS. The concentrated stresses of the HKS were approximately 21.2 MPa, based on finite element analysis. The stress risers in high-stress areas resulted from the structural design flaws of not having any supporting ribs.

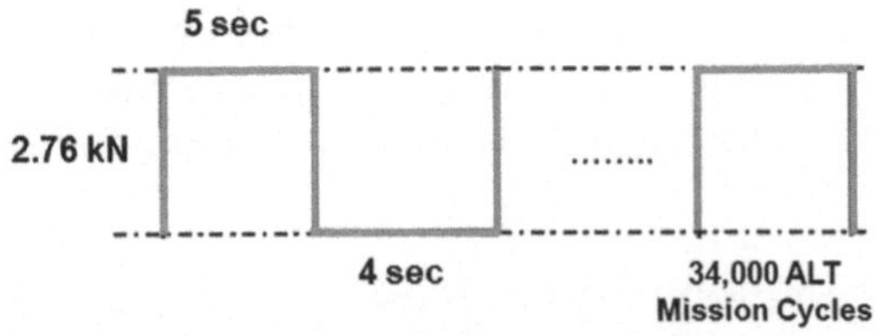

FIGURE 9.95 Duty cycles of the repetitive impact load *F* on HKS.

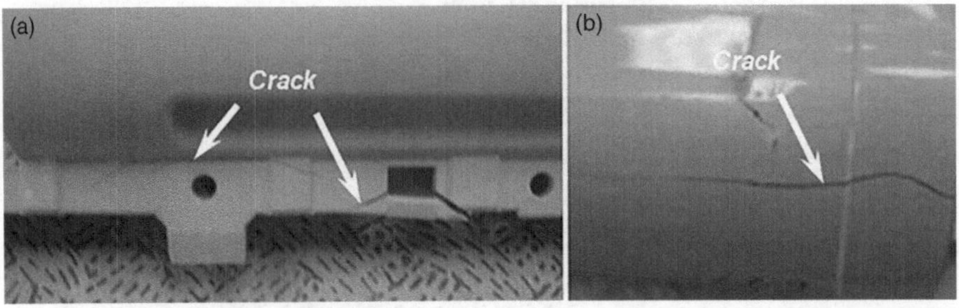

FIGURE 9.96 Failed products in the field and crack after the first ALT: (a) failed products in the field and (b) crack after the first ALT.

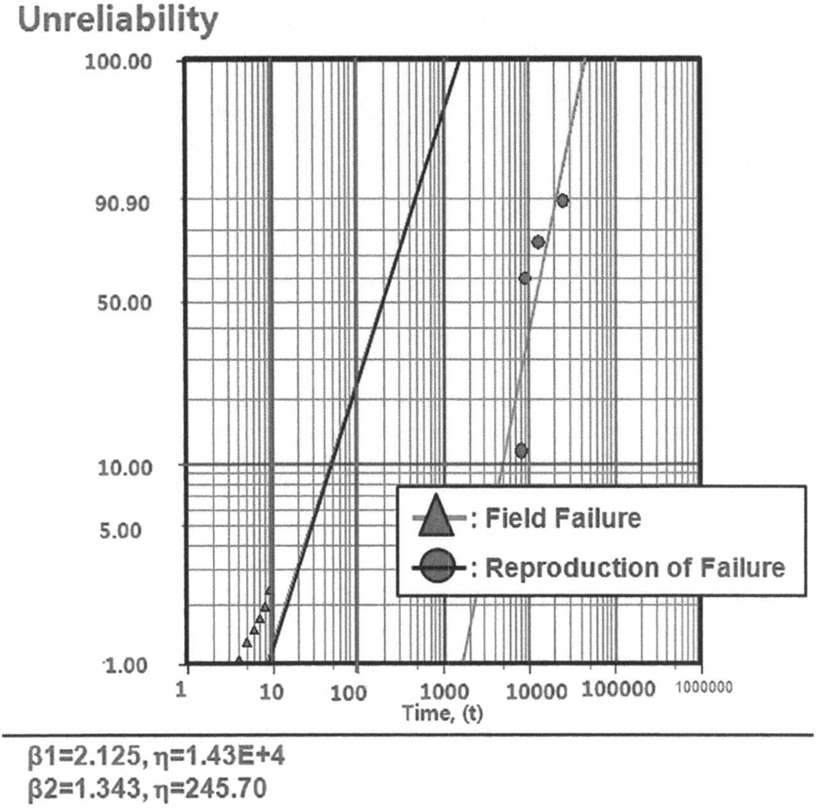

FIGURE 9.97 Field data and first ALT on a Weibull chart.

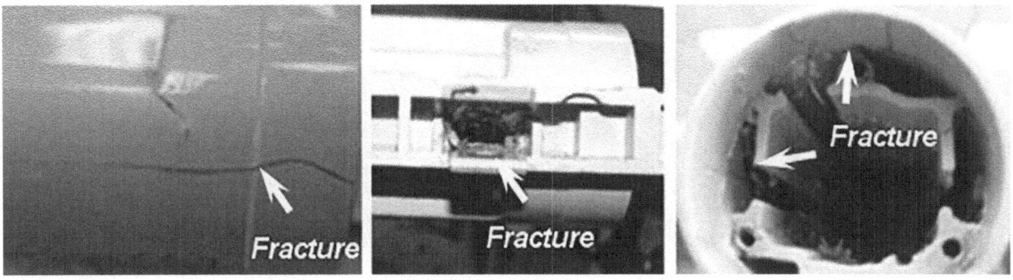

FIGURE 9.98 Structure of failing HKS in the first ALT.

The corrective action plan was to add the support ribs (Figure 9.99). Applying the new design parameters to the finite element analysis, the stress concentrations of the HKS decreased from 21.2 to 19.9 MPa. Therefore, the corrective action plan had to be made at the design stage before production.

The design target of the newly designed samples was more than the lifetime target – a B1 life of 10 years. The confirmed values of AF and β in Figure 9.98 were

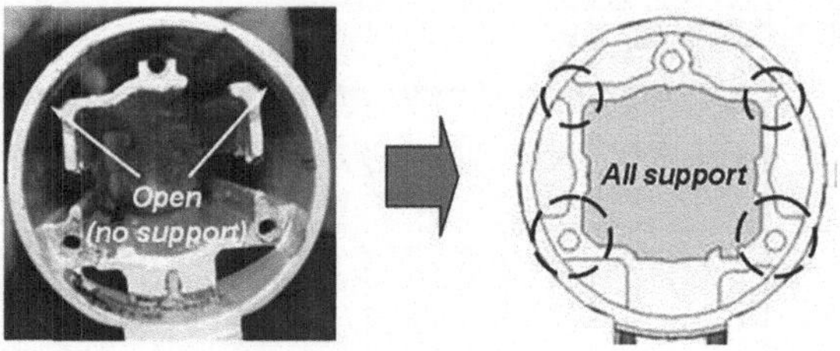

FIGURE 9.99 Redesigned HKS housing structure.

6.3 and 2.1, respectively. For the lifetime target – a B1 life of 10 years and six sample units, the mission cycles recalculated from Equation (8.29) were 41,000, respectively. To obtain the design parameters and their proper levels, second ALTs were performed. In the second ALTs, the crack of the torsional shaft occurred due to its sharp rounding and repetitive impact stresses (Figure 9.100).

The torsional shaft of the HKS was modified by giving it more roundness from R0.5 to R2.0 mm at the corner of the torsional shaft (Figure 9.101). Finally, the redesigned HKS could withstand the high impact force during the closure of the door. With this design change, the refrigerator could also be opened and closed more comfortably.

Table 9.19 shows the design parameters confirmed from a tailored set of ALTs and the summary of the results of the ALTs. With these modified parameters, the Kimchi refrigerator door could be smoothly closed for a longer period without failure. Throughout the three ALTs, the B1 life of the samples was guaranteed to be 10.0 years.

FIGURE 9.100 Problematic torsional shaft structure in HKS.

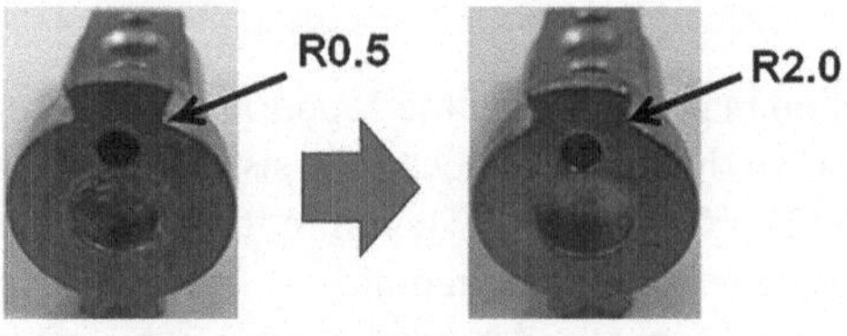

FIGURE 9.101 Redesigned torsional shaft of HKS.

TABLE 9.19
Results of ALTs in Improving the Lifetime of HKS in Refrigerator

	First ALT	Second ALT	Third ALT
	Initial Design	**Second Design**	**Final Design**
In 41,000 cycles, HKS has no crack	3,000 cycles: 2/6 crack (HKS housing)	12,000 cycles: 4/6 crack (torsional shaft)	41,000 cycles: 6/6 OK
Photo (HKS structure)		 	
Material and specification	Supporting Rib C1: No → 2 supports	Roundness corner of torsional shaft C2: R0.5 mm → R2.0 mm	

Index